Thermodynamic Theory of
STRUCTURE, STABILITY AND FLUCTUATIONS

Thermodynamic Theory of STRUCTURE, STABILITY AND FLUCTUATIONS

P. GLANSDORFF

and

I. PRIGOGINE

Université Libre de Bruxelles, Brussels, Belgium, and University of Texas, Austin, Texas

WILEY – INTERSCIENCE
a division of John Wiley & Sons, Ltd.
London – New York – Sydney – Toronto

Library of Congress Catalog Card No. 78–147070

ISBN 0 471 30280 5

Printed by Unwin Brothers Limited, Old Woking, Surrey

Contents

Introduction

The basic description of a mechanical system is in terms of the coordinates and the momenta of the molecules, or in terms of its wave function. However such a description when applied to systems of interest in chemical physics, hydrodynamics or biology leads to great practical and conceptual difficulties. Even if we could conceive computers big enough to study the molecular dynamics of say 10^{23} molecules in a macroscopic system, the knowledge of their positions and their velocities would be of little interest as we would never be able to repeat an experiment involving the same initial state.

The great importance of thermodynamic and hydrodynamic methods is that they provide us with a 'reduced description', a 'simplified language' with which to describe macroscopic systems. In many cases of interest such a reduced description is all that is needed. For instance, to predict the temperature evolution of some piece of metal it is sufficient to solve the Fourier equation with appropriate initial and boundary conditions. The temperature at every point is an average taken over a large number of molecules. The agreement between the predictions of the Fourier equation and experiment shows that a more detailed study of the evolution in terms of mechanical quantities is not required. It is not the purpose of this monograph to analyse the relation between the mechanical and the macroscopic descriptions. This can only be done with the help of statistical mechanics of many-body systems. Here we shall be concerned solely with macroscopic methods.

How far can we proceed with such methods? What is the class of phenomena which may be investigated? These are some of the problems we shall deal with in this book.

It is well known that once the second law is formulated, classical thermodynamics concentrates essentially on the study of equilibrium states. Classical thermodynamics concentrates on the properties of systems which have reached thermodynamic equilibrium (e.g. Schottky, 1929). It is mainly during the last twenty years that we have witnessed the rapid growth of thermodynamics of irreversible processes. The great importance of this development lies in the fact that it makes possible the application of macroscopic methods outside equilibrium (for a short history of this subject see I. Prigogine, 1947). However this development was essentially limited to the *near-equilibrium* region. In this region, the thermodynamic forces (such as temperature gradient, chemical affinities, . . .) and the thermodynamic flows (such as flow of heat, chemical reactions rates, . . .) are linked by linear relations.

The Onsager reciprocity relations (1931) and the Theorem of Minimum Entropy production (1945) both belong to this *Linear Non-Equilibrium Thermodynamics.*

Today this branch of Thermodynamics of irreversible processes, is a classical subject and is adequately treated in many monographs (especially de Groot and Mazur, 1961).

Is it necessary to go further? Some examples will suffice to stress the interest of an extension of thermodynamics into the non-linear region. Let us first consider the case of chemical reactions. It is well known that if the reaction rate is sufficiently slow not to perturb to an appreciable extent the Maxwell equilibrium distribution of each component, a macroscopic description in terms of average concentrations of the components is possible (for more details, see e.g. Prigogine, 1967). Still the relations between chemical rates and affinities are in general *non-linear*.

Another important field of research where macroscopic methods have been applied with success is *hydrodynamics*. Of special interest for us will be the theory of hydrodynamic stability. It is well known that some simple patterns of flow (such as the Poiseuille flow) are realized only for certain ranges of parameters. Beyond these ranges they become unstable.

As a simple example we may consider the thermal stability in horizontal layers of a fluid heated from below. This is the so-called Bénard problem which we shall study in detail in Chapters XI and XII of this monograph (Chandrasekhar, 1961). For some critical value of a dimensionless parameter called the Rayleigh number, the state of the fluid at rest becomes unstable and cellular convection sets in. Now both below and beyond this instability a macroscopic description of the fluid is possible. Thermodynamic considerations should then be of great importance to understand the location and the meaning of the instability.

Our central problem is thus the following: can we extend the methods of thermodynamics to treat the entire range of phenomena starting from equilibrium and including non-linear situations and instabilities?

We shall see that this extension is indeed possible for the whole class of situations for which the local entropy may be expressed in terms of the *same independent variables* as if the system were at equilibrium. This is the assumption of 'local equilibrium', the validity of which implies the dominance of collisional effects which tend to restore thermodynamic equilibrium. In other words, at no moment molecular distribution functions of velocities or of relative positions, may deviate strongly from their equilibrium form (Chapter II, §2). This condition should be considered here as a *sufficient* condition for the application of thermodynamic methods. It is quite possible that a unified thermodynamic approach could be set up under less restrictive conditions. However, we do not explore this possibility here.

Even so restricted, the extension referred to above, leads to a substantial increase of the power of macroscopic methods. Various problems treated till now by quite different methods may be approached in a new unified way; even some problems of classical equilibrium thermodynamics find their natural answer once formulated in the frame of a more general approach.

Any theory whose aim is to include the possibility of new organization of matter in far from equilibrium conditions such as those applying beyond an unstable transition, has to face the problem of fluctuations. A purely causal description is no longer sufficient even for systems involving a large number of degrees of freedom. As an illustration consider a typical problem in hydrodynamics: the stability of the laminar flow of a fluid. Suppose a small fluctuation δE_{kin} appears in the kinetic energy. To this fluctuation will correspond some small 'hump' in the velocity profile, as shown on the figure.

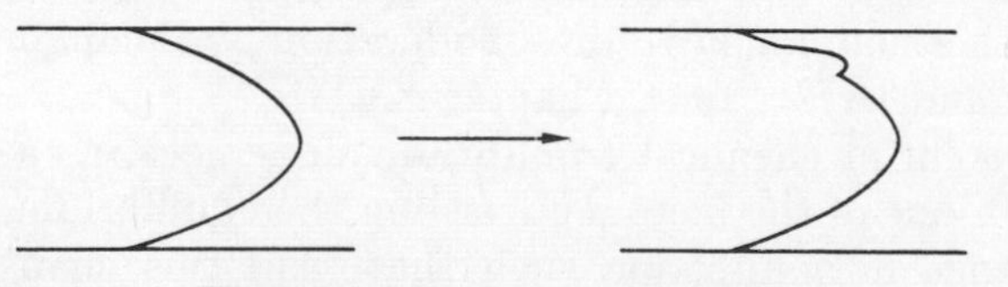

Fluctuation in the velocity profile

If δE_{kin} vanishes everywhere for $t \to \infty$, the flow is stable. On the contrary, if δE_{kin} increases with time, then a new state of flow will be reached. As is well known from classical hydrodynamics, this will be the case if the Reynolds number lies beyond a critical value corresponding to turbulence.

The main point is therefore the following: a new 'structure' is always the result of an instability. It originates from a fluctuation. Whereas a fluctuation is normally followed by a response that brings the system back to the unperturbed state, on the contrary, at the point of formation of a new structure, fluctuations are amplified. This idea is of course the basis of classical stability theory derived from normal mode analysis (e.g. Chandrasekhar, 1961). One considers small perturbations around a steady state which satisfy linear equations of evolution. The time dependence of each normal mode is of the form $exp\ \omega t$ where ω is in general a complex quantity $\omega_r + i\,\omega_i$. The stability condition implies then that for each normal mode:

$$\omega_r < 0 \tag{1}$$

One of our main objects will be to relate stability theory to the thermodynamics of irreversible processes in order to obtain as much information as possible, independently of a detailed normal mode analysis. Clearly, we must then in some way incorporate in our thermodynamic description, the response of the system to fluctuations. In other words we have to build a *generalized thermodynamics* which will also include a macroscopic theory of fluctuations.

Let us emphasize that the fluctuations may have either an external or an internal origin. They may result for example, from a temporary disturbance of the boundary conditions. However, the existence of many degrees of freedom in a macroscopic system automatically implies spontaneous *fluctuations*. The stability conditions of a given process then become the conditions for the *regression of fluctuations*.

The problem of the response to spontaneous fluctuations is also closely related to the famous 'Le Châtelier-Braun' principle of classical thermodynamics (or 'moderation' principle). It states (Prigogine and Defay, 1954, Chapter XVII):

'Any system in chemical equilibrium undergoes, as a result of a variation in one of the factors governing the equilibrium, a compensating change in a direction such that, had this change occurred alone it would have produced a variation of the factor considered in the *opposite* direction'.

For equilibrium situations, the moderation principle when applied to intensive variables (pressure, temperature, mole fractions) can easily be proved.

But what happens when we apply the moderation principle to non-equilibrium situations?

A step towards the discussion of such problems was the proof of the theorem of minimum entropy production (I. Prigogine, 1945). When a steady state is characterized by the minimum of entropy production, fluctuations will regress exactly as in thermodynamic equilibrium and the moderation principle will be satisfied.

The very existence of hydrodynamic instabilities shows that this is no longer necessarily so for situations *far* from equilibrium. We come to one of the most basic questions of macroscopic physics: *Under which conditions may we extrapolate results obtained by equilibrium thermodynamics or by linear non-equilibrium thermodynamics to far from equilibrium conditions*? More specifically: What is the generality of instability phenomena? What is the possibility of their occurrence in purely dissipative systems? And how is a system being organized beyond such a transition?

Classical thermodynamics had solved the problem of the competition between randomness and organization for equilibrium systems. But what happens far from equilibrium? Can we find there new organizations, new structures stabilized through the interaction with the outside world?

From a macroscopic point of view it is necessary to distinguish between two types of structure:

(a) equilibrium structures;

(b) dissipative structures.

Equilibrium structures may be formed and maintained through *reversible* transformations implying no appreciable deviation from equilibrium. A crystal is a typical example of an equilibrium structure. Dissipative structures have a quite different status: they are formed and maintained through the effect of exchange of energy and matter in non-equilibrium conditions. The formation of cell patterns at the onset of free convection (Chapter XI) is a typical example of a dissipative structure. We may consider a convection cell as a *giant fluctuation* stabilized by the flow of energy and matter prescribed by the boundary conditions.

As we shall see, such dissipative structures may, under well defined conditions, exist also for open systems involving chemical reactions (especially Chapters VII, XIV–XVI).

A hint towards a thermodynamic theory which would also include a macroscopic theory of fluctuations, is provided by the Einstein theory of fluctuations. More precisely a generalization of Einstein's theory which may be applied both to equilibrium and non-equilibrium macroscopic evolutions (Chapter VIII) indicates that the basic quantity to consider is the 'curvature' $\delta^2 S$ of the entropy. For isolated systems and small fluctuations, this quantity is identical to the entropy change considered by Einstein. But the important feature is that $\delta^2 S$ retains a simple physical meaning under much more general conditions.

In the whole range of macroscopic physics for which the local equilibrium assumption remains valid, $\delta^2 S$, or its straightforward generalization involving inertial effects, is a negative definite quadratic function.

The problem of the regression of fluctuations, or equivalently of the validity of a moderation principle, leads to the study of the time evolution of $\delta^2 S$. This approach corresponds clearly to the basic ideas of Liapounoff's stability theory (e.g. La Salle and Lefshetz, 1961, Pars, 1965).

As well known non-equilibrium thermodynamics is based on the balance equation for entropy:

$$\mathrm{d}S = \mathrm{d}_e S + \mathrm{d}_i S \tag{2}$$

with

$$\mathrm{d}_i S \geqslant 0 \tag{3}$$

Here $\mathrm{d}_e S$ denotes the contribution of the outside world (entropy flow) and $\mathrm{d}_i S$, the entropy production due to the irreversible processes inside the system. This term $\mathrm{d}_i S$, may be expressed in terms of the rates of the irreversible processes and the corresponding forces. We wish now to go beyond equation (2) and to establish *a new balance equation for* $\delta^2 S$, giving $\mathrm{d}\delta^2 S$. The corresponding source term, which we call the '*excess entropy production*', is of fundamental importance. Whenever its sign is positive, the system is stable. One finds that near equilibrium this condition is identically satisfied. The Le Châtelier–Braun principle is then also satisfied and fluctuations regress. However far from equilibrium this is no longer so. At the marginal state, corresponding to the transition between stability and instability, the excess entropy production vanishes. In this way the physical meaning of instabilities can be studied with great generality†.

† At least as long as both the unperturbed and perturbed states may be described macroscopically.

We see that the method followed here combines various points of view: the emphasis on balance equations (as in linear non-equilibrium thermodynamics), the classical thermodynamic stability theory, Liapounoff's stability theory and an extension of Einstein's fluctuation formula. All of them contribute to achieve a unified treatment of macroscopic physics, involving both reversible and irreversible processes in both the near and far from equilibrium situations.

It is worth noting that in a very interesting paper, G. N. Lewis (1931) proposed to unify fluctuation theory and thermodynamics. However, he was concerned only with equilibrium situations, where the effect of fluctuations is generally negligible (with the exception of critical phenomena).

Before we comment on the organization of this monograph, we would like to mention another major result of our approach. We derive a very general inequality valid for any evolution of a macroscopic system under fixed boundary conditions. Because of this high degree of generality, we call this inequality the '*universal*' evolution criterion (Chapter IX). Usually this criterion appears in the form of a non-exact differential which means that no thermodynamic potential in the classical sense, can be associated with this criterion. Still the criterion may be used to obtain a *generalization* of the concept of thermodynamic potential. This is 'local potential' (Chapter X). The main feature of the local potential is that each unknown function (e.g. the distribution of temperature in the non-linear heat conduction problem) appears *twice*: once as an average quantity and once as a fluctuating quantity. This then leads to a generalization of classical variational techniques valid for non-self-adjoint problems. The local potential presents a minimum (in the functional sense) when the average quantity coincides with the most probable one.

Applications of the local potential method to the convergence of successive approximations are presented in Chapter X, while a few examples of its use for the solution of stability problems are studied in Chapter XII.

In order to obtain a self-contained text we have reformulated in Chapters I–IV a number of important results of equilibrium thermodynamics as well as of linear non-equilibrium thermodynamics. This includes the conservation laws, the second law of thermodynamics, the basic theorems of linear non-equilibrium thermodynamics such as the Onsager relations, the theorem of minimum entropy production, and finally the classical Gibbs–Duhem stability theory. These elements are presented here in a form which should permit the reader to understand the more recent developments without having to refer elsewhere.

Chapters V–VII are devoted to an extension of classical thermodynamic stability theory to general equilibrium and non-equilibrium conditions. It is interesting to note that even for equilibrium conditions, the classical theory was restricted to the few cases where the minimum of a thermodynamic potential exists (for example systems of given volume and temperature). In many cases what is given are well defined *boundary conditions* and not the values of some thermodynamic variables *inside* the system. As a rule, no minimum property of a thermodynamic potential is then available and we had first to develop a new approach to the stability problem (Chapter V) which could then be extended to non-equilibrium situations. As already mentioned the essential result of this approach is the introduction of the so-called 'excess entropy production'. The sign of this quantity is directly related to the stability of a non-equilibrium process in respect to its fluctuations.

The extension of Einstein's fluctuation theory is discussed in Chapter VIII. In a macroscopic theory such as the one considered in this monograph, fluctuations are introduced in a somewhat *ad hoc* manner to test stability. This method of treatment has some serious draw-backs. For instance, it cannot lead to an estimate of the time delay which may be involved in the transition from one stable state to another. Also 'average equations' such as the equations of chemical kinetics may correspond only to a first approximation, in the vicinity of such a transition point, as the fluctuations are then likely to increase much beyond their normal level.

These are very interesting questions and we are actively involved in the study of some of these aspects. In the near future publications, we hope to go beyond the few preliminary results stated in Chapters VIII, XIV–XVI.

We have already mentioned in this introduction the concepts of 'universal evolution criterion' and of 'local potential' studied in Chapter IX and X.

Chapters XI–XVI are devoted to applications. Because of the variety of problems to which the theory may be applied we wished only to present a few examples to illustrate some characteristic features. We begin in Chapter XI with stability problems for fluid layers such as the problem of thermal instability (Bénard problem).

Our thermodynamic stability criterion then leads directly to the variational principles for the Bénard problem as derived from the normal mode analysis by Chandrasekhar and others (Chandrasekhar, 1961). In our opinion, this illustrates the degree of unification achieved in our approach between thermodynamic and hydrodynamic methods.

In Chapter XII, we deal with more complicated stability problems for fluid layers, such as laminar flow instability, and the mutual influence of flow and thermal gradients on stability. This also provides typical illustrations of the local potential technique.

A quite different type of problem is studied in Chapter XIII where we deal mainly with the stability of finite amplitude wave propagation in ideal fluids. The interesting point is that in general the excess entropy production (more exactly its generalization including inertial effects) appears as either a positive *definite* or a negative *definite* function. As a consequence the stability problem can then be solved without any reference to the properties of the marginal state. We have here examples of *time-dependent* evolutions which may be unstable (compression waves). There exist therefore situations where a solution of the partial differential equations of macroscopic physics (here of wave propagation) while correct, does not correspond to stable physical situations.

Chapters XIV–XVI are devoted to the investigation of open chemical systems. It seems to us that such a study presents a special interest due to a number of unexpected features and the direct relevance of the results to biological problems.

Far from equilibrium we may then have oscillations in time around the steady state. We may have also either instabilities, or multiple steady states each being stable in some range. The problem of oscillations is studied in Chapter XIV. The first models for chemical oscillations were introduced a long time ago initially by Lotka (1920) and Volterra (1931). But it is only in recent years, especially in the domain of biochemical reactions, that a considerable amount of data on low frequency chemical oscillations has become available.

As pointed out in Chapter XIV there are two types of chemical oscillations: the first corresponds to oscillations on the 'thermodynamic branch'. This is the situation realized in the Lotka–Volterra model. More precisely this model corresponds to the limit of infinite chemical affinity. The other type corresponds to oscillations beyond the marginal stability of the thermodynamic branch. This leads then to the concept of 'limit cycle' introduced by Poincaré in theoretical mechanics (1892). These limit cycles are of great interest, as they provide us with a beautiful example of time order generated by irreversible processes.

Chemical instabilities which lead to space organization are studied in Chapter XV. Such 'symmetry breaking instabilities' are of special interest as they lead to a spontaneous 'self organization' of the system both from the point of view of its *space order* and its *function*. We have here typical examples of what we have called dissipative structures corresponding to a low entropy value. Such situations may arise in systems which are able to use part of the energy or matter exchanged with the outside world to establish a macroscopic internal order.

The existence of these dissipative structures has now been confirmed both by computer and laboratory experiments (Büsse 1969, Herschkowitz, 1970). In far from equilibrium conditions, chemical reactions may compensate the effect of diffusion, and lead to organized structures on a macroscopic level. This is a fact of primary importance and is likely to open new perspectives in classical thermodynamics.

Moreover the requirements necessary to obtain an instability in far from equilibrium conditions are compatible with the mechanisms of some of the most important biochemical reactions responsible for the maintainance of biological activity (Prigogine, Lefever, Goldbeter and Herschkowitz, 1969).

Another interesting point is that the variety of steady states accessible to an open system may become much larger in far from equilibrium conditions. Examples are studied in Chapter XVI. Again, this enlarged variety of possibilities has important biological implications. As an illustration we study a model of membrane excitation due to Blumenthal, Changeux and Lefever (1969) in which co-operative behaviour together with irreversible processes far from equilibrium lead to a new type of 'dissipative' phase transition.

All these results indicate that dissipation may indeed be a source of order both in time and space. It is difficult to avoid the feeling that such considerations may ultimately contribute to narrow the gap which still exists today between biology and theoretical physics. Far from equilibrium and beyond the instability we have really a new state of matter induced by a prescribed flow of free energy. Do biological processes belong to this state? This is quite a challenging problem which still requires a considerable amount of thought and study. At least what seems henceforth certain is that important biological processes involve situations beyond the instability and consequently, can not be accounted for by mere extrapolation from thermodynamic equilibrium.

This monograph comprises three distinct parts, the first (Chapters I–IX), is devoted to the general theory, the second (Chapters X–XIII), to the variational techniques and hydrodynamic applications, and finally the third (Chapters XIV–XVII), to instabilities in chemical systems. Readers particularly interested in the last part, which contains a discussion about the possible application to biology, may leave out Part II.

The french philosopher Heuri Bergson (1907) called the second law of thermodynamics the most 'metaphysical' of all laws of nature. Whether a compliment or a criticism, this applies also to the 'generalized thermodynamics' we develop in this monograph.

Classical thermodynamics is essentially a theory of '*destruction of structure*'. One may even consider the entropy production as a measure of the 'rate' of this destruction. But in some way such a theory has to be completed by a theory of '*creation of structure*', lacking in classical thermodynamics.

We have seen that in addition to the entropy production this approach introduces the '*excess entropy production*' which seems to be the basic quantity whose behaviour characterizes the occurrence of new structures and their stability. In the case of chemical reactions we shall see that the stability is determined by a rather complex inter-play of both kinetic and thermodynamic quantities. No statements completely independent of kinetics can be made at present. Specific classes of chemical reactions have to be considered (e.g., systems of monomolecular reactions, cross-catalytic reactions). As a consequence there exists a whole wealth of possibilities. This contrasts with the *universal* character of the statements made by classical thermodynamics about systems approaching equilibrium.

But this multiplicity of possibilities seems precisely necessary to describe the various far from equilibrium situations. A flow of energy may organize systems and *decrease* their entropy (as in the case of symmetry breaking instabilities referred to before). In other cases it may *increase* their entropy. Likewise it may also increase their entropy production as in the Bénard instability by adding a new mechanism of dissipation, or it may decrease it. We shall study examples of all these situations in this monograph.

The importance of stability theory has been emphasized in a wide range of fields of research such as biology, economics, sociology. To quote P. Weiss (1968): 'Considering the cell as a population of parts of various magnitude, the rule of order is objectively described by the fact that the resultant behaviour of the population as a whole is infinitely less variant from moment to moment than are the momentary activities of its parts'.

This statement applies as well to the cell as to a human population. In spite of this stability property, a modification of the state variables may lead to a new pattern of organization.

Now in all these cases we basically deal with situations which correspond much more closely to the non-equilibrium conditions than to the situations studied by classical equilibrium thermodynamics. Whichever we consider, a cell or a society, it interacts with its medium and the exchange of energy and matter is an essential element of its very existence.

Therefore we may hope that the far from equilibrium approach we develop in this book may act as an element of unification bringing closer problems belonging to a wide range of disciplines.

It is well known that the most detailed analysis of 'order' made in physics refers to equilibrium situations. But here we have to extend this concept to non-equilibrium situations. To borrow an expression introduced by P. Weiss (1968), we must study 'molecular ecology', analyse the order in terms of population dynamics and compare it with the order in equilibrium systems. The relation between this order and probability is a completely different one than that of equilibrium. In the cell pattern corresponding to thermal instabilities, a macroscopic number of molecules has a coordinated motion over macroscopic times. This would correspond for equilibrium situations to a probability smaller than anything we could imagine.

But even the time evolution of such systems has to be described in new terms. We have already emphasized the relation between fluctuations and instabilities. Therefore the evolution of the system now involves both deterministic and statistical aspects and from the macroscopic point of view at least, contains some essential indeterministic features.

Already classical thermodynamics had added a new element to the concept of time through distinction between reversible and irreversible processes. Now still another element is added; the history of successive instabilities. In this way such systems acquire a 'historical' dimension. Their state can no longer be characterized by the value of variables at a given moment, but in addition we need to know the succession of instabilities which have occurred in the past. Biological systems carry with them their information. Is this information not at least related to the 'historical dimension'?

These are fascinating questions, and we feel that we are only at the very beginning. Still, as we shall see, specific examples may already be discussed along these lines.

Certainly one of the most attractive features of thermodynamics has always been its power of unification, its reduction of a large variety of phenomena to a few basic ideas. This is the tradition we have tried to follow in this monograph.

We would like to express our gratitude to: Professor R. Balescu, Brussels, Professor B. Baranowski, Warsaw, Professor J. P. Changeux, Paris, Professor J. Chanu, Paris, Professor A. Katchalsky, Rehovoth, Israël, Professor R. Mazo, Eugen, Oregon, Professor R. Narashima, Bangalore, Professor G. Nicolis, Brussels, Professor P. H. Roberts, Newcastle, Professor R. S. Schechter, Austin, Texas, Professor G. Thomaes, Brussels, Professor W. Strieder, Notre Dame, Indiana.

It is with pleasure that we acknowledge the efforts of Drs. A. Babloyantz, N. Banaï, M. Goche, A. Goldbeter, J. R. Hamm, M. Herschkowitz, R. Lefever, J. Legros and J. Platten from our group in Brussels and of J. Herschkowitz from Piccatiny Arsenal, New Jersey. Our grateful thanks for the typewriting are due to Mrs. L. Février.

These colleagues, friends and co-workers have in various ways contributed to the development of this work and to the preparation of the manuscript.

The work described here, was stimulated by our contacts with members of the General Motors Research Laboratories. We would like to express our appreciation to Mr. J. M. Campbell, former scientific director of the Laboratories, Dr. P. F. Chenea, vice-president of the Research Laboratories, Dr. R. Davies, D. Hayes and Dr. R. Herman, for their interest and support.

This research has been sponsored in part by the R. A. Welch Foundation of Houston, Texas and the Fonds National Belge de la Recherche Scientifique Collective.

P. Glansdorff
I. Prigogine

PART I
General Theory

CHAPTER I

Conservation Laws and Balance Equations

1. GENERAL FORM OF A BALANCE EQUATION

Let us consider a system of volume V limited by the surface Ω. We wish to follow the time change of the integral

$$I(t) = \int f \, \mathrm{d}V \tag{1.1}$$

This integral is extended over the volume V of the system we are interested in. The surface Ω is assumed to be at rest. In the terminology usually adopted in thermodynamics, $I(t)$ is an *extensive* quantity. For instance, it may be the mass or the energy of the system. On the contrary, $f(x, y, z, t)$ is an *intensive* quantity, which does not depend on the system as a whole. It corresponds to the volume density associated to I and may be represented by the functional derivative

$$f \equiv \frac{\delta I}{\delta V} \tag{1.2}$$

We may distinguish between two *mechanisms* for the change in time of $I(t)$

$$\frac{\partial I}{\partial t} = P[I] + \Phi[I] \tag{1.3}$$

The first term on the r.h.s. corresponds to the *production* per unit time of the quantity I inside the volume V. It can be written as a volume integral:†

$$P[I] = \int \sigma[I] \, \mathrm{d}V \tag{1.4}$$

where $\sigma[I]$ denotes the *source* of I per unit time and unit volume.

† In order to simplify the notation, the limit of a definite integral is not explicitly indicated when it corresponds either to the whole volume V of the system as in (1.1) and (1.4), or to the boundary surface, as in (1.5).

The second term in the r.h.s. of (1.3) represents the *flow* of the quantity I through the boundary surface Ω. It can be written as a surface integral

$$\Phi[I] = \int j_n[I] \, \mathrm{d}\Omega \tag{1.5}$$

This formula introduces the *density of flow* $\mathbf{j}[I]$ associated with I; $j_n[I]$ is its projection along the inside normal to the surface (Figure 1.1). For simplicity, we shall also use the term flow instead of density of flow (or current, or flux).

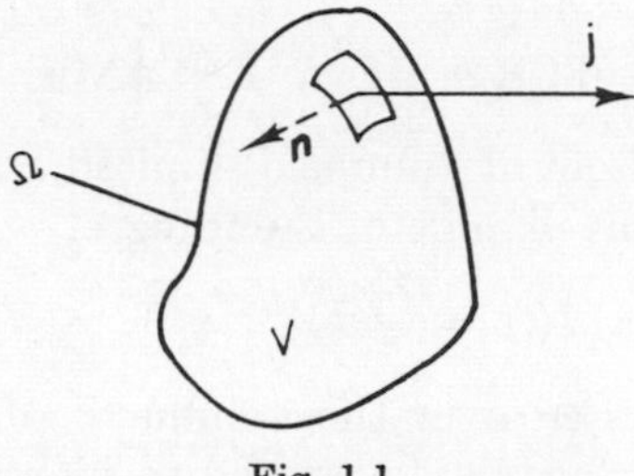

Fig. 1.1.

In general, both the source and the flow terms in (1.3) may be positive or negative quantities.

Using equations (1.4) and (1.5) we obtain the so-called *balance equation* corresponding to the extensive variable I as:

$$\frac{\partial I}{\partial t} = \int \sigma[I] \, \mathrm{d}V + \int j_n[I] \, \mathrm{d}\Omega \tag{1.6}$$

In this way, the time variation of I has been decomposed into two terms, which represent, one a volume integral and the other a surface integral. Let us also observe that in the case of a vectorial extensive variable $\mathbf{I}(t)$, as e.g. the momentum of the system, the balance equation may still be used in the form (1.6) for each projection I_x, I_y, I_z taken separately.

It is sometimes useful to write the balance equation (1.3) or (1.6) in the symbolic form

$$\mathrm{d}I = \mathrm{d}_i I + \mathrm{d}_e I \tag{1.7}$$

where $\mathrm{d}_i I$ represents the source term and $\mathrm{d}_e I$ the flow term. One may write as well

$$\mathrm{d}_i I = \mathrm{d}I + (-\mathrm{d}_e I) \tag{1.8}$$

In this form, we see that the source d_iI contributes on one side to the time change of I and on the other to the flow from the system to the outside world ($-d_eI$). It must be emphasized that only dI is in general a total differential of the state variables.

Equality (1.6) has to be valid whatever the volume V. Therefore, the application of Green's formula gives us directly the balance equation in *local* form:†

$$\frac{\partial f}{\partial t} = \sigma[I] - \operatorname{div} \mathbf{j}[I] \tag{1.9}$$

One of the advantages of this formulation is that all conservation equations can be expressed by the statement that the source term corresponding to a conserved quantity vanishes. For example, if I represents the total mass of the system one has

$$I = M$$

f becomes the mass density ρ (see (1.2))

$$f = \rho$$

and conservation of mass is expressed by the relation

$$\sigma[M] = 0 \tag{1.10}$$

The time change of the density ρ is then, apart from the sign, equal to the divergence of the mass flow. Moreover, the mass flow is clearly

$$\mathbf{j}[M] = \rho \mathbf{v} \tag{1.11}$$

where $\mathbf{v}$ is the velocity of matter. Using (1.10) and (1.11), we therefore obtain the classical continuity equation

$$\frac{\partial \rho}{\partial t} + \operatorname{div} \rho \mathbf{v} = 0 \tag{1.12}$$

Likewise, conservation of total energy U (first law of thermodynamics), and of total momentum $\mathbf{Q}$ (in the absence of external forces), may be expressed as

$$\sigma[U] = 0; \quad \sigma[\mathbf{Q}] = 0 \tag{1.13}$$

† We use the standard notation:

$$\operatorname{div} \mathbf{j} \equiv \frac{\partial j_x}{\partial x} + \frac{\partial j_y}{\partial y} + \frac{\partial j_z}{\partial z}$$

As we shall see in Chapter II, the source term for entropy S plays a special role as the second principle of thermodynamics postulates the inequality:

$$\sigma[S] \geqslant 0 \tag{1.14}$$

Entropy is not a conserved quantity but increases as the result of irreversible processes included in the source term. It is only for reversible processes that the change of entropy is entirely due to entropy exchanges with the outside world.

Let us go back to the density of flow $\mathbf{j}[I]$ in the balance equation (1.9). In general, we are concerned not only with a convection flow such as (1.11) but also with a conduction flow $\mathbf{j}_{cond}$ (say), which occurs even in a system at rest. We have therefore

$$\mathbf{j} = \mathbf{j}_{cond} + \mathbf{j}_{conv} = \mathbf{j}_{cond} + f\mathbf{v} \tag{1.15}$$

For example, the heat flow $\mathbf{W}$ is the conduction current associated with internal energy (cf. §4). We shall consider other examples later on in this chapter.

In many cases, it is useful to introduce the so-called *substantial* or *hydrodynamic* derivative

$$\frac{\mathrm{d}}{\mathrm{d}t} \equiv \frac{\partial}{\partial t} + \sum_i \mathrm{v}_i \frac{\partial}{\partial x_i} \tag{1.16}$$

The continuity equation (1.12) leads then to the equalities:

$$\rho \frac{\mathrm{d}\varphi}{\mathrm{d}t} = \rho \frac{\partial \varphi}{\partial t} + \sum_i \rho \mathrm{v}_i \frac{\partial \varphi}{\partial x_i} = \frac{\partial(\rho\varphi)}{\partial t} + \sum_i \frac{\partial}{\partial x_i} (\rho \mathrm{v}_i \varphi) \tag{1.17}$$

This relation may be applied to an arbitrary intensive variable $\varphi(x, y, z, t) \equiv \varphi(x_1, x_2, x_3, t)$.

Let us now apply this formalism to the most important quantities which occur in continuous media. We shall follow closely the procedure one of us outlined some years ago (Prigogine, 1947). Therefore, we shall limit ourselves to a minimum of detail. Supplementary information may also be found in other text books (e.g. de Groot and Mazur, 1962).

2. CONSERVATION OF MASS

Let us consider a continuous medium formed by c components $\gamma(\gamma = 1, 2, \ldots, c)$. The macroscopic velocity of component γ is

$\mathbf{v}_\gamma$. Moreover, the mass fraction N_γ and the partial density ρ_γ of component γ are defined by the relations (1.2)

$$N_\gamma = \frac{\delta m_\gamma}{\delta m}, \qquad \sum_\gamma N_\gamma = 1 \tag{1.18}$$

$$\rho_\gamma = \rho N_\gamma, \qquad \sum_\gamma \rho_\gamma = \rho \tag{1.19}$$

The centre of mass velocity (which we shall also call the barycentric velocity) is given by the equality

$$\rho \mathbf{v} = \sum_\gamma \rho_\gamma \mathbf{v}_\gamma \tag{1.20}$$

The diffusion current of component γ, is by definition:

$$\mathbf{\Delta}_\gamma = \mathbf{v}_\gamma - \mathbf{v} \tag{1.21}$$

As the result of (1.19) and (1.20) we have the identity

$$\sum_\gamma \rho_\gamma \mathbf{\Delta}_\gamma = 0 \tag{1.22}$$

The current $\mathbf{j}[M_\gamma] = \rho_\gamma \mathbf{v}_\gamma$ which appears in the balance equation (1.9) of component γ may therefore be split into a conduction current (here the diffusion current) $\rho_\gamma \mathbf{\Delta}_\gamma$ and a convection current $\rho_\gamma \mathbf{v}$. Moreover if component γ participates in some chemical reactions we have to introduce a source term. Let us write

$$0 = \sum_\gamma \nu_\gamma M_\gamma \tag{1.23}$$

This is the so-called stoichiometric equation corresponding to the chemical reaction. The coefficient ν_γ corresponds to the component γ with molar mass M_γ and is positive when the component is produced in the chemical reaction, negative in the opposite case and equal to zero if γ does not participate in the reaction.

The number of moles transformed by the chemical reaction per unit time and volume gives us the source term

$$\sigma[M_\gamma] = \nu_\gamma M_\gamma w \tag{1.24}$$

where w is called the chemical reaction rate.† Of course, if component γ participates in r simultaneous chemical reactions, the source term (1.24) becomes

$$\sigma[M_\gamma] = \sum_\rho \nu_{\gamma\rho} M_\gamma w_\rho \quad (\rho = 1, 2, \ldots, r) \tag{1.25}$$

† In formula (1.24) the reaction rate is expressed in *moles* per unit time and unit volume. To avoid the factor M_γ it may be convenient to introduce a rate of mass production instead of w.

The balance equation (1.9) therefore takes the explicit form

$$\sum_{\rho} \nu_{\gamma\rho} M_{\gamma} w_{\rho} = \frac{\partial \rho_{\gamma}}{\partial t} + \sum_{i} \frac{\partial}{\partial x_i} (\rho_{\gamma} \Delta_{\gamma i} + \rho_{\gamma} \mathrm{v}_i) \tag{1.26}$$

Adding these equations for all components $\gamma = 1, 2, \ldots, c$, we obtain the conservation equation (1.10) for the total mass. Indeed, (1.23) leads immediately to

$$\sum_{\gamma}\sum_{\rho} \nu_{\gamma\rho} M_{\gamma} w_{\rho} = \sum_{\rho} w_{\rho} (\sum_{\gamma} \nu_{\gamma\rho} M_{\gamma}) = 0 = \sigma[M]$$

To simplify the notations, we shall use the dummy indices i, j instead of explicit summation over the geometrical coordinates x, y, z. We shall also use the subscript $,_j$ for the derivation symbol $\partial/\partial x_j$ with respect to the coordinate x_j and the symbol ∂_t for the time derivative $\partial/\partial t$. With these notations, the balance equations (1.12) and (1.26) take the condensed forms:

$$0 = \partial_t \rho + (\rho \mathrm{v}_j),_j \tag{1.27}$$

and

$$\sum_{\rho} \nu_{\gamma\rho} M_{\gamma} w_{\rho} = \partial_t \rho_{\gamma} + [\rho_{\gamma} \Delta_{\gamma j} + \rho_{\gamma} \mathrm{v}_j],_j \tag{1.28}$$

3. CONSERVATION OF MOMENTUM AND EQUATION OF MOTION

Using the notations introduced above, the equations of motion of a continuous medium may be written as (e.g. Landau and Lifshitz, 1959):

$$\rho \frac{\mathrm{dv}_i}{\mathrm{d}t} = \rho F_i - (P_{ij}),_j \tag{1.29}$$

The derivative $\mathrm{dv}_i/\mathrm{d}t$ of the ith component of velocity is taken in accordance with the definition (1.16); F_i is the ith component of the external force per unit mass and P_{ij} represent the components of the pressure tensor. The definition (1.16) shows that (1.29) is a *non-linear* equation in the velocities. This is an essential feature of hydrodynamics. The whole study of hydrodynamic instabilities is based on this non-linearity. If hydrodynamics had a linear structure we should never observe transitions from laminar to turbulent flow.

Equation (1.29) may be considered as a postulate on which classical continuum mechanics is based. Its physical meaning becomes much clearer if it is written in the form of a balance equation. Upon applying equation (1.17) to (1.29) we obtain:

$$\partial_t \rho v_i = \rho F_i - [P_{ij} + \rho v_i v_j]_{,j} \tag{1.30}$$

By comparison with the general expression (1.9) of a balance equation, we see that the source associated to the ith component of the total momentum $\mathbf{Q}$ is given by

$$\sigma[Q_i] = \rho F_i \tag{1.31}$$

This is exactly what had to be expected. The expression (1.31) of the source of momentum shows the total momentum is *conserved* in the absence of external forces.

On the other hand, the flow of momentum contains both a convection current $(\rho v_i)\mathbf{v}$ and a conduction current which corresponds to the pressure tensor. This tensor coincides with the total *flow of momentum* as calculated with respect to the centre of mass motion. This is in agreement with the microscopic interpretation of the pressure tensor (Prigogine, 1947, Hirschfelder, Curtiss and Bird, 1954, Massignon, 1957).

The denomination here adopted and the separation between flow and source are somewhat different from the definition introduced in the theory of continuous media. The *conservation* properties we defined here, are however in agreement with the basic microscopic properties of the collisional invariants considered in the kinetic gas theory (cf. Chapman, Cowling, 1939). In order to avoid any misunderstanding, let us illustrate this difference for the total momentum conservation. In classical hydrodynamics a conserved momentum is often defined as a momentum which remains constant along the macroscopic motion defined in (1.16), that is

$$\frac{d\mathbf{Q}}{dt} = 0$$

Obviously, this is not in agreement with the conservation property as prescribed by the vanishing of the source in the second relation (1.13) (see 1.31).

As a rule, the tensor P_{ij} may be decomposed into an *elastic* contribution p^e_{ij} and a *dissipative* contribution (related e.g. to viscosity), as:

$$P_{ij} = p^e_{ij} + p_{ij} \tag{1.32}$$

For a fluid, the elastic tensor reduces to the scalar hydrostatic pressure p, and (1.32) takes the simpler form:

$$P_{ij} = p\delta_{ij} + p_{ij}; \qquad \delta_{ij} = \begin{cases} 0 & i \neq j \\ 1 & i = j \end{cases} \tag{1.33}$$

In what follows we shall limit ourselves to situations in which the pressure tensor is symmetric ($P_{ij} = P_{ji}$). At thermodynamic equilibrium the only contribution to (1.32) is the elastic term p^e_{ij}.

Let us also observe that in a system formed by several components, each of which is subjected to an external force $\mathbf{F}_\gamma$, the total force which occurs in the equation of motion (1.29) is defined by the relation

$$\rho\mathbf{F} = \sum_\gamma \rho_\gamma \mathbf{F}_\gamma \tag{1.34}$$

Let us now consider the balance equation for the energy.

4. CONSERVATION OF ENERGY

The total energy U of a system will generally include the kinetic (macroscopic) energy E_{kin}, potential energies Ω_γ corresponding to the external forces $\mathbf{F}_\gamma$ as well as the internal energy E. We have therefore

$$U = E_{kin} + \sum_\gamma \Omega_\gamma + E \tag{1.35}$$

where Ω_γ is the potential energy of component γ corresponding to $\mathbf{F}_\gamma$.

To derive the balance equation for the kinetic energy of the centre of mass motion it suffices to multiply the equation of motion (1.29) by v_i and to apply (1.17). The result is

$$\rho F_i \mathrm{v}_i + P_{ij}\mathrm{v}_{i,j} = \partial_t \frac{\rho \mathrm{v}^2}{2} + [P_{ij}\mathrm{v}_i + \rho \frac{\mathrm{v}^2}{2} \mathrm{v}_j]_{,j} \tag{1.36}$$

Again i and j are dummy indices over which a summation is performed and $\mathrm{v}_{i,j}$ stands for $\dfrac{\partial \mathrm{v}_i}{\partial x_j}$.

The source of kinetic energy is given by:

$$\sigma[E_{kin}] = \rho F_i \mathrm{v}_i + P_{ij}\mathrm{v}_{i,j} \tag{1.37}$$

It contains two contributions: one related to the work done per unit volume and unit time by the external forces, the other to the work per unit volume and unit time associated with the stress tensor.

Similarly, the source of the potential energy Ω_γ corresponds to the work done per unit volume and unit time by the force $\mathbf{F}_\gamma$ acting on component γ, along the motion $\mathbf{v}_\gamma$, that is:

$$\sigma[\Omega_\gamma] = -\rho_\gamma F_{\gamma i} v_{\gamma i} \tag{1.38}$$

The flow term corresponds to the convection of the potential energy along the same motion. Therefore, denoting by ω_γ the potential energy per unit mass of component γ, the balance equation for Ω_γ may be written as:

$$-\rho_\gamma F_{\gamma i} v_{\gamma i} = \partial_t \rho_\gamma \omega_\gamma + [\rho_\gamma \omega_\gamma v_{\gamma i}]_{,i} \tag{1.39}$$

In writing (1.39) we have assumed that the potential energy is a property of the matter moving along with it. To take into account other forms of energy exchange, as e.g. the energy of radiation, it would be necessary to introduce an additional conduction flow (the Poynting vector of the electromagnetic field), into the balance equation (1.39). In what follows we shall not discuss problems involving such supplementary terms.

Let us now establish the balance equation of internal energy. The source of internal energy has to be chosen in such a way as to satisfy the conservation principle for the total energy (1.13). This leads to

$$\begin{aligned} \sigma[E] &= -\sigma[E_{kin}] - \sum_\gamma \sigma[\Omega_\gamma] \\ &= \sum_\gamma F_{\gamma i} \rho_\gamma \Delta_{\gamma i} - P_{ij} v_{i,j} \end{aligned} \tag{1.40}$$

The work done by diffusion currents against external forces is therefore transformed into internal energy.

Again the flow of internal energy may be split into a convection flow $\rho e \mathbf{v}$ and a conduction flow. This conduction flow is by definition the heat flow $\mathbf{W}$. Therefore:

$$\mathbf{j}[E] = \rho e \mathbf{v} + \mathbf{W} \tag{1.41}$$

where e denotes the energy density per unit mass.

This definition is in agreement with the microscopic definition of the heat flow. Other equivalent forms of $\mathbf{j}[E]$ may be introduced,

(Prigogine, 1947, de Groot and Mazur, 1962) but we shall not use them in this monograph.

Combining (1.40) and (1.41), we obtain the balance equation for internal energy in the form:

$$\sum_\gamma F_{\gamma i}\rho_\gamma \Delta_{\gamma i} - P_{ij}\mathrm{v}_{i},_j = \partial_t(\rho e) + [W_j + \rho e \mathrm{v}_j],_j \tag{1.42}$$

In many problems the external force per unit mass is the same for all components γ. This is e.g. the case in an external gravitational field where

$$\mathbf{F}_\gamma = \mathbf{F} = \mathbf{g} \tag{1.43}$$

The first contribution to (1.42) then vanishes as a consequence of (1.22) and the balance equation (1.42) takes the same form as for a simple component. If moreover P_{ij} is replaced by the equilibrium pressure p (1.33) one obtains:

$$-p\mathrm{v}_j,_j = \partial_t(\rho e) + [W_j + \rho e \mathrm{v}_j],_j \tag{1.44}$$

or using (1.16) and (1.17)

$$-p\,\frac{\mathrm{d}v}{\mathrm{d}t} = \frac{\mathrm{d}e}{\mathrm{d}t} + \mathrm{div}\,\mathbf{W} \tag{1.45}$$

where

$$v = \rho^{-1}$$

Assuming now that the pressure is uniform, we get after an integration over the whole system:

$$\mathrm{d}Q = \mathrm{d}E + p\,\mathrm{d}V \tag{1.46}$$

where $\mathrm{d}Q$ denotes the heat *received* by the system during the time $\mathrm{d}t$. This relation corresponds to the usual formulation of the so called *first law of thermodynamics* given in all elementary textbooks.

Let us finally consider the balance equation for the total energy U. Taking into account (1.36), (1.39) and (1.42) we obtain:

$$0 = \partial_t[\tfrac{1}{2}\rho \mathrm{v}^2 + \rho e + \sum_\gamma \rho_\gamma \omega_\gamma] \\ + [W_j + P_{ij}\mathrm{v}_i + (\tfrac{1}{2}\mathrm{v}^2 + e)\rho \mathrm{v}_j + \sum_\gamma \rho_\gamma \mathrm{v}_{\gamma j}\omega_\gamma],_j \tag{1.47}$$

This relation expresses the conservation of total energy in local form. Again, when the forces per unit mass are the same for all components γ, e.g. in (1.43), the balance equation (1.47) takes the simpler form:

$$0 = \partial_t[\rho(\tfrac{1}{2}\mathrm{v}^2 + e + \omega)] + [W_j + P_{ij}\mathrm{v}_i + (\tfrac{1}{2}\mathrm{v}^2 + e + \omega)\rho\mathrm{v}_j]_{,j} \quad (1.48)$$

as for a single component.

The balance equations for mass, momentum and energy we have derived in this chapter, will be used repeatedly in this monograph. In Chapter II, we will use these equations in order to derive an explicit expression of the entropy flow and entropy production, appearing in the entropy balance equation. In Chapter VII, they will serve in the discussion of the stability conditions of equilibrium and non equilibrium processes. Finally, in Chapter IX and X, we will use them to formulate the evolution criterion and to introduce the concept of local potential.

CHAPTER II

The Second Law of Thermodynamics and the Entropy Balance Equation

1. THE SECOND LAW OF THERMODYNAMICS

Let us first introduce the phenomenological formulation of the second law. We postulate the existence of a state function S, the entropy, which has the following properties:

(*a*) The entropy is an extensive quantity. If a system consists of several parts the total entropy is equal to the sum of the entropies of each part. The change of entropy $\mathrm{d}S$ can therefore be split into two parts: the entropy production d_iS due to changes inside the system, and the flow of entropy d_eS due to the interaction with the outside world. We have in agreement with the general relation (1.7)

$$\mathrm{d}S = \mathrm{d}_iS + \mathrm{d}_eS \tag{2.1}$$

(*b*) The entropy production d_iS due to changes inside the system is never negative

$$\mathrm{d}_iS \geqslant 0 \tag{2.2}$$

The primary aim of the entropy concept is to introduce a clear distinction between two types of processes: reversible processes (or changes) and irreversible processes. The entropy production d_iS vanishes when the system undergoes only reversible changes, and is always positive if the system is subject to irreversible processes. Hence:

$$\mathrm{d}_iS = 0 \quad \text{(reversible processes)} \tag{2.3}$$

$$\mathrm{d}_iS > 0 \quad \text{(irreversible processes)} \tag{2.4}$$

In this chapter, we give explicit expressions for the entropy flow $\mathrm{d}_e S$ and the entropy production $\mathrm{d}_i S$. We also derive the basic thermodynamic relations linked to the Gibbs' formula. Most of these relations are well known and can be found in many textbooks (e.g. Prigogine and Defay, 1954, Guggenheim, 1949). They are included in this chapter in order to produce a self-contained text. We study in more detail the properties of the second order differential expressions of entropy, due to their importance in the general stability theory developed in the following chapters.

For an isolated system, the entropy flow is by definition equal to zero and equations (2.1), (2.2), reduce to the classical formulation of the second law

$$\mathrm{d}S \geqslant 0 \quad \text{(isolated system)} \tag{2.5}$$

This inequality states that the entropy of an *isolated* system can never decrease. In agreement with the definition (1.4), we write for the entropy production per unit time

$$\frac{\mathrm{d}_i S}{\mathrm{d}t} = P[S] = \int \sigma[S]\,\mathrm{d}V \geqslant 0 \tag{2.6}$$

where $\sigma[S]$ is the entropy source, that is the entropy production per unit time and volume. As this inequality has to remain valid for an arbitrary macroscopic volume V, we deduce from (2.6) the inequality

$$\sigma[S] \geqslant 0 \tag{2.7}$$

Likewise, we have for the entropy flow, according to (1.5)

$$\frac{\mathrm{d}_e S}{\mathrm{d}t} = \Phi[S] = \int \mathscr{S}_n\,\mathrm{d}\Omega \tag{2.8}$$

where $\mathscr{S}_n$ is the component of the entropy current along the interior normal n to the boundary surface Ω.

So far, the system of interest may be *open* or *closed*. A *closed system* may exchange energy but not matter with the outside world. A particularly simple case is that of a closed system at uniform temperature T (T is the so-called absolute temperature). Classical thermodynamics is based on the definition

$$\mathrm{d}S = \frac{\mathrm{d}Q}{T} \quad \text{(reversible process)}$$

In our notation this relationship takes the equivalent form

$$\mathrm{d}_e S = \frac{\mathrm{d}Q}{T} \tag{2.9}$$

The entropy flow is then simply given by the heat flow divided by T.

If no irreversible processes occur inside this system we deduce from (2.1), and (2.3):

$$\mathrm{d}S = \mathrm{d}_e S = \frac{\mathrm{d}Q}{T} \tag{2.10}$$

On the contrary, in the presence of irreversible processes, such as for example chemical reactions inside the system, the entropy production no longer vanishes and we recover, according to (2.1) and (2.9) the classical Carnot–Clausius inequality:

$$\mathrm{d}_i S = \mathrm{d}S - \frac{\mathrm{d}Q}{T} \geqslant 0 \tag{2.11}$$

Let us now consider more general situations.

2. LOCAL EQUILIBRIUM

In general, the evaluation of entropy production and entropy flow can only be performed using the methods of non-equilibrium statistical mechanics or of kinetic theory of gases. Even the problem of defining entropy outside the state of equilibrium is clearly a subject which goes beyond macroscopic thermodynamics. We shall however consider throughout this monograph situations for which a macroscopic evaluation of entropy production and entropy flow still remains possible. This will be the case when there exists within each small mass element of the medium a state of *local equilibrium* for which the local entropy s is the same function of the local macroscopic variables as at equilibrium state. This assumption of local equilibrium is not in contradiction with the fact that the system as a whole is out of equilibrium. As a simple example, expansion of a gas in a pipe corresponds to a non-homogeneous state, as well as to a non-equilibrium process. However, at each point the relation between temperature, pressure and density is still expressed by the same law, e.g. $pv = RT$, as for equilibrium. Likewise, entropy remains defined by Gibbs' formula. In other words, the local state is completely described by an equation of state independent of the gradients. Such situations occur in many problems of macroscopic physics. Situations, sometimes considered in rheology, where the equation of state involves the gradients, are beyond the scope of this theory.

Let us emphasize from the outset, that the local equilibrium assumption implies that dissipative processes are sufficiently dominant to exclude large deviations from statistical equilibrium. In many situations, the mechanism of dissipation which permits the system to reach thermodynamical equilibrium may be described simply in terms of 'collision' processes. Therefore we need enough collisions to compensate the effect of imposed gradients or of chemical affinities. This is likely to be the situation in moderately dense gases and *a fortiori* in liquids and solids. Ideal fluids are also included (Chapter XIII).

As a result of such considerations we may apply the local equilibrium assumption to transport processes described by linear laws, such as for example Fourier's law for heat conduction, as well as to not too-fast chemical reactions.

On the contrary, the assumption of local equilibrium will certainly fail in highly rarefied gases when collisions become too infrequent, or for interacting fermions at very low temperature where dissipative processes will become ineffective. A continuum hydrodynamic description is then impossible.

Such situations are explicitly excluded from our considerations. It is possible that some generalization of thermodynamics beyond the assumption of local equilibrium may prove in the future to be of interest (e.g. Truesdell, 1965). However, in this monograph we shall only explore the consequences of the simple local equilibrium assumption (cf. Nicolis, Velarde and Wallenborn, 1969).

The thermodynamical variables describing the state of a small element of a system may be chosen to be the energy density e per unit mass, the specific volume $v = \rho^{-1}$ (or the pressure p) and the mass fractions N_γ defined in (1.18).

The local equilibrium assumption is then expressed by the equation:

$$s = s(e, v, N_\gamma) \tag{2.12}$$

together with the relations familiar from equilibrium theory:

$$\left(\frac{\partial s}{\partial e}\right)_{v, N_\gamma} = T^{-1}; \quad \left(\frac{\partial s}{\partial v}\right)_{e, N_\gamma} = pT^{-1}; \quad \left(\frac{\partial s}{\partial N_\gamma}\right)_{e, v, (N_\gamma)} = -\mu_\gamma T^{-1} \tag{2.13}$$

The variables T, p as well as the chemical potentials μ_γ (per unit mass) are taken at local equilibrium and have therefore the same meaning as in equilibrium thermodynamics. In the last relation (2.13) the index (N_γ) means that all mass fractions except N_γ are maintained constant. The condition (1.18), $\sum_\gamma N_\gamma = 1$, between the mass fractions is considered as a supplementary condition in the thermodynamical variable space e, v, N_γ. Relations (2.12) and (2.13) give rise to the Gibbs relation for the total differential δs of the entropy s per unit mass:

$$T\delta s = \delta e + p\delta v - \sum_\gamma \mu_\gamma \delta N_\gamma \tag{2.14}$$

This is the basic relation we shall use to obtain the entropy balance equation.

3. ENTROPY BALANCE EQUATION

Let us start from the Gibbs formula (2.14) and introduce the balance equations for mass and internal energy established in Chapter I (see also de Groot and Mazur, 1962, Prigogine, 1947, as well as other textbooks on non-equilibrium thermodynamics). We first write (2.14) along the centre of mass motion as:

$$T\frac{\mathrm{d}s}{\mathrm{d}t} = \frac{\mathrm{d}e}{\mathrm{d}t} + p\frac{\mathrm{d}v}{\mathrm{d}t} - \sum_\gamma \mu_\gamma \frac{\mathrm{d}N_\gamma}{\mathrm{d}t} \tag{2.15}$$

We then multiply the two sides of this equation by the density ρ and use (1.17) and (1.19). We obtain in this way:

$$\begin{aligned} T\partial_t(\rho s) + T(\rho s \mathrm{v}_j)_{,j} &= \partial_t(\rho e) + (\rho e \mathrm{v}_j)_{,j} \\ &\quad + p\mathrm{v}_{j,j} - \sum_\gamma \mu_\gamma[\partial_t \rho_\gamma + (\rho_\gamma \mathrm{v}_j)_{,j}] \end{aligned} \tag{2.16}$$

The first three terms on the right hand side may be replaced by the explicit form deduced from the balance equation for internal energy (1.42), as well as from relation (1.33). Also the explicit form of the last term is given by the mass balance equation (1.28). We get in this way

$$\begin{aligned} \partial_t(\rho s) = &-(\rho s \mathrm{v}_j)_{,j} - W_{j,j}T^{-1} - p_{ij}T^{-1}\mathrm{v}_{i,j} \\ &+ \sum_\gamma (\rho_\gamma \Delta_{\gamma j})(T^{-1}F_{\gamma j}) + \sum (\rho_\gamma \Delta_{\gamma j})_{,j}(\mu_\gamma T^{-1}) \\ &+ \sum_\rho w_\rho A_\rho T^{-1} \end{aligned} \tag{2.17}$$

We have introduced the chemical affinity A_ρ of reaction ρ which is defined in terms of the chemical potentials through (e.g. Prigogine and Defay, 1954, p. 69):

$$A_\rho = -\sum_\gamma \nu_{\gamma\rho} M_\gamma \mu_\gamma \qquad (\rho = 1, \ldots, r) \tag{2.18}$$

Let us recall that the symbol μ_γ denotes here the chemical potential per unit mass.†

Applying the rules for the derivation of a product to the second and the fifth terms on the right hand side of (2.17) we obtain the entropy balance equation

$$\begin{aligned} W_j T^{-1}_{,j} - \sum_\gamma (\rho_\gamma \Delta_{\gamma j})[(\mu_\gamma T^{-1})_{,j} - T^{-1} F_{\gamma j}] \\ - T^{-1} p_{ij} \mathrm{v}_{i,j} + \sum_\rho w_\rho A_\rho T^{-1} \\ = \partial_t(\rho s) + [W_j T^{-1} - \sum_\gamma (\rho_\gamma \Delta_{\gamma j})(\mu_\gamma T^{-1}) + \rho s \mathrm{v}_j]_{,j} \end{aligned} \tag{2.19}$$

This equality has the form (1.9):

$$\sigma[S] = \partial_t(\rho s) + \operatorname{div} \mathscr{S} \geqslant 0 \tag{2.20}$$

The entropy production is

$$\begin{aligned} \sigma[S] = W_j T^{-1}_{,j} - \sum_\gamma (\rho_\gamma \Delta_{\gamma j})[(\mu_\gamma T^{-1})_{,j} - T^{-1} F_{\gamma j}] \\ - p_{ij} T^{-1} \mathrm{v}_{i,j} + \sum_\rho w_\rho A_\rho T^{-1} \geqslant 0 \end{aligned} \tag{2.21}$$

Also the entropy flow is given by

$$\mathscr{S} = \mathbf{W} T^{-1} - \sum_\gamma (\rho_\gamma \mathbf{\Delta}_\gamma)(\mu_\gamma T^{-1}) + \rho s \mathbf{v} \tag{2.22}$$

The separation between source and flow has been performed in such a way that the entropy production vanishes when the system is in thermodynamical equilibrium. The entropy flow (2.22) contains the convection term $\rho s \mathbf{v}$ as well as *conduction* terms related to transport of heat and matter through diffusion. It should be noticed that the separation between flow and production terms is not unique. A number of alternative formulations using different definitions of a heat flow, are possible (for more detail see de Groot and Mazur, 1962).

Let us also notice the structure of the entropy source. It is formed by the sum of contributions corresponding to the transport of heat,

† If we use a chemical potential per mole (or per molecule), (2.18) takes the simpler form:

$$A_\rho = -\sum_\gamma \nu_{\gamma\rho} \mu_\gamma$$

matter and momentum as well as by a contribution due to chemical reactions. Each of these contributions is a bilinear form which contains two types of factors: a flow or a rate of an irreversible process (from this point of view the pressure tensor p_{ij} is also a flow corresponding to transport of momentum, see Chapter I, §3), and a gradient or a chemical affinity. The latter are considered as 'generalized' thermodynamic forces denoted by X_α. In the case of chemical reactions, X_α will be, by definition, equal to $A_\rho T^{-1}$. Let us also use the notation J_α for the flows or rates. The local entropy production (2.21) takes then the remarkable form:

$$\sigma[S] = \sum_\alpha J_\alpha X_\alpha \geqslant 0 \tag{2.23}$$

Through integration we obtain for the entropy production of the whole system

$$P[S] = \int \sum_\alpha J_\alpha X_\alpha \, \mathrm{d}V \geqslant 0 \tag{2.24}$$

By its very meaning, the entropy production describes the dissipative irreversible processes inside the system. At equilibrium both the flows and the forces vanish:

$$J_\alpha = 0 \quad \text{and} \quad X_\alpha = 0 \tag{2.25}$$

As a result, near equilibrium, the entropy production becomes a second order quantity in the deviations from equilibrium.

The entropy production is both a *thermodynamic* quantity through the thermodynamic forces X_α and a *kinetic* quantity through the flows J_α.

It is clear that alternative expressions for the entropy production may be derived. Different sets of generalized flows J'_α and generalized forces X'_α may be introduced. However, no physical consequence of the theory can depend on such transformations and the entropy production has to remain invariant:

$$\sigma[S] = \sum_\alpha J_\alpha X_\alpha = \sum_\alpha J'_\alpha X'_\alpha \tag{2.26}$$

It should be noticed that the invariance of the entropy production may not be sufficient to ensure the *equivalent* character of the alternative descriptions $(J_\alpha X_\alpha)$ and $(J'_\alpha X'_\alpha)$. Additional conditions may be necessary (e.g. Prigogine, 1967, Chapter IV, also this monograph Chapter IX, §7). This restriction must be kept in mind to avoid misinterpretations.

The diffusion current $\mathbf{\Delta}_\gamma$ and the stress tensor p_{ij} are defined with respect to the barycentric motion. However, for problems in which the stress tensor does not enter, it is often more convenient to use another reference velocity for diffusion instead of (1.20) (for example, the mean molar velocity $c\mathbf{v} = \sum_\gamma c_\gamma \mathbf{v}_\gamma$ where c_γ is the molar concentration of component γ). These questions are studied elsewhere (de Groot and Mazur, 1962, Chapter V, §2; Chapter XI, §2).

Note also that for closed systems the entropy flow corresponding to the two last terms in (2.22) vanish on the boundary surface Ω. We then recover the classical expression for the entropy flow due to heat transport (see 2.9).

The entropy production is one of the two basic quantities which will play a fundamental role in our approach. The other is the second differential of the entropy we shall study starting from §5 of this chapter.

4. BASIC THERMODYNAMIC RELATIONS

For convenience, we collect in this paragraph a few classical relations we need later on.

Mass, energy, and entropy are extensive variables which may therefore be written in accordance with (1.1) in the form of volume integrals

$$M = \int \rho \, \mathrm{d}V$$
$$E = \int \rho e \, \mathrm{d}V$$
$$S = \int \rho s \, \mathrm{d}V \tag{2.27}$$

Similar formulae hold for enthalpy

$$H = \int \rho h \, \mathrm{d}V$$
$$h = e + pv \tag{2.28}$$

the Helmholtz free energy

$$F = \int \rho f \, \mathrm{d}V$$
$$f = e - Ts \tag{2.29}$$

and the Gibbs free energy

$$G = \int \rho g \, \mathrm{d}V$$
$$g = h - Ts = \sum_\gamma N_\gamma \mu_\gamma \tag{2.30}$$

Using the four thermodynamical potentials E, H, F, G, the basic Gibbs formula (2.14) may be written in one of the following equivalent forms

$$\begin{aligned}
\delta e &= T\delta s - p\delta v + \sum_\gamma \mu_\gamma \delta N_\gamma \\
\delta h &= T\delta s + v\delta p + \sum_\gamma \mu_\gamma \delta N_\gamma \\
\delta f &= -s\delta T - p\delta v + \sum_\gamma \mu_\gamma \delta N_\gamma \\
\delta g &= -s\delta T + v\delta p + \sum_\gamma \mu_\gamma \delta N_\gamma
\end{aligned} \tag{2.31}$$

The left hand side of (2.31) are exact differentials. This implies identities such as

$$\left(\frac{\partial g}{\partial T}\right)_{p,N_\gamma} = -s; \quad \left(\frac{\partial g}{\partial p}\right)_{T,N_\gamma} = v; \quad \left(\frac{\partial g}{\partial N_\gamma}\right)_{T,p,(N_\gamma)} = \mu_\gamma \tag{2.32}$$

as well as

$$\left(\frac{\partial v}{\partial T}\right)_{p,N_\gamma} = -\left(\frac{\partial s}{\partial p}\right)_{T,N_\gamma}; \quad \mu_{\gamma\gamma'} = \mu_{\gamma'\gamma} = \left(\frac{\partial \mu_\gamma}{\partial N_{\gamma'}}\right)_{T,p,(N_\gamma)} \tag{2.33}$$

$$\left(\frac{\partial \mu_\gamma}{\partial T}\right)_{p,N_\gamma} = -\left(\frac{\partial s}{\partial N_\gamma}\right)_{T,p,(N_\gamma)} = -s_\gamma \tag{2.34}$$

$$\left(\frac{\partial \mu_\gamma}{\partial p}\right)_{T,N_\gamma} = \left(\frac{\partial v}{\partial N_\gamma}\right)_{T,p,(N_\gamma)} = v_\gamma \tag{2.35}$$

Here v_γ and s_γ are the so-called specific partial volume and the specific partial entropy of component γ (i.e. per unit mass of γ). Many other identities can be derived in a similar way from (2.31). As an example, we have

$$\begin{aligned}
\mu_\gamma &= \left(\frac{\partial e}{\partial N_\gamma}\right)_{s,v,(N_\gamma)} = \left(\frac{\partial h}{\partial N_\gamma}\right)_{s,p,(N_\gamma)} \\
&= \left(\frac{\partial f}{\partial N_\gamma}\right)_{T,v,(N_\gamma)} = \left(\frac{\partial g}{\partial N_\gamma}\right)_{T,p,(N_\gamma)}
\end{aligned} \tag{2.36}$$

The most commonly used independent variables are either T, v, N_γ or T, p, N_γ. Introducing these variables into the expressions (2.31) of δe and δh we obtain the equalities:

$$c_v = \left(\frac{\partial e}{\partial T}\right)_{v,N_\gamma} = T\left(\frac{\partial s}{\partial T}\right)_{v,N_\gamma}$$

$$c_p = \left(\frac{\partial h}{\partial T}\right)_{p,N_\gamma} = T\left(\frac{\partial s}{\partial T}\right)_{p,N_\gamma} \tag{2.37}$$

These relations give us the specific heat c_v, c_p at constant volume or pressure in terms of thermodynamic potentials. We obtain also the Kelvin relations

$$\left(\frac{\partial e}{\partial v}\right)_{T,N_\gamma} = T\left(\frac{\partial p}{\partial T}\right)_{v,N_\gamma} - p$$

$$\left(\frac{\partial h}{\partial p}\right)_{T,N_\gamma} = v - T\left(\frac{\partial v}{\partial T}\right)_{p,N_\gamma} \tag{2.38}$$

Using the Kelvin relations and formula (2.37) for the specific heat, we obtain the difference between the specific heats at constant pressure and constant volume in terms of the isothermal compressibility and the thermal dilatation coefficient at constant pressure.

$$c_p - c_v = \frac{T}{\chi v}\left(\frac{\partial v}{\partial T}\right)^2_{p,N_\gamma}$$

$$\chi = -\frac{1}{v}\left(\frac{\partial v}{\partial p}\right)_{T,N_\gamma} \tag{2.39}$$

The second relation (2.30) combined with (2.34) gives us

$$h_\gamma \equiv \left(\frac{\partial h}{\partial N_\gamma}\right)_{T,p,(N_\gamma)} = Ts_\gamma + \mu_\gamma$$

$$= -T\left(\frac{\partial \mu_\gamma}{\partial T}\right)_{p,N_\gamma} + \mu_\gamma \tag{2.40}$$

or alternatively

$$\left[\frac{\partial(\mu_\gamma T^{-1})}{\partial T}\right]_{p,N_\gamma} = -h_\gamma T^{-2} \tag{2.41}$$

This important Gibbs–Helmholtz equality relates the specific partial enthalpy to the temperature derivative of the chemical potential.

Let us now go over to another group of relations which result from the fact that the extensive variables E, S, H, F, and G considered as functions of pressure, temperature, and masses, are homogeneous of the first degree in the masses, $m_1, m_2, \ldots, m_c$. Indeed, we have for the energy

$$E(T, p, km_1, km_2, \ldots, km_c) = kE(T, p, m_1, \ldots, m_c) \quad (2.42)$$

The application of Euler's theorem on homogeneous functions gives us† (for more details see Prigogine and Defay, 1954, p. 3):

$$E = \sum_\gamma m_\gamma \left(\frac{\partial E}{\partial m_\gamma}\right)_{T,p,(m_\gamma)} = \sum_\gamma m_\gamma e_\gamma \quad (2.43)$$

and therefore also

$$e = \sum_\gamma N_\gamma e_\gamma \quad (2.44)$$

Taking relation (1.18) as a separate condition, we may also write

$$\left(\frac{\partial E}{\partial m_\gamma}\right)_{T,p,(m_\gamma)} = \left(\frac{\partial e}{\partial N_\gamma}\right)_{T,p,(N_\gamma)}$$

Likewise, we have

$$h = \sum_\gamma N_\gamma h_\gamma$$

$$s = \sum_\gamma N_\gamma s_\gamma$$

$$g = \sum_\gamma N_\gamma \mu_\gamma \quad (2.45)$$

The last relation provides a justification for (2.30). Combined with the expression (2.31) of δg we obtain the Gibbs–Duhem formula

$$s\delta T - v\delta p + \sum_\gamma N_\gamma \delta\mu_\gamma = 0 \quad (2.46)$$

This equality relates the variations of the intensive variables T, p, μ_γ. Using (1.19) and (2.30) we may write the Gibbs–Duhem formula in the two equivalent forms

$$\rho h \delta T^{-1} + T^{-1}\delta p - \sum_\gamma \rho_\gamma \delta(\mu_\gamma T^{-1}) = 0$$

$$\rho e \delta T^{-1} + \delta(pT^{-1}) - \sum_\gamma \rho_\gamma \delta(\mu_\gamma T^{-1}) = 0 \quad (2.47)$$

† If $f(kx, ky, \ldots) = k^m f(x, y, \ldots)$ we have

$$\frac{\partial f}{\partial x} x + \frac{\partial f}{\partial y} y + \ldots = mf(x, y, \ldots).$$

This is Euler's theorem, and the function f is called homogeneous of the mth degree.

Let us now consider the relations which result from the fact that the chemical potential μ_γ considered as a function of pressure, temperature and masses is a homogeneous function of order zero in respect to the masses. We have indeed

$$\mu_\gamma(p, T, km_1, km_2, \ldots, km_c) = \mu_\gamma(p, T, m_1, \ldots, m_c) \quad (2.48)$$

Application of Euler's theorem (footnote after (2.42)) gives us the c equations

$$\sum_{\gamma'} N_{\gamma'} \left(\frac{\partial \mu_\gamma}{\partial N_{\gamma'}}\right)_{T, p, (N_{\gamma'})} = \sum_{\gamma'} N_{\gamma'} \mu_{\gamma\gamma'} = 0 \quad (2.49)$$

$$(\gamma = 1, 2, \ldots, c)$$

Using the second relation (2.33), we may also write (2.49) in the form:

$$\sum_{\gamma} N_\gamma \mu_{\gamma\gamma'} = 0 \quad (2.50)$$

The two last equations can be considered as special cases of the Gibbs–Duhem equation because they may be deduced from 2.46, keeping temperature and pressure constant.

This completes our survey of the basic thermodynamic relations we shall use in the subsequent chapters.

5. SECOND ORDER DIFFERENTIAL OF ENTROPY

We shall now derive a set of relations for the second order quantity $\delta^2 s$. These relations may be obtained straightforwardly using the well known equalities given in the preceding paragraph. We shall however go into more detail as these relations are of basic importance for the stability theory as developed in Chapter IV–VII.

Let us first calculate $\delta^2 s$ for a one component system, using e and v as independent variables. This quantity corresponds to twice the quadratic terms in the Taylor expansion of the increment Δs that is

$$\delta^2 s = \frac{\partial^2 s}{\partial e^2} (\Delta e)^2 + 2 \frac{\partial^2 s}{\partial e \partial v} \Delta e \Delta v + \frac{\partial^2 s}{\partial v^2} (\Delta v)^2 \quad (2.51)$$

Using the equalities (2.13), (2.51) becomes:

$$\delta^2 s = \left(\frac{\partial T^{-1}}{\partial e} \Delta e + \frac{\partial T^{-1}}{\partial v} \Delta v\right) \Delta e + \left(\frac{\partial (pT^{-1})}{\partial e} \Delta e + \frac{\partial (pT^{-1})}{\partial v} \Delta v\right) \Delta v \quad (2.52)$$

Hence:

$$\delta^2 s = \delta T^{-1} \Delta e + \delta(pT^{-1}) \Delta v \tag{2.53}$$

or:

$$\delta^2 s = \delta T^{-1} \delta e + \delta(pT^{-1}) \delta v \tag{2.54}$$

Indeed one has for the *independent* variables:†

$$\Delta e = \delta e; \; \Delta v = \delta v; \quad \delta^2 e = \delta^2 v = 0 \tag{2.55}$$

Accordingly, equality (2.54) could have been directly deduced from the Gibbs formula (2.14), by simple differentiation. For a multi-component system, we get in this way:

$$\delta^2 s = \delta T^{-1} \delta e + \delta(pT^{-1}) \delta v - \sum_\gamma \delta(\mu_\gamma T^{-1}) \delta N_\gamma \tag{2.56}$$

in variables e, v, N_γ. We have given a more detailed treatment in order to underline the importance of the choice adopted for the independent variables, when one proceeds with second order differentials. Indeed, for variables other than $\{e, v, N_\gamma\}$ the r.h.s. of (2.56) no longer represents $\delta^2 s$, as the differentiation of (2.14), then introduces additional quantites due to the non-vanishing values of $\delta^2 e$, $\delta^2 v$, $\delta^2 N_\gamma$. For instance, the second time derivative $\partial_t^2 s$ is by no means given by replacing simply the increments δ *by* ∂_t in (2.56).

It is clear that such a precaution is not necessary when one deals with a first order differential as it is invariant in respect to the choice of the independent variables. For instance, in the Gibbs formula (2.14), e, v, N_γ may be interpreted either as independent or dependent variables. On the contrary, in the second order relations, the transformations performed are strictly limited to the variables e, v, N_γ.

Using again Gibbs formula, equality (2.56) can also be written as

$$T\delta^2 s = -\delta T \delta s + \delta p \delta v - \sum_\gamma \delta \mu_\gamma \delta N_\gamma \tag{2.57}$$

Furthermore, let us now express the variation of the chemical potentials $\delta\mu_\gamma$ in the variables, T, p, N_γ. We obtain (see (2.34), (2.35)):

$$\begin{aligned} T\delta^2 s = &-\delta T[\delta s - \sum_\gamma s_\gamma \delta N_\gamma] \\ &+ \delta p[\delta v - \sum_\gamma v_\gamma \delta N_\gamma] \\ &- \sum_{\gamma\gamma'} \mu_{\gamma\gamma'} \delta N_\gamma \delta N_{\gamma'} \end{aligned}$$

† see e.g. de La Vallée Poussin (1926, Chap. I).

This formula can obviously be written as

$$T\delta^2 s = -\delta T\left[\left(\frac{\partial s}{\partial T}\right)_{p,N_\gamma}\delta T + \left(\frac{\partial s}{\partial p}\right)_{T,N_\gamma}\delta p\right]$$
$$+ \delta p\left[\left(\frac{\partial v}{\partial T}\right)_{p,N_\gamma}\delta T + \left(\frac{\partial v}{\partial p}\right)_{T,N_\gamma}\delta p\right]$$
$$- \sum_{\gamma\gamma'}\mu_{\gamma\gamma'}\delta N_\gamma \delta N_{\gamma'}$$

Using (2.33), (2.37) and (2.39) we then obtain the characteristic quadratic form:

$$\delta^2 s = -\frac{1}{T}\left[\frac{c_v}{T}(\delta T)^2 + \frac{\rho}{\chi}(\delta v)^2_{N_\gamma} + \sum_{\gamma\gamma'}\mu_{\gamma\gamma'}\delta N_\gamma \delta N_{\gamma'}\right] \quad (2.58)$$

By definition

$$(\delta v)_{N_\gamma} = \left(\frac{\partial v}{\partial T}\right)_{p,N_\gamma}\delta T + \left(\frac{\partial v}{\partial p}\right)_{T,N_\gamma}\delta p$$

One realizes immediately the importance of (2.58) in all questions related to stability of equilibrium: for an isolated system entropy has to be maximum at equilibrium. As a result the first order differential of entropy has to vanish and the second order differential, that is (2.58) has to be *negative*. We shall come back to these questions in Chapter IV. By an identical calculation (in which the operaor δ is replaced by ∂_t), we also obtain the important equality:

$$\partial_t T^{-1}\partial_t e + \partial_t(pT^{-1})\partial_t v - \sum_\gamma \partial_t(\mu_\gamma T^{-1})\partial_t N_\gamma$$
$$= -\frac{1}{T}\left[\frac{c_v}{T}(\partial_t T)^2 + \frac{\rho}{\chi}(\partial_t v)^2_{N_\gamma} + \sum_{\gamma\gamma'}\mu_{\gamma\gamma'}\partial_t N_\gamma \partial_t N_{\gamma'}\right] \quad (2.59)$$

To derive this relation, it is sufficient to compare the right hand sides of (2.56) and (2.58). As already underlined (2.59) is *not* equal to the second time derivative of the specific entropy.

For further applications, it will be often helpful to choose as independent variables, the $c + 1$ quantities ρe, $\rho N_\gamma (\rho v = 1)$ instead of the $c + 2$ quantities e, v, N_γ. The first group of variables represents intensive quantities taken per unit volume while the second group corresponds to variables taken per unit mass. Densities per unit volume appear directly in the definitions (2.27)–(2.30) of the extensive quantities. In the new variables, the Gibbs formula (2.14) may be written as

$$T\delta(\rho s) = \delta(\rho e) - \sum_{\gamma}\mu_{\gamma}\delta\rho_{\gamma} \tag{2.60}$$

Therefore, we now obtain instead of (2.56):

$$\delta^2(\rho s) = \delta T^{-1}\delta(\rho e) - \sum_{\gamma}\delta(\mu_{\gamma}T^{-1})\delta\rho_{\gamma} \tag{2.61}$$

Using the Gibbs–Duhem relation (2.47), we see that we have simply

$$v\delta^2(\rho s) = \delta^2 s \tag{2.62}$$

The two sides of this equality are therefore both given by the same fundamental quadratic form (2.58). However, let us again emphasize that in the left hand side of (2.62), the independent variables are ρe, ρ_{γ} while in the right hand side they are e, v, N_{γ}. Likewise, for different choices of the independent variables, the Gibbs formula (2.14) used together with the relation (2.57), leads directly to the following set of equalities:

$$\begin{aligned} \underset{[\rho e,\rho_{\gamma}]}{T\delta^2(\rho s)} &= \underset{[e,v,N_{\gamma}]}{T\rho\delta^2 s} = \underset{[s,v,N_{\gamma}]}{-\rho\delta^2 e} = \underset{[\rho s,\rho_{\gamma}]}{-\delta^2(\rho e)} \\ &= \underset{[\rho s,p,\rho_{\gamma}]}{-\delta^2(\rho h)} = \underset{[s,p,N_{\gamma}]}{-\rho\delta^2 h + 2\rho\delta v\delta p} \end{aligned} \tag{2.63}$$

where the corresponding independent variables are indicated between brackets. Using exactly the same arguments as in the derivation of (2.59), we now obtain:

$$\begin{aligned} &\partial_t T^{-1}\partial_t(\rho e) - \sum_{\gamma}\partial_t(\mu_{\gamma}T^{-1})\partial_t\rho_{\gamma} \\ &\quad = -\frac{\rho}{T}\left[\frac{c_v}{T}(\partial_t T^{-1})^2 + \frac{\rho}{\chi}(\partial_t v)^2_{N_{\gamma}} + \sum_{\gamma\gamma'}\mu_{\gamma\gamma'}\partial_t N_{\gamma}\partial_t N_{\gamma'}\right] \end{aligned} \tag{2.64}$$

On the other hand, the differentiation of the Gibbs–Duhem relation (2.47), used in conjunction with (2.61), leads to the alternative expression of $\delta^2(\rho s)$ in the form:

$$\delta^2(\rho s) = -[\rho e\delta^2 T^{-1} + \delta^2(pT^{-1}) - \sum_{\gamma}\rho_{\gamma}\delta^2(\mu_{\gamma}T^{-1})] \tag{2.65}$$

The following *reciprocity relation* is also useful:

$$\begin{aligned} &\delta_2 T^{-1}\delta_1(\rho e) - \sum_{\gamma}\delta_2(\mu_{\gamma}T^{-1})\delta_1\rho_{\gamma} \\ &\quad = \delta_1 T^{-1}\delta_2(\rho e) - \sum_{\gamma}\delta_1(\mu_{\gamma}T^{-1})\delta_2\rho_{\gamma} \end{aligned} \tag{2.66}$$

Here δ_1 denotes one type of change, as e.g. the increment δ used so far, while δ_2 concerns another type of change, which might be either the local time derivative, or any component of the gradient. Equality (2.66) is easily proved by expanding $\delta_2 T^{-1}$ and $\delta_2(\mu_\gamma T^{-1})$ in terms of $\delta_2(\rho e)$ and $\delta_2 \rho_\gamma$, taking into account the equalities derived from (2.60):

$$\frac{\partial T^{-1}}{\partial \rho_\gamma} = -\frac{\partial(\mu_\gamma T^{-1})}{\partial(\rho e)}; \quad \frac{\partial(\mu_\gamma T^{-1})}{\partial \rho_{\gamma'}} = \frac{\partial(\mu_{\gamma'} T^{-1})}{\partial \rho_\gamma} \tag{2.67}$$

Now choosing more specifically:

$$\delta_1 = \frac{\partial}{\partial x_j} \mathrm{d}x_j; \quad \delta_2 = \delta$$

equality (2.66) becomes:

$$T^{-1}_{,j}\delta(\rho e) - \sum_\gamma (\mu_\gamma T^{-1})_{,j}\, \delta\rho_\gamma = \delta T^{-1}(\rho e)_{,j} - \sum_\gamma \delta(\mu_\gamma T^{-1})\rho_{\gamma,j} \tag{2.68}$$

We multiply the two sides by the component a_j of an arbitrary vector **a**. Using again the Gibbs–Duhem relation (2.47) written as

$$a_{j,j}[\rho e \delta T^{-1} + \delta(pT^{-1}) - \sum_\gamma \rho_\gamma \delta(\mu_\gamma T^{-1})] = 0$$

we then obtain the useful identity:

$$a_j T^{-1}_{,j}\delta(\rho e) - \sum_\gamma a_j(\mu_\gamma T^{-1})_{,j}\delta\rho_\gamma$$
$$= (a_j \rho e)_{,j}\delta T^{-1} + a_{j,j}\delta(pT^{-1}) - \sum_\gamma (a_j \rho_\gamma)_{,j}\delta(\mu_\gamma T^{-1}) \tag{2.69}$$

In the subsequent applications of (2.69), the vector **a** will represent e.g. the barycentric velocity **v** or its increment δ**v**. As already pointed out the second order quantities we have calculated play an essential role in the stability theory of equilibrium and non-equilibrium states.

6. USE OF COMPLEX VARIABLES

In the problems we shall subsequently consider, the basic increments $\delta(\rho e)$ and $\delta\rho_\gamma$ correspond to perturbations varying in the course of time. As a rule, this time dependence is governed by the partial differential equations deduced from the excess balance equations (Chapter I), used together with the phenomenological laws giving the flows (Chapter III), and discussed in Chapter VII.

However, the solution of the perturbation equations is generally obtained in the form of complex quantities. For instance, in the simple case of a single normal mode, around a steady state, the time dependence occurs as a complex quantity exp ωt, where

$$\omega = \omega_r + i\omega_i \tag{2.70}$$

is a complex frequency. The imaginary part is related to oscillations.

As the coefficients of the perturbation equations are real quantities, the complex conjugate of a solution is also a solution, which may be considered simultaneously with the first one.

The corresponding complex values of the increments of the other variables considered in the foregoing paragraphs are then defined by the same relations as for real quantities. For example, the complex value of δs is still defined by Gibbs formula (2.60).

Let us also recall that for a complex increment $\delta\varphi$ of a real variable φ (e.g. e, v, ρ, N_γ . . .) one has:†

$$(\delta\varphi)_r = \tfrac{1}{2}(\delta\varphi + \delta\varphi^*) \tag{2.71}$$

where r denotes the real part. Hence:

$$[\delta(\rho s)]_r = \tfrac{1}{2}[\delta(\rho s) + \delta(\rho s)^*] \tag{2.72}$$

After substitution into equation (2.61) we get:

$$[\delta^2(\rho s)]_r = \tfrac{1}{4}[\delta^2(\rho s) + \delta^2(\rho s)^* + 2\delta^2_m(\rho s)] \tag{2.73}$$

where $\delta^2_m(\rho s)$ denotes the *mixed* second differential:

$$\delta^2_m(\rho s) = \tfrac{1}{2}\{\delta T^{-1}\delta(\rho e)^* + \delta T^{-1*}\delta(\rho e) - \sum_\gamma[\delta(\mu_\gamma T^{-1})\delta\rho_\gamma{}^* + \delta(\mu_\gamma T^{-1})^*\delta\rho_\gamma]\} \tag{2.74}$$

Using equalities (2.71), (2.58) and (2.62), we obtain in the same way for $\delta^2_m(\rho s)$:

$$\delta^2_m(\rho s) = -\frac{\rho}{T}\left[\frac{c_v}{T}\delta T\delta T^* + \frac{\rho}{\chi}(\delta v)_{N_\gamma}(\delta v)^*_{N_\gamma} + \tfrac{1}{2}\sum_\gamma \mu_{\gamma\gamma'}(\delta N_\gamma \delta N^*_{\gamma'} + \delta N_\gamma{}^*\delta N_{\gamma'})\right] \tag{2.75}$$

† To simplify the notation, we write $\delta\varphi^*$ instead of $(\delta\varphi)^*$ to denote the complex conjugate.

This mixed second differential has a number of useful properties. In this respect let us observe that $\delta_m^2(\rho s)$ is a *real quadratic* expression. Also, whenever $[\delta^2(\rho s)]_r$ is a *definite* or a *negative definite* quadratic form, so is $\delta_m^2(\rho s)$. Moreover, $\delta_m^2(\rho s)$ reduces to $[\delta^2(\rho s)]_r$ in the case of real increments.

Expressions (2.74) and (2.75) give us the formation rule of $\delta_m^2(\rho s)$:

$$\delta A \delta B \rightarrow \tfrac{1}{2}(\delta A \delta B^* + \delta B \delta A^*) \tag{2.76}$$

The reader may easily verify this rule for other mixed second differential expressions, as e.g for the time derivative of $\delta_m^2(\rho s)$. Let us also observe that it is sometimes useful to take into account equality (2.66), in the particular form:

$$\begin{aligned} \delta T^{-1*}\delta(\rho e) - \sum_\gamma \delta(\mu_\gamma T^{-1})^* \delta\rho_\gamma \\ = \delta T^{-1}\delta(\rho e)^* - \sum_\gamma \delta(\mu_\gamma T^{-1})\delta\rho_\gamma^* \end{aligned} \tag{2.77}$$

CHAPTER III

Linear Thermodynamics of Irreversible Processes

1. FLOWS AND FORCES

In this chapter we shall give a brief summary of irreversible thermodynamics *close to equilibrium.* As in this region the relations between the flows (or rates, fluxes, currents) J_α and the forces (thermodynamic or generalized forces) X_α may be expected to be linear, this part of non-equilibrium thermodynamics may also be called the *linear* thermodynamics of irreversible processes. We shall go into a minimum of detail as there exist many texts covering this subject (de Groot and Mazur, 1962, Prigogine, 1947).

We return to the general expression (2.23) for the entropy production. At thermodynamic equilibrium, we have simultaneously for *all* irreversible processes

$$J_\alpha = 0; \quad X_\alpha = 0 \tag{3.1}$$

It is therefore quite natural to assume, at least for conditions near equilibrium, linear homogeneous relations between the flows and the forces. Such a scheme automatically includes empirical laws such as Fourier's law for heat flow or Fick's law for diffusion. Linear laws of this kind are called the *phenomenological relations.* They may be written as

$$J_\alpha = \sum_\beta L_{\alpha\beta} X_\beta; \quad \text{where } \alpha, \beta = 1, 2, \ldots, n \tag{3.2}$$

for n flows and n forces. The coefficients $L_{\alpha\beta}$ are called the *phenomenological coefficients.* The diagonal coefficients of the matrix $\|L_{\alpha\beta}\|$ are the *self* or the *proper* coefficients. They may stand e.g. for the heat conductivity, the electrical conductivity, the chemical drag coefficient and so on. The other coefficients $L_{\alpha\beta} (\alpha \neq \beta)$ are the *mutual* coefficients. They describe the *interference* between the irreversible processes α and β.

Clearly, the existence of such phenomenological relations has to be understood as an extra-thermodynamic hypothesis. Later on we shall study many examples where the relations between J_α and X_α are much more involved. However, once the linear relations are adopted, the thermodynamic method yields important information about the coefficient $L_{\alpha\beta}$, without involving any particular kinetic model.

To illustrate this point let us consider the case of two irreversible processes, for which the phenomenological relations (3.2) may be written as

$$\begin{aligned} J_1 &= L_{11}X_1 + L_{12}X_2 \\ J_2 &= L_{21}X_1 + L_{22}X_2 \end{aligned} \tag{3.3}$$

If the two irreversible processes represent thermal conductivity and diffusion, the coefficient L_{12} is connected with thermal diffusion. This means that a concentration gradient appears in an initially homogeneous mixture under the influence of a temperature gradient. Substituting the flows by their values (3.3) in the expression (2.23) for the entropy source we obtain the quadratic form:

$$\sigma[S] = L_{11}X_1^2 + (L_{12} + L_{21})X_1X_2 + L_{22}X_2^2 > 0 \tag{3.4}$$

This quadratic form has to be positive for all positive or negative values of the variables X_1, X_2, except when $X_1 = X_2 = 0$, in which case the entropy production vanishes. As shown in elementary texts on algebra this requirement implies the following inequalities

$$L_{11} > 0; \qquad L_{22} > 0 \tag{3.5}$$

$$(L_{12} + L_{21})^2 > 4L_{11}L_{22} \tag{3.6}$$

Hence the *self* phenomenological coefficients L_{11}, L_{22} are positive. On the other hand, the *mutual* coefficients L_{21}, L_{12} may be positive or negative, their magnitude being limited only by the condition (3.6). This is in agreement with the experimental observations; coefficients, like thermal conductivity or electrical conductivity, are always positive while, for example, the thermal diffusion coefficient has no definite sign.

Let us introduce the notation

$$\begin{aligned} L_{\alpha\beta} &= L_{(\alpha\beta)} + L_{[\alpha\beta]} \\ L_{(\alpha\beta)} &= L_{(\beta\alpha)} \\ L_{[\alpha\beta]} &= -L_{[\beta\alpha]} \end{aligned} \tag{3.7}$$

Therefore the $L_{(\alpha\beta)}$ are the elements of a symmetric matrix and the $L_{[\alpha\beta]}$ those of an antisymmetric matrix.

In the entropy production (3.4), the contribution of the antisymmetric part vanishes and we obtain

$$\sigma[S] = \sum_{\alpha\beta} L_{(\alpha\beta)} X_\alpha X_\beta \geqslant 0 \tag{3.8}$$

Let us now discuss some general properties of the phenomenological coefficients.

2. ONSAGER'S RECIPROCITY RELATIONS

An important theorem due to Onsager (1931) states that

$$L_{\alpha\beta} = L_{\beta\alpha} \qquad (\beta \neq \alpha) \tag{3.9}$$

These Onsager reciprocity relations express the property that when the flow J_α, corresponding to the irreversible processes α, is influenced by the force X_β of the irreversible process β. then the flow J_β is also influenced by the force X_α through the same interference coefficient $L_{\alpha\beta}$. Therefore the antisymmetric part $L_{[\alpha\beta]}$ in (3.7) vanishes. This property supplements the second law of thermodynamics, since as pointed out in (3.8), the entropy production cannot give any information about the elements of the antisymmetric matrix.

Here we do not reproduce the proof of the Onsager theorem, nor shall we discuss its conditions of validity (ref. given above). One of the simplest examples refers to the Fourier law for heat conduction. In this case, the generalized forces are the three components of the thermal gradient $T^{-1}_{,j}$. In agreement with (3.2) the heat flow is given by the linear relations

$$W_i = L_{ij} T^{-1}_{,j} \tag{3.10}$$

In an isotropic medium we have simply

$$\mathbf{W} = L \nabla T^{-1} \tag{3.11}$$

The usual coefficient of thermal conductivity is defined by

$$\lambda = L T^{-2} \tag{3.12}$$

and the corresponding expression of Fourier's law is

$$\mathbf{W} = -\lambda \nabla T \tag{3.13}$$

For an anisotropic medium we would write

$$W_i = -\lambda_{ij} T_{,j} \tag{3.14}$$

The Onsager reciprocity relations 3.9 imply that the thermal conductivity tensor will be symmetric. Hence

$$\lambda_{ij} = \lambda_{ji} \tag{3.15}$$

in agreement with the experimental data (ref. above).

3. SYMMETRY REQUIREMENTS ON COUPLING OF IRREVERSIBLE PROCESSES

As pointed out in §1, the mutual phenomenological coefficients $L_{\alpha\beta}$ ($\alpha \neq \beta$) describe the interference between two irreversible processes α and β. We shall now specify which of the irreversible processes are capable of mutual interference. Let us consider a system in which a heat flow propagates along the geometrical coordinate x and which is subject at the same time to a single chemical reaction. According to (2.21) the entropy source takes the form:

$$\sigma[S] = -\frac{W_x}{T^2}\frac{\partial T}{\partial x} + \frac{A w}{T} > 0 \tag{3.16}$$

and the phenomenological relations (3.2) become:

$$\begin{aligned} W_x &= -\frac{L_h}{T^2}\frac{\partial T}{\partial x} + L_{12}\frac{A}{T} \\ w &= -\frac{L_{21}}{T^2}\frac{\partial T}{\partial x} + L_h\frac{A}{T} \end{aligned} \tag{3.17}$$

A first reduction of the number of phenomenological coefficients is obtained by the Onsager reciprocity relations (3.9), which give $L_{12} = L_{21}$. In addition however, we can show in this case that

$$L_{12} = L_{21} = 0 \tag{3.18}$$

Indeed, suppose that $\dfrac{\partial T}{\partial x} = 0$, equation (3.17) gives

$$W_x = L_{12}\frac{A}{T} \tag{3.19}$$

so that the scalar *cause* A/T would produce the vector *effect* W_x. This would be contrary to the general requirements of symmetry principles. Chemical affinity cannot produce a directed heat flow. Therefore the interference coefficient must necessarily be zero.

In such a case it is not only the total entropy source which is a positive quantity but there are some irreversible processes or groups of irreversible processes which give separate positive contributions to the total entropy source. In (3.16) we have separately

$$-\frac{W_x}{T^2}\frac{\partial T}{\partial x} > 0 \quad \text{and} \quad \frac{Aw}{T} > 0 \tag{3.20}$$

As a rule, owing to the impossibility of mutual interferences, the entropy production may occur as the sum of contributions which are *separately* positive. One group stands for *scalar* processes such as chemical reactions, a second, for *vector* phenomena (diffusion and thermal conduction), and finally a third for *tensor* processes (such as viscous dissipation). We can only have coupling between irreversible processes possessing the same tensor character.

These conditions were first formulated by one of us (I. Prigogine, 1947) as extensions of the Curie symmetry principle. However, as emphasized recently by Katchalsky and coworkers, these symmetry requirements are only valid in an *isotropic* medium. If we are dealing with an anisotropic medium, such as a biological membrane, new possibilities of coupling may appear in the phenomenological laws (Katschalsky, 1968), for which limitations such as (3.18) are no longer valid (e.g. active transports).

Coupling effects may also be generated by the balance equations themselves. Such situations may arise e.g. in many steady-state problems or for states beyond an instability point. A typical example is provided by the Bénard instability, discussed repeatedly in this monograph (Chapter XI). Beyond the critical point a temperature gradient induces convection which contributes to an effective coupling, not included in the phenomenological laws.

4. NON-EQUILIBRIUM STEADY STATES AND THEOREM OF MINIMUM ENTROPY PRODUCTION

In many situations the boundary conditions imposed on the system prevent it from reaching equilibrium. We may for example consider a system composed of two vessels at equilibrium, and connected by means of a capillary or a membrane. A temperature difference is maintained between the two vessels. We have here two forces say X_{th} and X_m corresponding to the difference of temperature and of chemical potential between the two vessels, and two corresponding flows J_{th} and J_m. The system reaches a state in which the transport of matter J_m vanishes, while the transport of energy between the two phases at different temperatures as well as the entropy production remain non-vanishing. The state variables tend asymptotically to time independent values. We have then reached a steady non-equilibrium state or more briefly a steady state. No confusion should arise between such states and equilibrium which is characterized by a zero entropy production. A similar situation occurs in single component systems. In the steady state a so called *thermomolecular pressure* difference is maintained between the two vessels.

Another example of a stationary state is given by a system which receives a component A from the outside environment and transforms it through a certain number of intermediate compounds into a final product F, which returns to the environment. A stationary state arises when the concentrations of the intermediate components no longer vary with time. In this case the conditions for the occurrence of a stationary state are expressed by relations between the reaction rates of the different processes which correspond to the formation or the destruction of the intermediate compounds. We shall study repeatedly such chemical non-equilibrium steady states.

It is easy to show that if the steady states occur *sufficiently close to equilibrium states* they may be characterized by an extremum principle according to which *the entropy production has its minimum value at the steady state compatible with the prescribed conditions* (constraints), to be specified in each case.

In our first example (thermal diffusion or thermomolecular pressure difference) the constraint corresponds to the difference of temperature between the two vessels. In the second example, the constraints may be the values of the concentrations of the initial and final products A and F in the external environment.

Let us prove this theorem for the typical case of a transfer of matter and energy, as in our first example. According to equation (2.23) the entropy source is given by

$$\sigma[S] = J_{th}X_{th} + J_m X_m > 0 \tag{3.21}$$

and the phenomenological laws are (cf. (3.3))

$$\begin{aligned} J_{th} &= L_{11}X_{th} + L_{12}X_m \\ J_m &= L_{21}X_{th} + L_{22}X_m \end{aligned} \tag{3.22}$$

For the stationary state the flow of matter vanishes

$$J_m = L_{21}X_{th} + L_{22}X_m = 0 \tag{3.23}$$

We shall now show that (3.23) is equivalent to the condition that the entropy production is minimum for a given value of the constraint X_{th}. Using (3.22) and Onsager's reciprocity relation $L_{12} = L_{21}$, the entropy production (3.21) becomes

$$\sigma[S] = L_{11}X_{th}^2 + 2L_{21}X_{th}X_m + L_{22}X_m^2 \tag{3.24}$$

Taking the derivative of (3.24) with respect to X_m at constant X_{th} we have:

$$\frac{\partial}{\partial X_m}\sigma[S] = 2(L_{21}X_{th} + L_{22}X_m) = 2J_m = 0 \tag{3.25}$$

Therefore, one sees that the two conditions:

$$J_m = 0 \quad \text{and} \quad \frac{\partial}{\partial X_m}\sigma[S] = 0 \tag{3.26}$$

are completely equivalent as long as the linear relations (3.22) are valid and the coefficients $L_{\alpha\beta}$ may be considered as constants satisfying equation (3.9).

If no subsidiary conditions such as the value of X_{th} are imposed (that is in absence of constraints), the entropy production vanishes, and we recover the equilibrium state as a special case of the steady state. The theorem of minimum entropy production (I. Prigogine, 1945) is very general as it applies to all non-equilibrium steady states whatever the nature of the constrains. On the other hand this theorem is also subject to severe restrictions as it is only valid in the range of *linear* thermodynamics of irreversible processes and when in addition the phenomenological coefficients may be considered as *constants satisfying the Onsager relations* (3.9).

However, in practice it often happens that the flows J_α are given by linear phenomenological laws with constant values of the coefficients, but in terms of forces X'_α, linked to the X_α appearing in the expression of the entropy production by a positive weighting function:

$$\varepsilon^2 = \frac{X_\alpha}{X'_\alpha} \tag{3.27}$$

independent of α. Denoting by $l_{\alpha\beta}$ these constant coefficients, the entropy source (2.23) takes the form (also (2.26)):

$$\sigma[S] = \sum_\alpha J_\alpha X_\alpha = \sum_{\alpha\beta} l_{\alpha\beta}X'_\beta X_\alpha = \varepsilon^2 \sum_{\alpha\beta} l_{\alpha\beta}X'_\alpha X'_\beta \tag{3.28}$$

Dividing the two sides by ε^2, we recover a similar situation as in (3.24), but this time the entropy production is replaced by a *weighted* entropy production. This leads to a slight generalization of the minimum entropy production theorem. Of course, this extension is only possible when a weighting function independent of α exists.

To illustrate this aspect of the theorem, let us consider the case of a non-homogeneous continuous medium. In this case, the constraints correspond to the boundary conditions, and the conservation equations lead to linear partial differential equations.

For example, let us focus our attention on the heat conduction problem in an isotropic medium and assume the thermal conductivity λ, and the specific heat c_v are constants. If we substitute in the balance equation for internal energy (1.44), the heat flow $\mathbf{W}$ by its value (3.13), we obtain the linear Fourier equation:

$$\lambda \nabla^2 T = \rho c_v \partial_t T \tag{3.29}$$

The coefficient λ plays here the same role as coefficient l in (3.28).

On the other hand, the entropy production (2.24) weighted by the factor $\varepsilon^2 = T^2$ takes the form

$$\int \varepsilon^2(-\lambda T_{,j}) T^{-1}_{,j} \, \mathrm{d}V = \lambda \int (T_{,j})^2 \, \mathrm{d}V \tag{3.30}$$

For fixed boundary conditions and according to the variational method (Hilbert and Courant, 1953), the minimum of the functional (3.30) is reached for the value of T, satisfying the Euler–Lagrange equation:

$$\nabla^2 T = 0 \tag{3.31}$$

Here again the minimum of (3.30) corresponds to the steady state, as given by (3.29) ($\partial_t T = 0$, $\lambda =$ constant).

When the thermal conductivity λ is no longer constant, equation (3.29) has to be replaced by the *non-linear* partial differential equation:

$$\lambda \nabla^2 T + \lambda'_T (\nabla T)^2 = \rho c_v \partial_t T \qquad \left(\lambda'_T = \frac{\mathrm{d}\lambda}{\mathrm{d}T}\right) \tag{3.32}$$

At the same time the theorem of minimum entropy production weighted or not, is no longer valid.

For anisotropic media, the heat equation (3.32) becomes:

$$\lambda_{ij}(T_{,i})_{,j} + (\lambda'_T)_{ij}(T_{,i})(T_{,j}) = \rho c_v \partial_t T \tag{3.33}$$

Clearly, the solution of such type of equations is a problem of much greater difficulty. However, in the isotropic case (3.32) the change of the function

$$\Theta = \int_{T1}^{T} \lambda(T)\,\mathrm{d}T \quad (T_1 = \text{constant}) \tag{3.34}$$

leads again to a linear equation (Carslaw and Jaeger, 1959):

$$\nabla^2\Theta = \frac{\rho c_v}{\lambda}\frac{\partial\Theta}{\partial t} \tag{3.35}$$

But such an *ad hoc* transformation is valid only for isotropic media.

In conclusion, the theorem of minimum entropy production at the steady state applies only to the *strictly linear* case described by (3.31). The cases described by equation (3.32) and (3.33) correspond to phenomenological laws which are linear in the forces but contain phenomenological coefficients $L_{\alpha\beta}$ which depend themselves on the thermodynamic variables. We shall then speak of linearity in an *extended sense*.

Certainly the validity of Gibbs formula (2.14) used to derive the explicit form of the entropy flow and entropy production, extends *beyond* the domain of strict linearity. It is even possible to include in the theory a number of important non-linear problems.

To illustrate this point we shall briefly consider the case of chemical reactions.

5. CHEMICAL REACTIONS

Chemical reactions provide us with simple situations in which the flows are no longer linear in terms of the generalized forces.

In order to compare the linear phenomenological laws (3.2) with the usual kinetic expressions for reaction rates, we shall consider the simple case of the synthesis of hydroiodic acid in the gaseous phase. The chemical reaction is

$$H_2 + I_2 \rightleftharpoons 2HI \tag{3.36}$$

and the corresponding affinity (2.18):

$$A = \mu_{H_2} + \mu_{I_2} - 2\mu_{HI} \tag{3.37}$$

For convenience, the chemical potentials are here defined in terms of numbers of moles n_γ, namely:

$$\mu_\gamma = \left(\frac{\partial G}{\partial n_\gamma}\right)_{p,T,(n_\gamma)} \tag{3.38}$$

On the contrary in equality (3.39) the chemical potentials are defined in terms of masses:

$$\mu_\gamma = \left(\frac{\partial G}{\partial m_\gamma}\right)_{p,T,(m_\gamma)} \tag{3.39}$$

Nevertheless, as it is usual, we use the same notation for (3.38) and (3.39) (cf. chap. II).

Since by definition

$$m_\gamma = M_\gamma n_\gamma \tag{3.40}$$

the masses M_γ in the definition of the affinities (2.18) disappear when molar chemical potentials are used. Now for perfect gases the molar chemical potentials may be written in the form

$$\mu_\gamma = \eta_\gamma(T) + RT \log c_\gamma \tag{3.41}$$

Introducing the equilibrium constant K given by (e.g. Prigogine and Defay, 1954):

$$RT \log K(T) = -\sum_\gamma \nu_\gamma \eta_\gamma(T) \tag{3.42}$$

the affinity becomes

$$\begin{aligned} A &= -\sum_\gamma \nu_\gamma \eta_\gamma(T) - RT \sum_\gamma \nu_\gamma \log c_\gamma \\ &= RT \log \frac{K(T)}{c_1^{\nu_1} \ldots c_c^{\nu_c}} \end{aligned} \tag{3.43}$$

For the synthesis of HI, described by the chemical reaction (3.36) it follows from (3.43) that

$$A = RT \log \frac{K(T)}{c_{I_2}^{-1} c_{H_2}^{-1} c_{HI}^{2}} \tag{3.44}$$

On the other hand, the usual kinetic expression for the reaction rate of (3.36) is given by the difference between two *partial* reaction rates as

$$
\begin{aligned}
w &= \overrightarrow{w} - \overleftarrow{w} = k_+ c_{I_2} c_{H_2} - k_- c_{HI}^2 \\
&= k_+ c_{I_2} c_{H_2} \left(1 - \frac{k_-}{k_+} \frac{c_{HI}^2}{c_{I_2} c_{H_2}}\right)
\end{aligned}
\tag{3.45}
$$

It is well known that the ratio of the kinetic constants k_+/k_- is equal to the equilibrium constant K, so that (3.45) can be written as (cf. (3.44))

$$
w = \overrightarrow{w}\left[1 - \exp\left(-\frac{A}{RT}\right)\right] \tag{3.46}
$$

This equation expresses the relation between reaction rate and affinity. It is valid for large classes of chemical reactions. Close to equilibrium one has:

$$
\left|\frac{A}{RT}\right| \ll 1 \tag{3.47}
$$

and the formula (3.46) reduces in first approximation to:

$$
w = \frac{\overrightarrow{w}_e}{R} \frac{A}{T} \tag{3.48}
$$

where $\overrightarrow{w}_e$ is the value of the partial rate $\overrightarrow{w}$ at equilibrium:

$$
\overrightarrow{w}_e = \overleftarrow{w}_e, \quad \text{for} \quad A = 0.
$$

It can now be seen that inequality (3.47) gives the condition for the validity of the linear law:

$$
w = LX \tag{3.49}
$$

with

$$
L = \frac{\overrightarrow{w}_e}{R}; \quad X = \frac{A}{T} \tag{3.50}
$$

The value of L depends only on the partial rate $\overrightarrow{w}$ at equilibrium.

Let us consider the other extreme case, corresponding to

$$
\frac{A}{RT} \to \infty \tag{3.51}
$$

In a more explicit form, this condition implies that, (cf. (3.44))

$$\frac{c_{I_2}c_{H_2}}{c_{HI}^2} \to \infty \quad \text{or} \quad c_{HI} \to 0 \tag{3.52}$$

which, for a closed system, corresponds to the initial stage of the reaction. The corresponding value of w is simply (cf. (3.46)):

$$w = \vec{w} \tag{3.53}$$

and thus becomes independent of the affinity. This situation corresponds to a sort of *saturation effect* with respect to the affinity and, in the region in which it appears, the entropy production becomes a linear function of the affinity.

If the chemical reaction under consideration is not too fast, elastic collisions will always maintain the molecular distribution functions near equilibrium. We may then again apply the Gibbs formula (2.14) based on the local equilibrium assumption and use our macroscopic approach *in spite of the non-linear character* of the relations (3.46) between rates and affinities.

6. CONCLUDING REMARKS

When we increase the value of the thermodynamic constraints and go beyond the region studied by linear thermodynamics of irreversible processes we may meet essentially three types of situation:

(*a*) The local equilibrium assumption breaks down. For example, this is the situation for a gas when the relative variation of temperature is no longer small within a length equal to the mean free path. These situations are explicitly excluded from our treatment. It should however be mentioned that most of the techniques we shall introduce in the later chapters of this monograph may be generalized to apply even to such situations (Nicolis, 1968). Nevertheless the starting point has to be non-equilibrium statistical mechanics or kinetic theory of gases.

(*b*) One remains in the domain of validity of the local equilibrium description with the properties of the system varying continuously when we increase the deviation from equilibrium.

For example, in (3.46) w is a continuous function of the affinity and reduces to the linear relation (3.48) in the neighbourhood of equilibrium. In such cases we may expect that the far from equilibrium situations will *share many properties of linear systems*. As another illustration we may mention that even when the heat conductivity coefficient is an arbitrary function of temperature, the average heat flow per unit volume

$$\langle W^2 \rangle = \frac{1}{V}\int W^2 \, \mathrm{d}V \tag{3.54}$$

is minimum at the steady state.

Indeed using the heat potential Θ of (3.34) we may also write:

$$\langle W^2 \rangle = \frac{1}{V}\int (\Theta, j)^2 \, \mathrm{d}V \tag{3.55}$$

and the minimum condition of this functional is (cf. 3.31):

$$\nabla^2 \Theta = 0 \tag{3.56}$$

This is precisely the condition of the steady state according to (3.35). Similar theorems also exist for some classes of chemical reactions (Prigogine, 1955). In the strictly linear region as defined in §4, we recover the theorem of minimum entropy production.

(*c*) One remains in the domain of validity of the macroscopic description, but far from equilibrium, new types of organization of matter appear. This situation familiar in hydrodynamics is by far the most interesting one. In the Bénard problem (see Chapter XI), we have to reach a critical temperature gradient before the convection sets in. This critical value is expressed in terms of a dimensionless parameter, the Rayleigh number. Similarly the Reynolds number has to reach a critical value for the laminar flow to become unstable.

In Chapter VII, §3, we shall derive Helmholtz's theorem according to which laminar flow corresponds to a minimum of energy dissipation. Under isothermal conditions, this is equivalent to minimum entropy production. Laminar flow corresponds to the state of the system near thermodynamic equilibrium while turbulent flow occurs only for conditions sufficiently distant from it when non-linearity due to inertial effects becomes dominant. In all such cases, the properties of the system beyond the instability point cannot be reached continuously starting from the neighbourhood of equilibrium.

As we shall see in Chapter VII, such instability phenomena are *not limited to hydrodynamics* and may also occur in purely dissipative systems.

When such instabilities occur, entirely new properties arise. The structure and properties of the solution below the instability point cannot be extrapolated even in a first approximation. Still we remain in the range where a macroscopic description should be possible. Indeed all these phenomena may occur in dense media where the number of collisions is sufficient to maintain the equilibrium conditions on the *microscopic level*.

Our immediate objective is therefore to investigate the kind of information which may be derived from the macroscopic theory about the occurrence of instabilities. This will be done in Chapters IV–VII. As an introduction to the study of non-equilibrium situations, we shall first summarize the classical theory of equilibrium thermodynamic stability.

CHAPTER IV

Gibbs–Duhem Stability Theory of Thermodynamic Equilibrium

1. INTRODUCTION

Before we approach the problem of stability of non-equilibrium states, it is useful to summarize the well known stability theory of thermodynamic equilibrium. The original theory is due to Gibbs (1876–1878, see Gibbs, 1931). Later on it was extended by many others and notably by Duhem (1897, 1911) (for a more recent account see Prigogine and Defay, 1954). This approach is based on the properties of the thermodynamic potentials such as E, H, F, G introduced in Chapter II, §4.

While the Gibbs method is well adapted to investigate most of the stability problems which arise in equilibrium theory, it does not provide us with a convenient starting point for the study of stability of non-equilibrium situations such as for example the steady states discussed in Chapter III. As a rule, the boundary conditions used in such problems are incompatible with the minimum properties of thermodynamic potentials. For this reason we introduce in Chapter V a more general approach to stability of equilibrium valid for *all types of boundary conditions* compatible with the maintenance of the equilibrium state. In Chapter VI, this theory will then be extended to non-equilibrium situations.

2. GIBBS–DUHEM STABILITY CRITERION

Let us introduce (1.46) into the inequality (2.11). We obtain in this way a formulation of the second law of thermodynamics valid for closed systems at uniform pressure and temperature in the form

$$T\,\mathrm{d}_iS = T\,\mathrm{d}S - \mathrm{d}E - p\mathrm{d}V \geqslant 0 \tag{4.1}$$

This relation leads to a stability criterion for thermodynamic equilibrium: indeed if no perturbation starting with equilibrium can satisfy inequality (4.1) the system has to remain in equilibrium.

It is important to realize that the perturbation refered to above, is not necessarily generated by an external action on the system. Molecular fluctuations lead inevitably to small spontaneous deviations of the macroscopic quantities from their average values.

Let us use the symbol δ for small but otherwise arbitrary increments. The stability criterion is then

$$\delta E + p\delta V - T\delta S \geqslant 0 \tag{4.2}$$

Note that the signs in (4.2) are opposite to those in (4.1). We kept the equality sign in (4.2) as, strictly speaking, there exists no reversible transformation. Therefore a process which would satisfy the equality sign in (4.2) would not affect the stability.

In the special case of systems at constant entropy and volume, (4.2) leads to the stability condition

$$\delta E \geqslant 0 \quad (S,\ V \text{ constant}) \tag{4.3}$$

This important inequality expresses that the internal energy is a *minimum* for stable equilibrium, that is

$$(\delta E)_{eq} = 0 \quad \text{(equilibrium)} \tag{4.4}$$

$$(\Delta E)_{eq} > 0 \quad \text{(stability)} \tag{4.5}$$

In (4.5) we used the symbol Δ to emphasize the possibility of finite amplitude perturbations. In the case of infinitesimal perturbations (4.5) reduces to the second order condition

$$(\delta^2 E)_{eq} > 0 \tag{4.6}$$

Alternatively, for systems at constant energy and volume (4.2) leads to the stability condition

$$\delta S \leqslant 0 \quad (E,\ V \text{ constant}) \tag{4.7}$$

By definition, a system maintained at constant energy and

volume is an *isolated* system. For such systems entropy is a *maximum* at stable equilibrium, and one has:

$$(\delta S)_{eq} = 0 \quad \text{(equilibrium)} \tag{4.8}$$

$$(\Delta S)_{eq} < 0 \quad \text{(stability)} \tag{4.9}$$

Again for infinitesimal perturbations (4.9) reduces to

$$(\delta^2 S)_{eq} < 0 \tag{4.10}$$

There exist situations where (4.6) or (4.10) may be satisfied while respectively (4.5) and (4.9) are not, at least for certain classes of perturbations. In such cases the equilibrium is *metastable* (e.g. §4). We shall often describe both stable and metastable systems as *stable* since both have properties in common which distinguish them from unstable systems.

Using the definition (2.30) of thermodynamic potentials we may derive from (4.2) three other alternative forms of the stability conditions. We recover in this way the well known conditions

$$\begin{aligned} \delta H &\geqslant 0 \quad (S, p \text{ constant}) \\ \delta F &\geqslant 0 \quad (T, V \text{ constant}) \\ \delta G &\geqslant 0 \quad (T, p \text{ constant}) \end{aligned} \tag{4.11}$$

Let us now deduce the explicit form of the stability conditions from these basic inequalities.

3. EXPLICIT FORM OF THE STABILITY CONDITIONS

We introduce (2.58) into the stability criterion (4.10):

$$(\delta^2 S)_{eq} = \int \rho \delta^2 s \mathrm{d}V = -\int \frac{\rho}{T} \left[\frac{c_v}{T} (\delta T)^2 + \frac{\rho}{\chi} (\delta v)^2_{N_\gamma} + \sum_{\gamma\gamma'} \mu_{\gamma\gamma'} \delta N_\gamma \delta N_{\gamma'} \right] \mathrm{d}V \leqslant 0 \tag{4.12}$$

The expression under the integral sign is a quadratic function which has to be negative definite. Therefore, as already pointed out for (3.4), the stability conditions are linked to the signs of the coefficients. It turns out that these conditions are independent of the requirement that the extensive quantities E and V have to be kept constant. As we shall prove in Chapter V, §2, this condition is not required as long as we deal with small perturbations (see also Chapter V).

From 4.12 we obtain as stability conditions:

$$c_v > 0 \quad \text{(thermal stability)} \tag{4.13}$$

$$\chi > 0 \quad \text{(mechanical stability)} \tag{4.14}$$

Thus the specific heat (at constant volume) and the isothermal compressibility have to be positive quantities. In addition we have to impose a positive sign to the quadratic form:

$$\sum_{\gamma\gamma'} \mu_{\gamma\gamma'} x_\gamma x_{\gamma'} > 0 \quad \text{(stability with respect to diffusion)} \tag{4.15}$$

for arbitrary values of x_γ ($\gamma = 1, 2, \ldots, c$).

The physical meaning of these conditions is simple (Prigogine and Defay, 1954, Chapter XV). Indeed, let us consider for example a perturbation which occurs as a heterogeneity in the composition of a binary system initially at uniform equilibrium. The inequality (4.15) means that the response of the system will be to restore the initial homogeneity. That is the reason why inequality (4.15) is called the stability condition in respect to *diffusion*. Similarly, (4.13) and (4.14) express the stability in respect to thermal and mechanical disturbances.

In §5, we show that the stability of chemical equilibrium is also secured by inequality (4.15).

4. PHASE SEPARATION IN BINARY MIXTURES

As a simple illustration of the stability condition (4.15), let us consider phase separation in binary mixtures. We obtain the inequalities

$$\mu_{11} > 0; \qquad \mu_{22} > 0$$

$$\begin{vmatrix} \mu_{11} & \mu_{21} \\ \mu_{12} & \mu_{22} \end{vmatrix} \geqslant 0 \tag{4.16}$$

Relations (2.33) and (2.49) allow us to simplify these conditions, since

$$\mu_{12} = \mu_{21}$$

$$n_1\mu_{11} + n_2\mu_{21} = 0$$

$$n_1\mu_{12} + n_2\mu_{22} = 0 \tag{4.17}$$

Therefore, the determinant in (4.16) vanishes and it remains only to consider the first two relations. Moreover according to (4.17) one has:

$$n_1^2\mu_{11} = n_2^2\mu_{22} \tag{4.18}$$

The first two inequalities in (4.16) are therefore equivalent. They imply also

$$\mu_{12} = \mu_{21} < 0 \tag{4.19}$$

For many systems known as *ideal systems* (such as mixtures of perfect gases or *perfect solutions* formed by components of nearly identical molecules) the chemical potentials take the form

$$\mu_\gamma = \eta_\gamma(p, T) + RT \log N_\gamma \tag{4.20}$$

For convenience, we use here *molar* chemical potentials as in Chapter III, §5.

In equation 4.20, $\eta_\gamma(p, T)$ is independent of composition while N_γ denotes the *mole fraction* ($N_\gamma = n_\gamma/n$). The expression (3.41) for perfect gases is a special case of (4.20).

It is easy to verify that the stability conditions (4.16) and (4.19) are satisfied. This is however no longer necessarily so with the so-called *regular solutions* (e.g. Guggenheim, 1952) for which (4.20) has to be replaced by:

$$\begin{aligned} \mu_1 &= \eta_1(T, p) + RT \log (1 - N_2) + \alpha N_2^2 \\ \mu_2 &= \eta_2(T, p) + RT \log N_2 + \alpha(1 - N_2)^2 \\ & \qquad (N_1 = 1 - N_2) \end{aligned} \tag{4.21}$$

Indeed we now have

$$\mu_{12} = \frac{\partial \mu_1}{\partial N_2} = -\frac{RT}{1 - N_2} + 2\alpha N_2 \tag{4.22}$$

Discussion of the sign on the r.h.s. shows that for

$$\frac{2\alpha}{RT} > 4 \tag{4.23}$$

there exists a range of mole fractions where the stability condition will be violated. We then obtain phase separation. The phase diagram is represented schematically in Figure 4.1.

One observes a critical point C corresponding to:

$$N_2 = 0{\cdot}5, \quad T_c = \frac{\alpha}{2R}$$

Above this critical point, the two components are miscible in all proportions. Below O C 1 we find two coexisting phases. For example at $T = T_1$, we have two phases corresponding to $N_2 = \gamma$ and

$N_2 = \beta$. Inside the region a C b the stability condition 4.19 is violated. The curve a C b is called the *spinodal*. It separates unstable states from the *metastable states*, which are located in the regions of coexistence O a C and b C 1. In these regions the mixture remains homogeneous.

For metastable states the stability conditions as derived in §3 are satisfied; however the Gibbs free energy (at constant p and T) is higher for the homogeneous mixture than for a system formed by

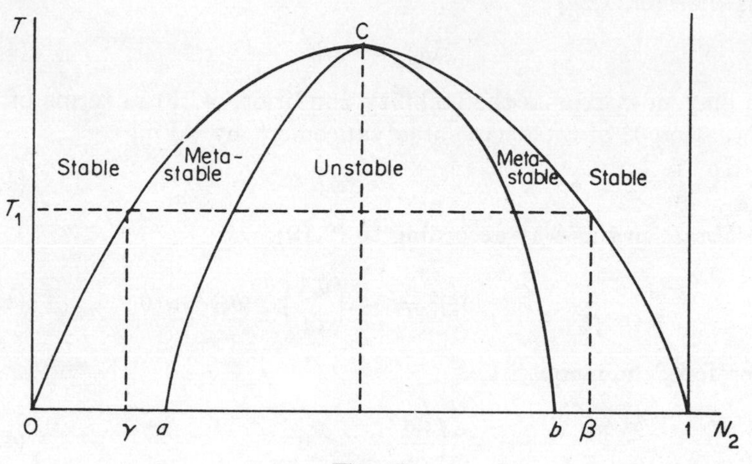

Fig. 4.1

two coexisting phases. Metastable systems are stable in respect to small perturbations (the second order stability conditions (4.10) as well as (4.19) are satisfied) but *unstable* in respect at least to some *finite perturbations* (the complete stability condition (4.9) is not satisfied).

The form (4.21) of the chemical potentials appears to be a first approximation for mixtures of molecules of the same size but differing in their molecular interactions. (Prigogine, Bellemans and Mathot, 1957).

5. STABILITY OF CHEMICAL REACTIONS

Let us now show how the stability of chemical equilibrium is deduced from the general stability criterion (4.12).

It is easily proved that if a system is stable with respect to diffusion all chemical equilibria are automatically stable (Duhem, 1893,

Jouguet, 1921; Prigogine and Defay, 1954, Chapter XV). Indeed, let us express the reaction rate (1.24) as

$$w = \frac{\mathrm{d}\xi}{\mathrm{d}t} \tag{4.24}$$

where ξ denotes the so-called degree of advancement (or the extent of reaction) introduced by De Donder (1936).† The change in the number of moles n_γ due to a single chemical reaction during the time $\mathrm{d}t$ is then (cf. 1.24)

$$\mathrm{d}n_\gamma = \nu_\gamma \,\mathrm{d}\xi \tag{4.25}$$

We may now express the stability condition (4.15) in terms of the fluctuation $\delta\xi$ of the degree of advancement by taking

$$x_\gamma = \nu_\gamma \delta\xi \tag{4.26}$$

We obtain in this way according to (2.18):

$$\sum_{\gamma\gamma'} \mu_{\gamma\gamma'} \nu_\gamma \nu_{\gamma'} (\delta\xi)^2 = -\left(\frac{\partial A}{\partial \xi}\right)_{eq} (\delta\xi)^2 > 0 \tag{4.27}$$

Therefore, the inequality

$$\left(\frac{\partial A}{\partial \xi}\right)_{eq} < 0 \tag{4.28}$$

expresses the condition of *chemical stability*, which is satisfied as a consequence of the stability condition with respect to diffusion.

In the general case of several chemical reactions, the *chemical stability criterion* occurs as a negative definite quadratic form

$$\sum_{\rho\rho'} \left(\frac{\partial A_\rho}{\partial \xi_{\rho'}}\right)_{eq} \delta\xi_\rho \delta\xi_{\rho'} < 0 \quad (\rho, \rho' = 1, 2, \ldots, r) \tag{4.29}$$

which implies relations between the coefficients as in (3.4), (4.12) and (4.16).

Again these conditions are consequences of the stability condition (4.15).

In Chapters VI and VII, we show that the situation is radically different for non-equilibrium steady states involving chemical reactions. Such steady states may be unstable even if the system is stable with respect to diffusion.

† Also Duhem (1911).

6. LIMITATION OF THE GIBBS–DUHEM THEORY

The Gibbs–Duhem theory leads to necessary and sufficient conditions for the stability of thermodynamic equilibrium both for infinitesimal and finite amplitude perturbations. However the theory is limited to groups of variables for which a thermodynamic potential can be defined. Unfortunately, this is a very restrictive condition. In general it is not possible to construct a thermodynamic potential corresponding to arbitrary physical variables x, y, . . . rather than the commonly used variables (T, V), (S, p) etc. . . . for which we have the potentials F, H, etc. . . . (Van Rysselberghe, 1935; De Donder, 1936).

Moreover in many circumstances the equilibrium state is defined *through boundary conditions* and *not* by imposing given values to some variables such as p and V.

For example, the conditions of thermal equilibrium in a solid may correspond to prescribed values of the temperature on the *boundaries*. One may also consider other types of conditions such as the vanishing of the heat flow on the boundaries. In all such situations the stability problem is studied starting from phenomenological laws such as the Fourier law for heat conduction. We then obtain partial differential equations which together with the boundary conditions determine the evolution of the system. The stability theory at equilibrium refers to the asymptotic state reached after a sufficiently long period of time. As already mentioned, in such situations there exists generally no thermodynamic potential which will reach a minimum at the steady state.

The limitations of the classical thermodynamic stability theory have been clearly recognized by the founders of the theory themselves (Duhem, 1911). Since then a few special cases have been studied, where a potential may be constructed. For example, we may consider the potential:

$$E + p_0 V - T_0 S$$

A theorem due to Maxwell and Gouy shows that the change of this potential, also called the *available energy*, gives us the maximum work performed by monothermal systems (e.g. Jouguet, 1909). Here, T_0 and p_0 are the uniform temperature and the uniform pressure of the *external medium* and are maintained constant. No condition is imposed on the temperature and the pressure of the system itself. In this case, the requirement is that sufficiently far from the system we have

$$T \rightarrow T_0 \quad \text{and} \quad p \rightarrow p_0$$

This is the potential used by Landau and Lifshitz (Landau and Lifshitz, 1958) to derive the stability conditions. In this case the boundary conditions are replaced by conditions at infinity. Again this method is not applicable to boundary value problems.

There exists another special case which can be treated by means of potentials (Gibbs, 1873, Duhem, 1904). The system is surrounded by membranes with specified properties (for example, such as to isolate the system in an adiabatic way from the surrounding). The introduction of such membranes brings us already nearer to the study of stability with well defined boundary conditions which will be undertaken in the next chapter.

CHAPTER V

General Stability Theory of Thermodynamic Equilibrium

1. THERMODYNAMIC STABILITY AND ENTROPY BALANCE EQUATION

We shall now use the physical ideas involved in the Gibbs–Duhem stability theory (Chapter IV) together with the entropy balance equation (Chapter II, §3) to obtain a new formulation of the stability theory of equilibrium, for given boundary conditions. We consider in this Chapter, only purely dissipative system, i.e. without convection.

For the system as a whole (2.20) and (2.23) give us

$$\int \sum_{\alpha} J_{\alpha} X_{\alpha} \, \mathrm{d}V = \frac{\partial S}{\partial t} + \Phi[S] \geqslant 0 \tag{5.1}$$

The left hand side is the entropy production (2.24). As pointed out in Chapter II, the entropy production is a quantity of second order with respect to the deviations from equilibrium. Let us now separate on the right hand side of (5.1) the terms of first and second order. We expand the entropy S about its equilibrium value S_e, retaining terms up to second order

$$S = S_e + (\delta S)_e + \tfrac{1}{2}(\delta^2 S)_e \tag{5.2}$$

Since S_e is time independent, (5.2) leads to

$$\partial_t S = \partial_t(\delta S)_e + \tfrac{1}{2}\, \partial_t(\delta^2 S)_e \tag{5.3}$$

Let us decompose in a similar way the entropy flow $\Phi[S]$. In accordance with (2.22) Φ is equal to

$$\Phi[S] = \int [W_n T^{-1} - \sum_{\gamma} \rho_{\gamma} \Delta_{\gamma n}(\mu_{\gamma} T^{-1}) + \mathrm{v}_n \rho s] \, \mathrm{d}\Omega \tag{5.4}$$

In the neighborhood of equilibrium we have both *first order terms* corresponding to the equilibrium flow, that is

$$\Phi[S]_e = \int [W_n T_e^{-1} - \sum_\gamma \rho_\gamma \Delta_{\gamma n} (\mu_\gamma T^{-1})_e] \, d\Omega \tag{5.5}$$

and *second order terms* as

$$\begin{gathered} \Delta\Phi[S] = \int [W_n \Delta T^{-1} - \sum_\gamma \rho_\gamma \Delta_{\gamma n} \Delta(\mu_\gamma T^{-1})] \, d\Omega \\ [T^{-1} = T_e^{-1} + \Delta T^{-1}; \, \mu_\gamma T^{-1} = (\mu_\gamma T^{-1})_e + \Delta(\mu_\gamma T^{-1})] \end{gathered} \tag{5.6}$$

provided there are no velocity fluctuations around the state at rest, on the boundary surface ($v_n = 0$ on Ω).

We now introduce (5.3), (5.5), and (5.6) into the balance equation (5.1), and identify *separately* first and second order terms. We obtain in this way the two equations:

$$\partial_t(\delta S)_e = -\Phi[S]_e \quad \text{(1st order)} \tag{5.7}$$

and

$$\tfrac{1}{2}\partial_t(\delta^2 S)_e = P[S] - \Delta\Phi[S] \quad \text{(2nd order)} \tag{5.8}$$

It should be kept in mind that the separation of the entropy balance equation (5.1) into two separate relations such as (5.7) and (5.8) is not always valid. This point will be developed in more detail in the discussion of the general theory (Chapter VII, §10). A detailed analysis based on the properties of the kinetic equations for the perturbations shows indeed that the terms $\delta_t(\delta S)_e$ and $\Phi[S]_e$, still contain additional second order quantities, that is, terms of the same order of magnitude as $\partial_t(\delta^2 S)_e$ and $\Delta\Phi[S]$. Nevertheless in the case discussed in this section, equation (5.8) remains valid because the additional second order quantities in $\partial_t(\delta S)_e$ and $\Phi[S]_e$ are equal in the absence of velocity fluctuations and cancel in equation (5.7). However when velocity fluctuations occur, equation (5.8) is no longer valid. For this reason, the equilibrium stability theory developed in the present chapter, is limited to systems at rest, as was also the theory discussed in Chapter IV.

Let us now consider separately each of the equalities (5.7) and (5.8).

1°/*First order conditions—Equilibrium*

We start from a state initially at equilibrium and integrate equation (5.7) over time. We obtain in this way

$$(\delta S)_e = -\int_0^t \Phi[S]_e \, dt \tag{5.9}$$

This equality may be considered as a generalized equilibrium condition. For an isolated system the right hand side vanishes and we come back to the classical equilibrium condition

$$[\delta S]_e = 0 \tag{5.10}$$

However, if the system is not isolated, a small change of entropy has to be compensated by the entropy flow which appears on the right hand side of (5.9). If such a compensation cannot occur, irreversible processes will take place and the initial reference system could not be in equilibrium.

It may also be observed that (5.9) is an alternative expression of the Gibbs formula (2.15) as applied to the equilibrium state. We can go back from (5.9) to the Gibbs formula, by using (5.5), the balance equations (1.42), (1.28) and neglecting all second order terms in deviations from equilibrium. These calculations present no difficulty and are left as an exercise to the reader.

2°/*Stability*

We assume that a set of well defined boundary conditions are given on the surface of the system. We do not exclude the possibility of fluctuations on this surface but we suppose that we can re-establish the given boundary conditions whenever they are altered. We therefore assume that the second order surface term (5.6) vanishes

$$\Delta\Phi[S] = 0 \tag{5.11}$$

This may be achieved either because the deviations ΔT^{-1}, $\Delta(\mu_\gamma T^{-1})$ vanish on the surface, or because the flows W_n, $\rho_\gamma \Delta_{\gamma n}$, vanish. Mixed situations in which some of the deviations and some of the flows vanish can also be considered. As before, we consider systems at rest ($\mathbf{v} = 0$). The expression (5.8) of the entropy balance reduces then to

$$\tfrac{1}{2}\, \partial_t(\delta^2 S)_e = P[S] \geqslant 0 \tag{5.12}$$

This inequality will be our starting point for the study of stability based on the entropy balance equation.

2. THERMODYNAMIC STABILITY CONDITIONS

The inequality 5.12 gives us an evolution criterion for states near equilibrium. Indeed it relates the time derivative of the curvature $(\delta^2 S)_e$ to the entropy production, i.e. to the irreversible processes inside the system. It also describes a kind of minimum property. Both sides vanish at equilibrium and are positive for all perturbed states.

Now, if no perturbation can satisfy the evolution criterion (5.12) and therefore cannot lead to a positive value for the entropy production, the system will remain in equilibrium. Our stability condition is therefore

$$\int_e^f P[S]\mathrm{d}t < 0 \tag{5.13}$$

where e denotes the initial equilibrium state and f the final state. The inverse process, that is the change from f to e, is then a spontaneous process accompanied by a positive entropy production

$$\int_f^e P[S]\,\mathrm{d}t > 0 \tag{5.14}$$

It is clear that the physical idea behind inequality (5.13) is the same as in the classical theory based on inequality (4.2). In both cases we express that the system will be stable when no evolution starting from the unperturbed state can satisfy the requirements of the second law. The only difference is that these requirements are expressed here by the definite positive character of the entropy production, independently of the boundary conditions imposed upon the system.

Let us introduce the condition (5.13) into (5.12). Through integration from e to the final state f we obtain using also (2.6):

$$\tfrac{1}{2}(\delta^2 S)_e = \int_0^t P[S]\,\mathrm{d}t = \int_0^t \mathrm{d}_i S = \Delta_i S < 0 \tag{5.15}$$

The stability *depends therefore only on the sign of the curvature* $(\delta^2 S)_e$ *taken at the equilibrium state.* It is not necessary to calculate explicitly the entropy production (5.13) associated with the fluctuation to discuss the stability problem.

Taking into account (2.27) this inequality can also be written

$$\int[\delta^2(\rho s)]_e\,\mathrm{d}V < 0 \tag{5.16}$$

It has to be satisfied for all arbitrary perturbations $\delta(\rho e)$, $\delta\rho_\gamma$, distributed in the volume V. Therefore (5.16) implies

$$[\delta^2(\rho s)]_e < 0 \tag{5.17}$$

Alternatively using (2.62) we have also

$$(\delta^2 s)_e < 0 \tag{5.18}$$

These inequalities express the stability in a *local* form. In (5.17) the independent variables are ρe, ρ_γ, while in (5.18) they are e, v, N_γ. We have seen that $\delta^2 s$ is given by the quadratic form (2.58). This quadratic form has therefore to be negative to ensure the stability of the system. This leads us immediately back to the Gibbs–Duhem stability conditions (4.13)–(4.15):

$$c_v > 0; \qquad \chi > 0; \qquad \sum_{\gamma\gamma'} \mu_{\gamma\gamma'} x_\gamma x_{\gamma'} > 0 \tag{5.19}$$

It is interesting to notice that in spite of the greater generality of our stability problem characterized by the boundary conditions (5.11) compatible with the existence of equilibrium, we do not obtain any new stability condition. Therefore the Gibbs–Duhem conditions (5.19) remain the necessary and sufficient conditions of stability for thermodynamic equilibrium even when no thermodynamic potential exists.

However for more general boundary conditions, supplementary *surface* stability conditions have to be introduced (see equation (7.81)).

As we have already observed, relation (5.15) shows that the entropy production due to a fluctuation in the system is directly related to $(\delta^2 s)_e$. As a consequence the total entropy production $\Delta_i S$ does not depend on the type of evolution which follows the initial perturbation, at least as long as terms of higher order than the second, are neglected. It is specially noteworthy that $\Delta_i S$ does not depend on the exchange of matter or energy with the outside world during this evolution. The stability condition (5.18) depends only on the sign of $(\delta^2 s)_e$ that is, finally on the conditions (5.19) which the reference equilibrium state has to satisfy.

In a few situations which are precisely those considered in the Gibbs–Duhem theory, the entropy production can itself be expressed in terms of a thermodynamic potential. Indeed, according to (4.1) as well as (2.28)–(2.30), we have the equalities:

$$-T_e \Delta_i S = (\Delta E)_{S,V} = (\Delta H)_{S,p} = (\Delta F)_{T,V} = (\Delta G)_{T,p} \tag{5.20}$$

Using (5.15) we obtain for small perturbations

$$\left.\begin{array}{r} -T_e(\delta^2 S)_e \\ [\rho e, \rho_\gamma] \\ (\delta^2 E)_e \\ [\rho s, \rho\gamma] \end{array}\right\} \begin{array}{l} = T_e(\delta^2 S)_{e;E,V} = (\delta^2 E)_{e;S,V} \\ \\ = (\delta^2 H)_{e;Sp} = (\delta^2 F)_{e;T,V} = (\delta^2 G)_{e;Tp} \end{array} \Bigg\} > 0 \tag{5.21}$$

We see again that it is the sign of the single quantity $(\delta^2 S)_e$ which determines in all cases the stability of thermodynamic equilibrium.

Let us stress again that the stability condition (5.15) refers to small perturbations. To decide if a system is stable with respect to finite amplitude perturbations, that is if it is metastable or stable, the only criterion we have presently at our disposal is based on thermodynamic potentials. It may well happen that for situations where no thermodynamic potential exists, the very concept of metastability looses its simple meaning. The stability of a system in state A as compared to a state B, at finite distance from A, could then depend not only on the properties of the states A and B but also on the path followed to go from A to B. This path itself may depend on the type of the initial finite fluctuation.

We shall not however go into more detail about this question which would deserve further investigation.

3. COMPARISON WITH KINETIC STABILITY THEORY

In this section, we show that the *thermodynamic theory of stability* when applied to equilibrium states, leads to conclusions in agreement with the *kinetic theory of stability*. We consider a closed system where the components $\gamma(\gamma = 1, \ldots, c)$ are involved in r chemical reactions characterized by chemical variables $\{\xi_\rho\}$ and by rates $\{w_\rho\}$ defined in (4.24). As in Chapter IV, §5, we consider only stability in respect to chemical equilibrium. Temperature and pressure are supposed to be constant. Let the values of the chemical variables be perturbed initially by arbitrarily small amounts $\{\delta\xi_\rho\}_{in}$. The kinetic stability of the system implies that

$$\delta\xi_\rho \to 0 \quad \text{for} \quad t \to \infty$$

In the neighbourhood of the equilibrium the variables $\{\delta\xi_\rho\}$ satisfy the linear equations:

$$\frac{\mathrm{d}(\delta\xi_\rho)}{\mathrm{d}t} = \sum_{\rho'=1}^{r} \alpha_{\rho\rho'}\delta\xi_{\rho'} \quad (\rho, \rho' = 1, \ldots, r) \tag{5.22}$$

where the $\alpha_{\rho\rho'}$ are constant coefficients related to the equilibrium state of interest. Equation (5.22) admits solutions of the form:

$$\delta\xi_\rho = (\delta\xi_\rho)_i e^{\omega t} \qquad (\rho = 1, \ldots, r) \tag{5.23}$$

Such a solution is called a normal mode.

Stability implies that the eigenvalue $\omega = \omega_r + i\omega_i$ corresponding to each normal mode has a negative real part. From (5.23) and according to the foregoing definition of the stability, it follows that all roots of the determinantal equation:

$$\left| \omega\delta_{\rho\rho'} - \alpha_{\rho\rho'} \right| = 0; \qquad \delta_{\rho\rho'} = \begin{cases} 0 & (\rho \neq \rho') \\ 1 & (\rho = \rho') \end{cases} \tag{5.24}$$

have to satisfy the condition:

$$\omega_r < 0 \tag{5.25}$$

Equation (5.24) is called the dispersion equation.

It is well known that the explicit stability conditions corresponding to (5.25) are given by the Hurwitz criterion (Hurwitz, 1895, cf. Cesari, 1963). If one writes the dispersion equation in the form

$$\mathscr{A}_0\omega^r + \mathscr{A}_1\omega^{r-1} + \ldots + \mathscr{A}_r = 0 \tag{5.26}$$

the Hurwitz criterion implies that the $r + 1$ determinants

$$\mathscr{A}_0, \mathscr{A}_1, \begin{vmatrix} \mathscr{A}_1 & \mathscr{A}_3 \\ \mathscr{A}_0 & \mathscr{A}_2 \end{vmatrix}, \begin{vmatrix} \mathscr{A}_1 & \mathscr{A}_3 & \mathscr{A}_5 \\ \mathscr{A}_0 & \mathscr{A}_2 & \mathscr{A}_4 \\ 0 & \mathscr{A}_1 & \mathscr{A}_3 \end{vmatrix}, \ldots \tag{5.27}$$

are separately positive quantities. These relations constitute the *kinetic stability conditions of equilibrium.*

Let us now observe that, around an equilibrium state, one may also write:

$$A_\rho = \sum_{\rho'=1}^{r} \beta_{\rho\rho'}\delta\xi_{\rho'}; \quad \beta_{\rho\rho'} = \frac{\partial A_\rho}{\partial \xi_{\rho'}} \tag{5.28}$$

since the affinity vanishes at equilibrium. Again, the subscript T, p, (ξ_ρ), is everywhere implied in the second equality (5.28). According to (2.18), (2.33) and (4.25) the matrix $\|\beta_{\rho\rho'}\|$ is symmetric. On the other hand, the rates w_ρ, i.e. here δw_ρ, are linked to the affinities by linear relations such as (3.2), with a symmetric matrix $\|L_{\rho\rho'}\|$ as a consequence of the Onsager relations (3.9).

It follows that the coefficients $\alpha_{\rho\rho'}$, of (5.22) occur as the elements of a symmetric matrix

$$\|\alpha_{\rho\rho'}\| = \|L_{\rho\rho'}\| \times \|\beta_{\rho\rho'}\| \tag{5.29}$$

since the product of two symmetric matrices is again a symmetric matrix. Therefore, as is well known, all the imaginary parts ω_i vanish identically. In other words, oscillating chemical disturbances never exist around an equilibrium state. In this case, the dispersion equation (5.24) is usually called the *secular* equation. This is a typical feature of the equilibrium stability theory, which is no longer valid in the kinetic theory of stability around non-equilibrium states (Chapter VII).

Let us finally show that the thermodynamic stability conditions (4.29) for chemical equilibrium are equivalent to (5.25) or (5.27). Indeed, according to (2.19) the entropy source as well as the entropy production due to the chemical reaction in a homogeneous system may be written as:

$$P = \sum_{\rho=1}^{r} w_\rho \frac{A_\rho}{T} = \sum_\rho \frac{d\xi_\rho}{dt} \frac{A_\rho}{T} = \sum_\rho \frac{d(\delta\xi_\rho)}{dt} \frac{A_\rho}{T} \geqslant 0 \tag{5.30}$$

Using equalities (5.23) and (5.28), the entropy production becomes:

$$P = \omega_r \sum_{\rho\rho'} \left(\frac{\partial A_\rho}{\partial \xi_{\rho'}}\right) \delta\xi_\rho \delta\xi_{\rho'} > 0 \tag{5.31}$$

for each normal mode taken separately.

Clearly, the *thermodynamic stability condition* (4.29) and the *kinetic stability conditions* (5.25) and (5.27) are mutually connected by the inequality (5.31).

In conclusion, we wish to emphasize the great simplicity and generality of the thermodynamic method for the study of stability versus the kinetic method. In the latter one always has to estimate the signs of the determinants (5.27) in terms of the coefficients $\alpha_{\rho\rho'}$ of the secular equation. This is often a very cumbersome task.

CHAPTER VI

Thermodynamic and Hydrodynamic Stability Conditions for Non-equilibrium States

1. INTRODUCTION

We now leave the equilibrium stability theory to start the much more difficult problem of stability of non-equilibrium states. From the kinetic point of view the problem is quite similar to that considered in Chapter V, §3. In the linear stability theory of steady states with respect to small perturbations one has to show that inequality (5.25) is satisfied by each normal mode.

We wish to supplement this kinetic criterion by a thermodynamic criterion which gives us a sufficient (and in some cases also a necessary) condition for the stability of stationary states as well as time dependent processes. The basic advantage we see in this extension of thermodynamic stability theory to non equilibrium situations is a greater physical insight into the mechanism of stability.

The main restriction is that we shall only consider small perturbations (linear stability).

Even so this domain is already very wide. It contains all hydrodynamic stability problems, such as the onset of convection or turbulence. Also stability problems of purely dissipative processes such as systems of chemical reactions, thermodiffusion and so on, are included. It is precisely one of the attractive features of this method to present a unified theory for a large variety of macroscopic systems.

In this chapter we derive the general conditions for the stability of non-equilibrium processes such as they arise from the very definition of stability. These conditions may be considered as complementing those of equilibrium theory studied in Chapter V. We then discuss briefly the link with the Liapounoff method, with the Le Châtelier–Braun principle and with the kinetic theory of stability.

2. THE DEFINITION OF STABILITY—LIAPOUNOFF FUNCTIONS

We first introduce a precise definition of stability applicable under wide conditions including equilibrium and non-equilibrium situations.

We consider an arbitrary point inside the volume V of the system. Using the symbol x for the set of the independent variables, which characterize the state of the system at this point, the local evolution is described by†

$$x = \varphi(t, t_0, x_0) \tag{6.1}$$

where x_0 denotes the state at time t_0.

Here φ is assumed to be a continuously differentiable function with respect to $t(t \geqslant t_0)$. Moreover we suppose that (6.1) remains valid in the neighbourhood of x_0, characterized by some deviation δ. The quantity:

$$y(t) = \varphi(t; x_0 + \delta) - \varphi(t; x_0) \tag{6.2}$$

represents the value reached by the initial perturbation δ at time t. Because of continuity $|y(t)|$ is small if $|\delta|$ is small and t not too large. Here $|y(t)|$ stands for the *distance* $\sqrt{(\Sigma y_i^2)}$ in the space of states.

This leads to the following definition for the stability of the evolution (6.1):

If to all $\varepsilon > 0$ there corresponds a $k(\varepsilon) > 0$, such that

$$|\varphi(t; x_0 + \delta) - \varphi(t; x_0)| < \varepsilon \tag{6.3}$$

for all values of t, provided that $|\delta| < k$, the evolution (6.1) is said to be *stable in the sense of Liapounoff* (for more details about Liapounoff stability, cf. Pars, 1965).

Besides, the stability is *asymptotic* (*or complete*), if for all δ in the domain, we have:

$$\lim_{t \to \infty} |\varphi(t; x_0 + \delta) - \varphi(t; x_0)| = 0 \tag{6.4}$$

† The function φ in equation (6.1) is such that in the space of states, cycles or crossings may occur.

In this case the disturbed motion tends to coincide with the original one as $t \to \infty$.

As a result, if the positive definite sum y^2 (square of the distance) is non-increasing, that is, if its time derivative satisfies the condition

$$(\dot{y^2}) \leqslant 0 \quad (<0) \tag{6.5}$$

for all values of t, evolution (6.1) is stable ($\leqslant 0$), or asymptotically stable (<0). However, in these two cases, (6.5) provides only a *sufficient condition*, since oscillating perturbations of y^2 consistent with the basic definition of stability are here excluded.

Likewise using, instead of y^2, other positive (or negative) definite quadratic forms with constant values of the coefficients, we may obtain different sufficient stability conditions. A definite function such as y^2 which leads to the stability conditions (6.5) is called a '*Liapounoff function*'.

3. STABILITY OF DISSIPATIVE SYSTEMS

The entropy balance method used in Chapter V cannot be extended without modification to the stability problem of non-equilibrium states (either stationary or time dependent). Indeed, at arbitrary distance from thermodynamic equilibrium we can no longer split the entropy balance equation (5.1) into the separate (5.7) and (5.8), since the entropy production is no longer a second order quantity. Therefore, inequality 5.12, which was the starting point of the equlibrium stability theory is not valid in the present case.

However, we still use our fundamental assumption of local equilibrium. In addition we suppose, once and for all, that the local equilibrium state is stable. According to (5.17) and (5.18) this implies:

$$\delta^2 s < 0 \quad \text{or} \quad \delta^2(\rho s) < 0 \tag{6.6}$$

Moreover it follows from (2.58) and (2.62) that these quantities are negative definite expressions in the increments of the independent variables, resp. e, v, N_γ and ρe, ρ_γ, which characterize the local state of a dissipative system (i.e. without convective effects). This remark leads to an approach to the stability problem based on the use of either $\delta^2 s$ or $\delta^2(\rho s)$ as Liapounoff functions in the sense defined in the preceding paragraph.

We obtain in this way the stability conditions

$$(\dot{\delta^2} s)_{t_0} \geqslant 0; \quad [\dot{\delta^2}(\rho s)]_{t_0} \geqslant 0 \tag{6.7}$$

for all values of time ($t \geqslant t_0$)

The subscript t_0 indicates that in the time derivation, the coefficients of the quadratic forms (6.6) are maintained constant, i.e. with the same values as at time t_0.

This subscript will be tacitly implied in the subsequent formulae, as well as the separate asymptotic condition included into (6.5).

Let us first compare in more detail the conditions (6.7) with an alternative definition of stability based on inequality (6.5) as well as on an extension of the classical Le Châtelier–Braun principle of equilibrium thermodynamics.

4. THEOREMS OF MODERATION AND THE LE CHÂTELIER–BRAUN PRINCIPLE

Let us apply inequality (6.5) to the asymptotic stability in variables e, v, N_γ. We have more explicitly

$$\delta e(\delta \dot{e}) + \delta v(\delta \dot{v}) + \delta N_\gamma(\delta \dot{N}_\gamma) < 0 \tag{6.8}$$

Near equilibrium one has:

$$(\delta \dot{e}) = \dot{e}; \quad (\delta \dot{v}) = \dot{v}; \quad (\delta \dot{N}_\gamma) = \dot{N}_\gamma \tag{6.9}$$

and (6.8) leads therefore to the inequalities

$$\begin{aligned} \dot{e}\delta e &< 0 \quad (v, N_\gamma \text{ constant}) \\ \dot{v}\delta v &< 0 \quad (e, N_\gamma \text{ constant}) \\ \dot{N}_\gamma \delta N_\gamma &< 0 \quad (e, v \text{ constant}) \end{aligned} \tag{6.10}$$

Such inequalities are well known in equilibrium thermodynamics. They correspond to the so-called Le Châtelier–Braun principle. This principle may be stated as follows (Le Châtelier, 1888): Any system in chemical equilibrium undergoes, as a result of a variation in one of the factors governing the equilibrium, a compensating change in a direction such that, had this change occurred alone it would have produced a variation of the factor considered in the *opposite* direction.

This principle implies therefore a *moderation* (or damping) of the changes of the variables characterizing the equilibrium state. For this reason inequalities such as (6.10) are also called *moderation theorems* (Prigogine and Defay, 1954). It should be kept in mind that inequalities (6.8) or (6.10) refer to intensive variables. For extensive variables the Le Châtelier–Braun principle is not always valid even at equilibrium. This question is studied elsewhere (Prigogine and Defay *loc. cit.*)

We may conclude that (6.8) gives us an extended form of the Le Châtelier–Braun principle which reduces to the usual form near equilibrium. Conversely, the extended Le Châtelier–Braun principle expressed by inequality (6.8) could be adopted as a starting point for the stability problem, in the same way as inequalities (6.7) based on the use of $\delta^2 s$, or $\delta^2(\rho s)$, as Liapounoff functions. However for reasons discussed in §6, the inequalities (6.7) are more convenient and will be used in the subsequent chapters. If one can show, using $\delta^2 s$ as the Liapounoff function, that the system is stable, the Le Châtelier–Braun principle will of course be automatically satisfied.

5. GLOBAL STABILITY CONDITIONS

In the general stability theory of dissipative processes we deal with boundary value problems described by partial differential equations. More specifically, we have to consider the time changes of perturbations $\delta(\rho e)$, $\delta\rho_\gamma$, . . . such as $\partial_t\delta(\rho e)$, $\partial_t\delta\rho_\gamma$, . . . The latter are described by the balance equations for the perturbed motion, and are given explicitly in Chapter VII.

Clearly the local formulation of the stability conditions such as (6.7) or (6.8) is no longer appropriate in this case since we have to take into account the boundary conditions. We need then a global (or integral) formulation. As the stability definition adopted in §2 refers to a fixed point of the system, the symbol 'dot' used in (6.7) for the space of states, stands here for the partial time derivative ∂_t taken at constant values of the coordinates x_j. Likewise δ is a localized variation ($\delta x_j = 0$). Let us now integrate (6.7) over an arbitrary volume V' inside the whole volume V of the system. Denoting by S' the corresponding entropy we obtain:

$$\partial_t\delta^2 S' \geqslant 0 \quad (>0) \quad (t \geqslant t_0) \tag{6.11}$$

since (6.7) has to be fulfilled at each point of V for all t. Conversely, (6.11) has to be satisfied for all arbitrary small V'. This in turn implies (6.7), as a consequence of the continuity of the integrand. Applying now (6.11) to the whole volume V of the system we finally get the sufficient stability condition in the *global* form:

$$\partial_t\delta^2 S \geqslant 0 \quad (>0) \tag{6.12}$$

together with [cf. (6.6)]

$$\delta^2 S < 0 \tag{6.13}$$

Let us emphasize that we recover here the two basic inequalities (5.12) and (5.15) of the equilibrium stability theory. However the situation here is actually reversed. Indeed, in the case of equilibrium, inequality (6.12) represents the starting point arising from the second law of thermodynamics, whereas (6.13) is deduced as a stability condition. On the contrary, in the case of non-equilibrium, inequality (6.13) becomes the starting point (local equilibrium) while (6.12) then represents the stability condition.

Therefore, in the stability theory of dissipative processes the set of inequalities (6.12) and (6.13) appears as a kind of *complementary principle*, when compared to the situation at equilibrium.

6. CHARACTERISTIC PROPERTIES OF $\delta^2 s$ CONSIDERED AS A LIAPOUNOFF FUNCTION

As already pointed out in §2, a number of definite quadratic forms other than $\delta^2 s$ could be considered at least in principle, as Liapounoff functions to be used in stability theory. Let us now state the main reasons for which the choice of $\delta^2 s$ finally prevails.

The set of relations (6.12), (6.13), already makes this choice quite attractive as it leads both to the equilibrium and non-equilibrium stability theory. Still this does not provide by itself a sufficient justification.

Let us consider the quadratic expressions of $\delta^2 s$ and $\delta^2(\rho s)$ respectively in variables δe, δv, δN_γ and $\delta(\rho e)$, $\delta \rho_\gamma$. According to Euler's theorem on homogeneous functions of the second degree we have

$$2f(u, v, w) = uf'_u + vf'_v + wf'_w \tag{6.14}$$

By comparison with (2.56) and (2.61), their derivatives for constant values of the coefficients may be written as:

$$\tfrac{1}{2}\, \partial_t \delta^2 s = \delta T^{-1}\, \partial_t \delta e + \delta(pT^{-1})\, \partial_t \delta v - \sum_\gamma \delta(\mu_\gamma T^{-1})\, \partial_t \delta N_\gamma \tag{6.15}$$

$$\tfrac{1}{2}\, \partial_t \delta^2(\rho s) = \delta T^{-1}\, \partial_t \delta(\rho e) - \sum_\gamma \delta(\mu_\gamma T^{-1})\, \partial_t \delta \rho_\gamma \tag{6.16}$$

One sees that the only time derivatives contained in the r.h.s. are precisely the quantities which are given by the mass and energy balance equations of the perturbed motion (excess balance equations,

Chapter VII). Moreover, the other factors δT^{-1}, $\delta(pT^{-1})$ and $\delta(\mu_\gamma T^{-1})$ are directly linked to the boundary conditions. Indeed, the partial differential equations deduced from these balance equations and from the phenomenological laws, contain as gradients, precisely these quantities, or quantities directly connected to them (e.g. heat equation, diffusion equation).

These two properties are of fundamental importance for the general problem of stability as studied in detail in the next chapter.

On the other hand, definite functions which would contain terms such as e.g. $\delta T\, \partial_t\, \delta T$ or $\delta(\rho e)\, \partial_t \delta(\rho e)$, would be very difficult to handle in the general theory. In the first case, because $\partial_t \delta T$ is not directly given by the balance equations, and in the second case because $\delta(\rho e)$, is not directly linked to the boundary conditions.

For the same reason, the other definite quadratic forms occuring in (2.63) are not suitable as Liapounoff functions, since they would lead to very heavy expressions which are unlikely to be useful. It is precisely for this reason that we do not start our investigation of stability with the extension of the Le Châtelier–Braun principle introduced in §4. A full appreciation of these remarks requires the preliminary reading of the next chapter where the explicit stability conditions are derived. Finally only $\delta^2 s$ or $\delta^2(\rho s)$ and some other definite functions closely connected to these expressions† seem to be of interest for stability theory.

Besides, there is a direct physical reason for choosing $\delta^2 s$ as the starting point for stability theory. It is indeed this quantity which is, through Einstein's formula, directly linked to the statistical macroscopic fluctuation theory (Chapter VIII). Therefore, our approach leads to a unified theory applicable both to equilibrium and non-equilibrium situations in agreement with the statistical meaning of stability.

7. STABILITY INVOLVING CONVECTIVE EFFECTS

So far, fluctuations of the mean velocity at each point of the macroscopic system (local centre-of-mass motion, barycentric velocity in multicomponent systems, see (1.20)) have been disregarded, except of course for fluctuations generated by density perturbations. To take into account such convective effects in addition to the dissipative effects already considered, the stability problem requires

† As e.g. $\varepsilon^2 \delta^2 s$ where ε^2 denotes a simple positive weighting function ($\delta \varepsilon^2 \equiv 0$) sometimes useful in computations [see e.g. (3.27)].

a larger set of independent variables, as e.g.: e, v, N_γ, v_i instead of e, v, N_γ or ρe, ρ_γ, $\rho \mathrm{v}_i$, instead of ρe, ρ_γ; here v_i denotes the velocity along the x_i axis. Therefore $\delta^2 s$ and $\delta^2(\rho s)$, remain no longer definite functions with respect to the increments, resp. δe, δv, δN_γ, $\delta \mathrm{v}_i$ and $\delta(\rho e)$, $\delta\rho_\gamma$, $\delta(\rho \mathrm{v}_i)$. Instead they become *semi-definite* functions, since they may vanish for non-vanishing values of velocity perturbations (according to the definition of a definite and semi-definite function). As a result, $\delta^2 s$ and $\delta^2(\rho s)$, can no longer be adopted as Liapounoff functions. However, setting:

$$z = s - \tfrac{1}{2} T_0{}^{-1} \mathrm{v}^2 \tag{6.17}$$

One may adopt as a suitable Liapounoff function the negative definite quadratic expression:

$$\delta^2 z = \delta^2(s - \tfrac{1}{2} T_0^{-1} \mathrm{v}^2) = \delta^2 s - T^{-1}(\delta \mathrm{v})^2 \tag{6.18}$$

Here T_0 denotes the temperature of the reference state at time t_0, that is a non-varied quantity. The first stability condition (6.7) then becomes

$$(\dot{\delta^2 z}) \geqslant 0 \tag{6.19}$$

More explicitly, due to (2.56) and (6.15), we get respectively for (6.18) and (6.19),

$$\delta^2 z = \delta T^{-1} \delta e + \delta(p T^{-1}) \delta v - \sum_\gamma \delta(\mu_\gamma T^{-1}) \delta N_\gamma - T^{-1}(\delta \mathrm{v})^2 < 0 \tag{6.20}$$

and

$$\begin{aligned} \tfrac{1}{2}\, \partial_t \delta^2 z = {} & \delta T^{-1}\, \partial_t \delta e + \delta(p T^{-1})\, \partial_t \delta v \\ & - \sum_\gamma \delta(\mu_\gamma T^{-1})\, \partial_t \delta N_\gamma - \tfrac{1}{2} T^{-1}\, \partial_t (\delta \mathrm{v})^2 \geqslant 0 \quad (>0) \end{aligned} \tag{6.21}$$

We may also derive similar relations in variables ρe, ρ_γ, $\rho \mathrm{v}_i$, which are, as already mentioned, more suitable for the global formulation of the stability conditions. We first observe that, in variables ρ, $\rho \mathrm{v}_i$ one has:

$$\delta(\rho \mathrm{v}_i^2) = 2 \mathrm{v}_i \delta(\rho \mathrm{v}_i) - \mathrm{v}_i^2 \delta\rho \tag{6.22}$$

This gives in turn

$$\delta^2(\tfrac{1}{2} \rho \mathrm{v}_i^2) = \delta \mathrm{v}_i \delta(\rho \mathrm{v}_i) - \delta(\tfrac{1}{2} \mathrm{v}_i^2) \delta\rho \tag{6.23}$$

or

$$\underset{[\rho \mathrm{v}_i, \rho]}{\delta^2(\tfrac{1}{2} \rho \mathrm{v}^2)} = \rho(\delta \mathrm{v})^2 = \underset{[\mathrm{v}_i]}{\rho \delta^2(\tfrac{1}{2} \mathrm{v}^2)} \tag{6.24}$$

This is a positive quadratic quantity. The square brackets indicate the corresponding independent variables. Taking into account (2.58) and (2.62) as well as (6.23) and (6.24) we now obtain as a Liapounoff function, the negative definite quadratic expression:

$$\delta^2(\rho z) = \delta^2(\rho s - \tfrac{1}{2}\rho T_0^{-1}\mathrm{v}^2) < 0 \tag{6.25}$$

The corresponding stability condition is now

$$\partial_t \delta^2(\rho z) \geqslant 0 \quad (>0) \tag{6.26}$$

Therefore, in the same way as in (6.20) and (6.21), the local stability in respect to small perturbations around non equilibrium processes is governed by the set of inequalities [cf. (2.61), (6.23)].

$$\delta^2(\rho z) = \delta T^{-1}\delta(\rho e) - \sum_\gamma \delta(\mu_\gamma T^{-1} - \tfrac{1}{2}T_0^{-1}\mathrm{v}^2)\delta\rho_\gamma - T^{-1}\delta \mathrm{v}_i \delta(\rho \mathrm{v}_i) < 0 \tag{6.27}$$

and

$$\tfrac{1}{2}\,\partial_t\delta^2(\rho z) = \delta T^{-1}\,\partial_t\delta(\rho e) - \sum_\gamma \delta(\mu_\gamma T^{-1} - \tfrac{1}{2}T_0^{-1}\mathrm{v}^2)\,\partial_t\delta\rho_\gamma - T^{-1}\delta\mathrm{v}_i\,\partial_t\delta(\rho\mathrm{v}_i) \geqslant 0 \quad (>0) \tag{6.28}$$

in variables ρe, ρ_γ, $\rho\mathrm{v}_i$ $(\delta\rho = \sum_\gamma \delta\rho_\gamma)$.

Here again, the time derivatives on the r.h.s. are directly given by the balance equations, including now the momentum equation associated with the perturbation. Also the other factors are connected with the boundary conditions.

The corresponding global formulation is deduced following the same line of reasoning as in §5. We write:

$$Z = \int \rho z \,\mathrm{d}V \tag{6.29}$$

where the function z introduced by equation (6.17) here replaces the specific entropy s. We then obtain by direct integration of (6.27) and (6.28),

$$\delta^2 Z < 0 \tag{6.30}$$

together with

$$\partial_t \delta^2 Z \geqslant 0 \quad (>0) \tag{6.31}$$

These relations generalize respectively (6.13) and (6.12) to the case of non-equilibrium processes, including convective as well as dissipative effects. Therefore they represent a *general sufficient thermodynamic and hydrodynamic condition of stability* ($\geqslant 0$), *or asymptotic stability* (>0). We shall use these inequalities in the next chapter to obtain explicit stability criteria for specific situations.

Let us also observe that for systems at rest, that is in mechanical equilibrium, one has $v_i = 0$ and therefore also $Z = S$ according to (6.17). Still $\delta^2 Z$ generally differs from $\delta^2 S$ due to the velocity fluctuations around the state at rest ($\delta v_i \neq 0$).

8. COMPARISON WITH KINETIC STABILITY THEORY

We have already shown in a simple case that the thermodynamic stability conditions for equilibrium imply kinetic stability (Chapter V, §3). We wish now to extend this comparison to the more general case of non equilibrium steady states. We again consider a single *normal mode*. The time variation of a quantity, say $\delta\varphi$, will then be given by

$$\partial_t \delta\varphi = \omega \delta\varphi \tag{6.32}$$

However here, the frequency ω contains both a real and an imaginary part as in (2.70). It is then convenient to introduce complex variables following the method developed in Chapter II, §6. We obtain, starting from (6.27):

$$\begin{aligned} \delta_m^2(\rho z) = \tfrac{1}{2}\{ & \delta T^{-1}\delta(\rho e)^* + \delta T^{-1*}\delta(\rho e) \\ & - \sum_\gamma [\delta(\mu_\gamma T^{-1} - \tfrac{1}{2}T_0^{-1}v^2)\delta\rho_\gamma^* \\ & + \delta(\mu_\gamma T^{-1} - \tfrac{1}{2}T_0^{-1}v^2)^*\delta\rho_\gamma] \\ & - T^{-1}[\delta v_i \delta(\rho v_i)^* \\ & + \delta v_i^* \delta(\rho v_i)]\} < 0 \end{aligned} \tag{6.33}$$

Applying the same operation to the time derivative (6.28), we obtain directly

$$\partial_t \delta_m^2(\rho z) = 2\omega_r \delta_m^2(\rho z) \tag{6.34}$$

The important point is that this expression contains only the real part ω_r of the frequency. This is of course a direct consequence of the definition of $\delta_m^2(\rho z)$ or $\delta_m^2 z$.† We see immediately that our stability conditions (6.20), (6.21), as well as (6.25), (6.26), or the

† For a state of reference, corresponding to a steady state, the requirement that the time derivatives be taken for constant values of the coefficients, is identically fulfilled.

global forms (6.30)–(6.31), rewritten with the operator δ_m^2, imply for each normal mode:

$$\omega_r \leqslant 0 \tag{6.35}$$

In the linear stability theory, it is generally assumed that the most general perturbation can be represented as a complete set of normal modes (Chandrasekhar, 1961).

Inversely, if we assume $\omega_r \leqslant 0$ together with $\delta^2(\rho z) < 0$ (which implies $\delta_m^2(\rho z) < 0$, as pointed out in Chapter II, §6), relation (6.34) shows that:

$$\partial_t \delta_m^2(\rho z) \geqslant 0 \quad (>0) \tag{6.36}$$

for *each normal mode*. Therefore, (6.36), or the corresponding global form:

$$\partial_t \delta_m^2 Z \geqslant 0 \quad (>0) \tag{6.37}$$

becomes then the *necessary and sufficient* condition of stability.

If we consider now an arbitrary perturbation, which is formed by the superposition of two or more normal modes, the r.h.s. of (6.34) will depend also on the imaginary parts ω_i of the frequencies involved. Its sign could then be either positive or negative even for stable systems. However, for long times, $t \to \infty$, the damping terms $\exp - |\omega_r| t$ become dominant and $\partial_t \delta_m^2(\rho z)$ reaches positive values.

We see now why our basic stability condition (6.31) or (6.37) is too strong in the general case of an arbitrary small perturbation, and provides us only with a sufficient condition of stability.†

In this respect, it should be noticed that even for a single complex normal mode together with its complex conjugate, the sign of

$$\partial_t [\delta^2 Z]_{real}$$

depends also on ω_i. The proof is straightforward by applying (2.71) to each increment included in the r.h.s. of (6.28). Therefore, the property expressed by (6.34) represents a typical feature of the mixed second differential defined in Chapter II.

In conclusion, the set of inequalities

$$\delta^2 Z < 0 \quad \partial_t \delta^2 Z \geqslant 0 \quad (t \geqslant t_0) \tag{6.38}$$

provides us with a *sufficient condition of stability* ($\geqslant 0$) or of *asymptotic stability* (>0). Likewise, the set of inequalities:

$$\delta_m^2 Z < 0 \quad \partial_t \delta_m^2 Z \geqslant 0 \quad (t \geqslant t_0) \tag{6.39}$$

† A less severe necessary and sufficient condition would be to prescribe that after a period Δt sufficiently large, the corresponding $\Delta \delta_m^2 Z$, becomes a positive quantity. Such a possibility will not be explored in this monograph.

provides us with a *necessary and sufficient condition of stability* ($\geqslant 0$) or of *asymptotic stability* (>0), for *each normal* mode taken separately, together with its complex conjugate. In each case the time derivative is taken, keeping the coefficients c_v, χ, $\mu_{\gamma\gamma'}$, ρ, T_0^{-1} as constants in the quadratic form $\delta^2(\rho z)$.

Finally, if the equality sign prevails for all values of t in the second relation (6.38), while the first remains satisfied, then $\delta^2_m Z$ becomes a constant. The perturbed motion then corresponds to a state of *marginal stability*.

9. SEPARATE THERMODYNAMIC AND HYDRODYNAMIC STABILITY CONDITIONS

Various other expressions of the stability conditions (6.38) and (6.39), may be obtained by introducing suitable weighting functions into the basic quadratic form (6.25). For instance, one may proceed with the Liapounoff function

$$\delta^2(\rho\zeta) = \varepsilon^2\delta^2(\rho s) - \tfrac{1}{2}T^{-1}\tau^2\delta^2(\rho \mathrm{v}^2) < 0 \tag{6.40}$$

Here, ε^2 and τ^2 denote positive smooth functions (for all values of x_i and t inside the volume V), related to the reference state ($\delta\varepsilon^2 = \delta\tau^2 = 0$) and having of course the same physical dimensions. Therefore, the quantity:

$$\zeta = \varepsilon^2 s - \tau^2 T_0^{-1}\frac{\mathrm{v}^2}{2} \tag{6.41}$$

is a simple generalization of (6.17). These requirements excepted, ε^2 and τ^2 represent arbitrary quantities which may be chosen independently. As a result, the stability criterion (6.38) takes the form of a set of two separate conditions:

$$\int \varepsilon^2\delta^2(\rho s)\,\mathrm{d}V < 0; \quad \partial_t\int \varepsilon^2\delta^2(\rho s)\,\mathrm{d}V \geqslant 0 \quad (>0); \qquad (t \geqslant t_0) \tag{6.42}$$

$$\int \tau^2 T_0^{-1}\tfrac{1}{2}\delta^2(\rho \mathrm{v}^2)\,\mathrm{d}V > 0; \quad \partial_t\int \tau^2 T_0^{-1}\tfrac{1}{2}\delta^2(\rho \mathrm{v}^2)\,\mathrm{d}V \leqslant 0 \quad (<0) \quad (t \geqslant t_0) \tag{6.43}$$

It should be emphasized that only the set of conditions (6.42), (6.43), considered as a whole represents here the stability criterion. Indeed, in the general case each integrand considered separately occurs as a *semi-definite* quadratic form (§7).

The new conditions (6.42) and (6.43) stand respectively for the *thermodynamic* and the *hydrodynamic* stability conditions (see e.g. Chapter XI).

CHAPTER VII

Explicit Form of the Stability Conditions for Non-equilibrium States

1. INTRODUCTION

We shall now give an explicit form to the criteria established above. This involves the study of the balance equation for $\delta^2 S$ and $\delta^2 Z$. The method we follow is very similar to that used in Chapter II to obtain the balance equation for entropy S. Exactly as in that problem, we have to introduce the balance equations for mass, momentum and energy. But as we deal here with *perturbations*, the corresponding balance equations become *excess balance equations*. They govern the behaviour of the perturbations of mass, momentum and energy. We then obtain, instead of the entropy production, the *excess entropy production* (or the generalized excess entropy production when convection effects occur). Let us first investigate a few simple situations.

It is of interest to notice that in the vicinity of thermodynamic equilibrium, we obtain a stability criterion which is identically satisfied. As might be expected, the stability of thermodynamic equilibrium implies the stability of states *near* equilibrium. This is the reason why all non-trivial stability problems cannot be approached by linear thermodynamics of irreversible processes. The possibility of new types of organization of matter past an instability point under the influence of non-equilibrium conditions, occurs only when the system is sufficiently far from equilibrium. The study of such a new organization, the so-called *dissipative structure*, arising from the exchange of matter and energy with the outside world, appears as one of the most fascinating subjects of macroscopic physics (cf. Chapters XI and XIV).

On the other hand, the method followed provides us with a separate boundary stability condition. As pointed out in §10, this condition is also applicable to equilibrium states. Therefore the stability theory of equilibrium is no longer restricted by condition (5.11) and henceforth many other types of boundary conditions may be considered as well (e.g. (7.20)). Besides, the excitation of convection starting from equilibrium is now included in the stability condition. In what follows we shall deal with two situations: (i) simple cases, for which $P[\delta S]$ (or more generally $P[\delta Z]$), is a *definite quadratic form*, which indicates if the reference state is stable or not (e.g. the near equilibrium situation of §8). For more interesting general cases for which $P[\delta S]$ (or $P[\delta Z]$) is not a definite form (see Chapter XI, XIV and XV). (6.37), will still be used to determine the *marginal stability*, corresponding to the appearance of the *first critical normal mode.*

2. THERMAL STABILITY

Let us consider the stability of heat conduction in a solid where the temperature is the unique variable. Relation (6.15) gives us first

$$\tfrac{1}{2}\,\partial_t\delta^2 s = \delta T^{-1}\,\partial_t\delta e \tag{7.1}$$

Then using the balance equation (1.44) we get†

$$\rho\,\partial_t\delta e = -[\delta W_j]_{,j} \tag{7.2}$$

This is a very simple example of an 'excess balance equation'. Introducing (7.2) into (7.1) we obtain

$$\tfrac{1}{2}\rho\,\partial_t\delta^2 s = -\delta T^{-1}[\delta W_j]_{,j} = \delta W_j\delta T^{-1}_{,j} - [\delta W_j\delta T^{-1}]_{,j} \tag{7.3}$$

Let us first consider the following boundary conditions on the surface Ω of the solid

$$(\delta T^{-1})_\Omega = 0 \quad \text{or} \quad [\delta W_j]_\Omega = 0 \tag{7.4}$$

These are the most commonly used boundary conditions for which either the temperature or the heat flow is fixed on the boundaries. The stability criterion (6.12) takes then the explicit form:

$$\tfrac{1}{2}\int\rho\,\partial_t\delta^2 s\,\mathrm{d}V = \int\delta W_j\delta T^{-1}_{,j}\,\mathrm{d}V > 0 \tag{7.5}$$

† Note the interchangeability of the symbols δ and ∂_x, as well as δ and ∂_t. Indeed the symbol δ used here denotes a *local* change and therefore may also be interpreted as a partial derivative with respect to some arbitrary parameter τ. The above interchangeability means simply:

$$\partial^2_{x\tau} = \partial^2_{\tau x} \quad \text{and} \quad \partial^2_{t\tau} = \partial^2_{\tau t}$$

We see that the fundamental quantity which determines the stability is the '*excess entropy production*':

$$P[\delta S] \equiv \int \delta W_j \delta T_{,j}^{-1} \, \mathrm{d}V \tag{7.6}$$

This excess entropy production has to be positive for all small perturbations. The reader should not confuse this excess entropy production with the *variation* $\delta P[S]$ of the entropy production:

$$P[S] = \int W_j T_{,j}^{-1} \, \mathrm{d}V > 0 \tag{7.7}$$

While the excess entropy production (7.6) is a second order quantity in the deviations from the reference state, the variation $\delta P[S]$ may involve both first and second order quantities. It has no simple connection with the stability problem except for strictly linear situations when the theorem of minimum entropy production holds (§9).

Also it should be noticed that the sign of the excess entropy production (7.6) is not prescribed once and for all by the second law of thermodynamics. Therefore we have to discuss its sign according to the phenomenological laws adopted.

The simplest case corresponds of course to the linear law (3.10), with constant phenomenological coefficients ($L_{ij} = L_{ij}^0$). We then obtain:

$$\tfrac{1}{2}\int \rho \, \partial_t \delta^2 s \, \mathrm{d}V = \int L_{ij}^0 \delta T_{,i}^{-1} \delta T_{,j}^{-1} \, \mathrm{d}V \tag{7.8}$$

We see that the stability condition is identically satisfied, since the quantity on the right hand side is precisely the same quadratic form as the entropy production (3.8). This is a particular case of a very general result, which will be derived later in §8. In the domain of validity of strictly linear thermodynamics of irreversible processes, the stability conditions are identically satisfied.

Let us now consider the situation where the L_{ij} are no longer constants. We first suppose that we may reduce them to constants by an appropriate choice of the generalized forces and the use of a positive weighting function ε^2 (Chapter III, §§2, 4). We then obtain instead of (7.3) the relation:

$$\tfrac{1}{2}\rho\varepsilon^2 \, \partial_t \delta^2 s = \delta W_j [\varepsilon^2 \delta T^{-1}]_{,j} - [\varepsilon^2 \delta W_j \delta T^{-1}]_{,j} \tag{7.9}$$

Together with the boundary conditions (7.4) this leads to the stability condition:

$$\tfrac{1}{2}\int \rho\varepsilon^2 \, \partial_t \delta^2 s \, \mathrm{d}V = \int \delta W_j [\varepsilon^2 \delta T^{-1}]_{,j} \, \mathrm{d}V > 0 \tag{7.10}$$

As an example, let us consider Fourier's law in the form [cf. (3.14)]:

$$W_i = -\lambda^0_{ij} T_{,j} \tag{7.11}$$

with temperature independent thermal conductivity coefficients λ^0_{ij}. We have:

$$L_{ij} = -\lambda^0_{ij} T^2 \tag{7.12}$$

Using as weighting function:

$$\varepsilon^2 = T^2 \tag{7.13}$$

the stability condition (7.10) takes the form:

$$\tfrac{1}{2}\int \rho T^2 \, \partial_t \delta^2 s \, \mathrm{d}V = \int \lambda^0_{(ij)} \delta T_{,i} \delta T_{,j} \, \mathrm{d}V > 0 \tag{7.14}$$

Again comparison with the corresponding expression for entropy production shows that this stability condition is identically satisfied.

A similar method still works for a thermal conductivity which is an arbitrary function of temperature, at least as long as the medium is isotropic. We have then to introduce the new function (3.34)

$$\Theta = \int_{T_1}^{T} \lambda(T) \mathrm{d}T \qquad (T_1 = \text{constant}) \tag{7.15}$$

As a consequence the heat flow is given by

$$W_j = -\Theta_{,j} \tag{7.16}$$

The new function $\Theta(T)$, is therefore a potential for the heat flow. Let us now take as the weighting function:

$$\varepsilon^2 = \lambda T^2 \tag{7.17}$$

We then have:

$$[\varepsilon^2 \delta T^{-1}]_{,j} = -(\delta\Theta)_{,j} = -\delta\Theta_{,j} = \delta W_j \tag{7.18}$$

Therefore in conjunction with (7.10) we now obtain, instead of (7.14) the stability condition:

$$\tfrac{1}{2}\int \rho \lambda T^2 \, \partial_t \delta^2 s \, \mathrm{d}V = \int (\delta W_j)^2 \, \mathrm{d}V > 0 \tag{7.19}$$

Again this inequality is identically satisfied.

Let us now consider more general boundary conditions. A situation frequently met in practice corresponds to the so-called Newton's law for heat exchange on an isothermal portion Ω_1 of the surface Ω:

$$W_n = \alpha(T - T_{ex}) \tag{7.20}$$

We suppose that the external temperature is prescribed ($\delta T_{ex} = 0$) and that α is a constant. The index n corresponds to the outward normal. We then have on Ω_1:

$$(\delta W_n)_{\Omega_1} = \alpha(\delta T)_{\Omega_1} \tag{7.21}$$

In the remaining part $\Omega - \Omega_1$ of the boundary surface, the usual conditions (7.4) are assumed. Relation (7.9) gives us now the stability criterion:

$$\tfrac{1}{2}\int \rho\varepsilon^2\, \partial_t \delta^2 s\, \mathrm{d}V = \int \delta W_j [\varepsilon^2 \delta T^{-1}]_{,j}\, \mathrm{d}V + \int_{\Omega_1} \varepsilon^2 \alpha T^{-2} (\delta T)^2\, \mathrm{d}\Omega > 0 \tag{7.22}$$

This condition contains two parts: the first reduces to the criterion (7.10) for the problem with fixed boundary conditions, while the second introduces the surface stability condition

$$\int_{\Omega_1} \varepsilon^2 T^{-2} \alpha (\delta T)^2\, \mathrm{d}\Omega > 0 \tag{7.23}$$

This splitting of (7.22) into two independent terms is justified by the fact that we are only dealing with sufficient conditions of stability. We see immediately that in this simple case the surface stability conditions leads to the inequality:

$$\alpha > 0 \tag{7.24}$$

which has an obvious physical meaning. It indicates that when (7.24) is fulfilled the stability of the reference state is secured even against fluctuations of temperature on the boundary.

More general boundary conditions may be studied in the same way. For example we may consider the case in which, through some 'feedback', we introduce a well defined relation between T_{ex} and the boundary temperature on Ω_1. In this case the variation of the external temperature does not vanish and (7.21) is replaced by

$$(\delta W_n)_{\Omega_1} = \alpha(\delta T - \delta T_{ex})_{\Omega_1} \tag{7.25}$$

We then obtain the stability condition

$$\int_{\Omega_1} \varepsilon^2 T^{-2} \alpha (\delta T - \delta T_{ex}) \delta T\, \mathrm{d}\Omega > 0 \tag{7.26}$$

which together with (7.24) leads to

$$(\delta T - \delta T_{ex}) \delta T > 0 \tag{7.27}$$

or alternatively

$$|\delta T_{ex}| < |\delta T| \tag{7.28}$$

If the variation of the external temperature (in absolute magnitude) is smaller than the variation of the boundary temperature, the stability is secured.

Boundary stability conditions such as (7.24) or (7.28), may also arise in the special case of equilibrium (§1).

In conclusion, referring to (7.19), we may say that no instability phenomena for thermal conduction is to be expected as long as linear equations of the Fourier type are valid. Of course outside the range of Fourier's law, instabilities are possible but one has then to be certain that a macroscopic description is still possible. The situation changes completely when both thermal conduction and convection occur. In such case instabilities on a macroscopic scale are often observed. This type of phenomena will be repeatedly discussed in this monograph (see especially Chapter XI).

3. HELMHOLTZ'S THEOREM ON THE MOTION OF VISCOUS FLUIDS

As a second example of our general approach, we study in this paragraph the slow steady motion of a viscous incompressible fluid in an external force field which derives from a potential. Moreover we suppose the fluid to be at uniform temperature. For sufficiently slow motion, this is a consistent assumption. As a result the entropy $s(\rho, T)$, will remain constant. We shall also restrict ourselves to perturbations satisfying the same conditions ($\delta T = \delta\rho = 0$, therefore $\delta s = \delta^2 s = 0$). Finally, the velocity field will be imposed on the boundary ($\delta v_i = 0$ on Ω). According to (6.18), the stability criterion in this limit is expressed in terms of the perturbations of kinetic energy.

Since the motion is slow we may neglect the inertial terms $\rho v_j v_{i,j}$ in (1.29), as second order quantities, and use the simplified equation of motion

$$\rho\, \partial_t v_i = \rho F_i - P_{ij,j} \tag{7.29}$$

Therefore for small perturbations:

$$\rho\, \partial_t \delta v_i = -\delta(P_{ij,j}) \tag{7.30}$$

and

$$\tfrac{1}{2}\rho\, \partial_t (\delta v)^2 = -\delta v_i \delta(P_{ij,j}) \tag{7.31}$$

By integration over the whole volume we obtain (cf. 1.32):

$$\tfrac{1}{2}\int \rho\, \partial_t(\delta \mathrm{v})^2\, \mathrm{d}V = \int \delta p_{ij} \delta \mathrm{v}_{i,j}\, \mathrm{d}V \tag{7.32}$$

To derive this formula we have taken into account the boundary conditions as well as the condition of incompressibility:

$$\mathrm{v}_{j,j} = 0 \quad (\rho = \text{constant}) \tag{7.33}$$

Exactly as in the case of heat conduction (7.5), the stability is determined by the *excess entropy production* which appears on the right hand side of (7.32) (apart from the constant factor T_0^{-1}). Also, as in the case of heat conduction, we have to introduce into (7.32) a phenomenological law and then discuss its sign.

Let us adopt the linear law of Newtonian fluid corresponding to the approximation of linear thermodynamics of irreversible processes (Chapter III), i.e.:

$$p_{ij} = -\eta(\mathrm{v}_{i,j} + \mathrm{v}_{j,i}) \tag{7.34}$$

where η is the viscosity coefficient. We also assume that η is a constant (for more general cases see Chapter XII). It follows that:

$$\delta p_{ij} = -\eta \delta(\mathrm{v}_{i,j} + \mathrm{v}_{j,i}) \tag{7.35}$$

Therefore, 7.32 gives rise to the stability criterion:

$$\int \rho\, \partial_t(\delta \mathrm{v})^2\, \mathrm{d}V = -\int \eta[\delta(\mathrm{v}_{i,j} + \mathrm{v}_{j,i})]^2\, \mathrm{d}V < 0 \tag{7.36}$$

Once more we see that this stability criterion is identically satisfied. Indeed the viscosity coefficient is necessarily positive as a consequence of the positive definite character of the entropy production (2.21). The expression:

$$2F = \int \eta(\mathrm{v}_{i,j} + \mathrm{v}_{j,i})^2\, \mathrm{d}V \geqslant 0 \tag{7.37}$$

is the so-called *dissipation function* introduced by Lord Rayleigh. Therefore the inequality (7.36) can also be written as:

$$\delta^2 F > 0 \tag{7.38}$$

Moreover, if the state of reference is stationary, one has also

$$\delta F = 0 \tag{7.39}$$

together with (7.38). This may be easily shown by using (7.37) to calculate δF together with (7.29) rewritten for the steady state.

The dissipation function is therefore minimum at the steady state. This is a classical theorem due to Helmholtz and Korteweg (e.g.

Lamb, 1960). In this case our stability condition leads therefore to a well known theorem of hydrodynamics, similar to the theorem of minimum entropy production since one has:

$$P[S] = FT^{-1}$$

We also observe that even for a non-steady reference state inequality (7.38) establishes the stability with respect to small perturbations.

It is interesting to compare more closely the thermal stability problem studied in (§2) with the problem treated in this paragraph. Stability appears to have rather different meanings in the two cases: in the case of thermal stability it is the entropy fluctuation $|\delta^2 s|$ which decreases in time, while in the case of Helmholtz's theorem it is the perturbation of kinetic energy $(\delta \mathrm{v})^2$ i.e. $\delta^2(\mathrm{v}^2/2)$ (cf. 6.24). The hydrodynamic case is the more intuitive one: if the perturbation of kinetic energy $(\delta \mathrm{v})^2$ increases at the expense of thermal energy we clearly leave the region of laminar slow motion (Orr 1907; see e.g. Lin 1955). We would have then the onset of instability. Also, in contrast to the thermal case where no instability is to be expected within the framework of a macroscopic description, Helmholtz's theorem is only valid for *slow* motion. Indeed, we have neglected in this section the non-linear terms in the equations of motion (7.29). If we relax this condition, the stability of laminar motion can no longer be proved. In fact, as discussed in detail in Chapter XII, the laminar flow will become unstable and the system will finally enter the turbulent regime.

4. CHEMICAL REACTIONS

As a third example we consider in this section the case of chemical reactions in uniform systems at rest. Relations (6.15) and (1.28) lead directly to the stability condition:

$$\tfrac{1}{2}\, \partial_t \delta^2 s = T^{-1} \sum_{\rho} \delta w_\rho \delta A_\rho > 0 \tag{7.40}$$

Here again the fundamental quantity which determines the stability is the excess entropy production in the sense of (7.6) due to the perturbation of the chemical affinities and of the corresponding reaction rates, that is [cf. (2.21)]:

$$\sigma[\delta S] = T^{-1} \sum_{\rho} \delta w_\rho \delta A_\rho \tag{7.41}$$

Problems of special interest are the chemical non-equilibrium steady states described in Chapter III, §4. Once more in the strictly linear case the sign of (7.41) is positive and all the steady states are stable. However as already discussed in Chapter III, §5, chemical reactions provide us with situations in which the rates are generally no longer linear in terms of the affinities. It is then easily seen that there exist mechanisms which may give a *negative* contribution to the sum (7.41). Indeed let us consider the chemical reaction

$$X + Y \rightarrow C + D \tag{7.42}$$

Since we are mainly interested in far from equilibrium situations we neglect the reverse reactions and write for the reaction rate (3.45):†

$$w = XY \tag{7.43}$$

According to (3.43), the affinity is given by

$$A = \log \frac{XY}{CD} \tag{7.44}$$

A fluctuation in the concentration X around some steady state value gives rise to the excess entropy production

$$\delta w \delta A = \frac{Y}{X} (\delta X)^2 > 0 \tag{7.45}$$

Such a fluctuation could therefore not violate the stability condition (7.40).

Let us now consider instead of (7.42) the autocatalytic reaction

$$X + Y \rightleftharpoons 2X \tag{7.46}$$

The reaction rate is still assumed to be given by (7.43) but the affinity is now

$$A = \log \frac{XY}{X^2} = \log \frac{Y}{X} \tag{7.47}$$

We have now the 'dangerous' contribution to the excess entropy production

$$\delta w \delta A = -\frac{Y}{X} (\delta X)^2 < 0 \tag{7.48}$$

† To simplify the notation, we take all kinetic and equilibrium constants as well as RT, equal to unity; also X is written for C_X. . . .

In Chapter XIV, we consider more realistic examples and show that indeed such systems may become unstable.

Here we wish only to make a few qualitative remarks. Both in hydrodynamics and in chemical kinetics, instabilities due to non-linear effects may occur far from equilibrium. In hydrodynamics, non-linear effects are generated by the inertial terms (critical Reynolds number). However the chemical kinetic problem may correspond to a practically infinite variety of possible mathematical structures. Indeed in the chemical case we are concerned with an arbitrary number of steps, each of which involves usually a monomolecular or a bimolecular mechanism. One of our main aims will be to investigate under what conditions instabilities have to be expected in such schemes. It will appear from this study, that some autocatalytic effect, in the general sense of the term, is always required: the same compound has to fulfill at least two different functions in the reaction scheme.

Finally, we see that in conditions far from equilibrium the chemical stability is no longer a consequence of the stability with respect to the diffusion, as is the case near equilibrium (Chapter IV, Duhem–Jouguet theorem).

5. EXCESS BALANCE EQUATIONS

Let us now study the general case. First, we have to introduce the explicit expressions of the time derivatives involved on the r.h.s. of the local stability condition (6.28). To this end we need the conservation laws of mass, momentum and energy which govern the evolution of small perturbations. In other words, we need the *linearized excess balance equations*. These equations are deduced straightforwardly starting from the general conservation laws for mass, momentum and energy given in (1.27), (1.28), (1.30) and (1.42). We obtain in this way:

$$\partial_t \delta\rho_\gamma = \sum \nu_{\gamma\rho} M_\gamma \delta w_\rho - [\delta(\rho_\gamma \Delta_{\gamma j} + \rho_\gamma \mathrm{v}_j)]_{,j} \tag{7.49}$$

$$\partial_t \delta\rho = -[\delta(\rho \mathrm{v}_j)]_{,j} \tag{7.50}$$

$$\partial_t \delta(\rho \mathrm{v}_i) = F_i \delta\rho - [\delta P_{ij} + \delta(\rho \mathrm{v}_i \mathrm{v}_j)]_{,j} \tag{7.51}$$

$$\partial_t \delta(\rho e) = \sum_\gamma F_{\gamma j} \delta(\rho_\gamma \Delta_{\gamma j}) - \delta(P_{ij} \mathrm{v}_{i,j}) - [\delta W_j + \delta(\rho e \mathrm{v}_j)]_{,j} \tag{7.52}$$

The external forces are assumed to be non-perturbed ($\delta F_i = \delta F_{\gamma i} = 0$). Again, as pointed out in Chapter I and II, the r.h.s. of each equation consists of a flow term between brackets (excess flow) and a source term (excess source). The excess source terms are respectively:

$$\sigma[\delta M_\gamma] = \sum_\rho \nu_{\gamma\rho} M_\gamma \delta w_\rho \quad \underset{(\gamma=1,\ldots,c)}{\text{(mass)}} \tag{7.53}$$

$$\sigma[\delta Q_i] = F_i \delta\rho \quad \underset{(i=1,2,3)}{\text{(momentum)}} \tag{7.54}$$

$$\sigma[\delta E] = \sum_\gamma F_{\gamma j} \delta(\rho_\gamma \Delta_{\gamma j}) - \delta(P_{ij} \mathrm{v}_{i,j}) \quad \text{(energy)} \tag{7.55}$$

and of course $\sigma[\delta M] = 0$.

The notation used on the l.h.s. indicates that we are concerned here with source terms due to perturbations around a given reference state (stationary or time dependent).

It should be noticed that the excess balance equations in the form (7.49)–(7.52), are also valid for finite values of the increments. However the symbol δ will be used as an infinitesimal operator in the following paragraphs (except in § 11). As a result, (7.49)–(7.52) reduce to the linearized excess balance equations.

6. EXCESS ENTROPY BALANCE EQUATION

We have now to introduce the expressions of the time derivatives given by equations (7.49), (7.51), and (7.52) into our local condition of stability (6.28). We proceed exactly as in Chapter II, § 3, in order to obtain the entropy balance equation.

We recall that in Chapter II, the mass, momentum and energy balance equations were used to deduce the entropy balance equation, starting from the Gibbs formula (2.14). We obtained in this way a formulation of the second law of thermodynamics in terms of the explicit form of the entropy production.

Likewise here, we shall use the excess balance equations to deduce the *excess entropy balance equation from* (6.28). The only difference is that the Gibbs equation (2.14) gives $\partial_t s$ while (6.28) contains $\partial_t \delta^2(\rho s)$, or $\partial_t \delta^2(\rho z)$ when convection occurs.

From this analogy we may expect to derive the general form of the stability conditions from the *excess entropy production*.

We now introduce (7.49), (7.51) and (7.52) into (6.28). After elementary manipulations and regrouping of terms, we obtain the excess entropy balance equation for small perturbations in the form:

$$\tfrac{1}{2}\,\partial_t \delta^2(\rho z) = \sigma[\delta Z] - \{\delta W_j \delta T^{-1} - \sum_\gamma \delta(\rho_\gamma \Delta_{\gamma j})\delta(\mu_\gamma T^{-1}) - T^{-1}[\delta P_{ij}\delta \mathrm{v}_i + \tfrac{1}{2}\rho \mathrm{v}_j(\delta \mathrm{v})^2] + \mathrm{v}_j \delta^2(\rho s)\}_{,j} \tag{7.56}$$

The term between braces is the *excess flow* term while the *excess source term* is given by:

$$\begin{aligned}\sigma[\delta Z] &= \sum_\alpha \delta J_\alpha \delta X_\alpha \\ &\quad - [\delta(\rho e)\delta \mathrm{v}_j + \delta P_{ij}\delta \mathrm{v}_i + \tfrac{1}{2}\rho \mathrm{v}_j(\delta \mathrm{v})^2]T^{-1}_{,j} \\ &\quad - \sum_\gamma \delta\rho_\gamma \delta \mathrm{v}_j[F_{\gamma j}T^{-1} - (\mu_\gamma T^{-1})_{,j}] - [\delta P_{ij}\delta T^{-1} \\ &\quad - T^{-1}\delta(\rho \mathrm{v}_j)\delta \mathrm{v}_i]\mathrm{v}_{i,j} + \tfrac{1}{2}T^{-1}(\delta \mathrm{v})^2(\rho \mathrm{v}_j)_{,j} \\ &\quad + [\delta(\rho e)\delta T^{-1}_{,j} - \sum_\gamma \delta\rho_\gamma \delta(\mu_\gamma T^{-1})_{,j}]\mathrm{v}^0_j\end{aligned} \tag{7.57}$$

To get (7.56) and (7.57) one has to use the following equalities:

1°/One deduces from (2.69), for $a_j = \delta \mathrm{v}_j$

$$\begin{aligned}&- \delta T^{-1}[\delta(\rho e \mathrm{v}_j)]_{,j} + \sum_\gamma \delta(\mu_\gamma T^{-1})[\delta(\rho_\gamma \mathrm{v}_j)]_{,j} \\ &= -\delta T^{-1}[\mathrm{v}_j \delta(\rho e)]_{,j} + \sum_\gamma \delta(\mu_\gamma T^{-1})[\mathrm{v}_j \delta\rho_\gamma]_{,j} \\ &\quad - \delta \mathrm{v}_j \delta(\rho e) T^{-1}_{,j} + \sum_\gamma \delta \mathrm{v}_j \delta\rho_\gamma (\mu_\gamma T^{-1})_{,j} \\ &\quad + \delta(pT^{-1})\delta \mathrm{v}_{j,j}\end{aligned} \tag{7.58}$$

2°/One has:

$$\begin{aligned}&- \delta T^{-1}\delta(P_{ij}\mathrm{v}_{i,j}) + \delta(pT^{-1})\delta \mathrm{v}_{j,j} \\ &= -\delta(P_{ij}T^{-1})\delta \mathrm{v}_{i,j} - \mathrm{v}_{i,j}\delta T^{-1}\delta P_{ij} \\ &\quad + \delta(pT^{-1})\delta \mathrm{v}_{j,j} + T^{-1}\delta P_{ij}\delta \mathrm{v}_{i,j} \\ &= -\delta(p_{ij}T^{-1})\delta \mathrm{v}_{i,j} - \mathrm{v}_{i,j}\delta T^{-1}\delta P_{ij} + T^{-1}\delta P_{ij}\delta \mathrm{v}_{i,j}\end{aligned} \tag{7.59}$$

The first term on the r.h.s. is then included in the sum over α in (7.57). In this respect let us observe that the factor T^{-1} is included in the flow J_α and not in the generalized force X_α.

We see that the source term (7.57) contains two types of contributions:

1°/The excess entropy source:†

$$\sum_{\alpha} \delta J_{\alpha} \delta X_{\alpha}$$

where J_{α} and X_{α} are given by (2.21).

2°/Terms involving convection, either through v_j or through δv_j.

It is this second type of terms which makes the source term $\sigma[\delta Z]$ so important for the understanding of situations in which *both* dissipative and mechanical phenomena play a role. On the contrary, the usual entropy production, given by (2.21) as well as the excess entropy production involve *only* dissipative terms. Since the excess source term (7.57) is directly connected to our stability criterion (6.31), we see that the presence of both types of terms will permit us to treat stability problems for very general situations. The first type of terms, is responsible for instabilities in purely dissipative systems as are the chemical systems considered in Chapter XIV. On the contrary, the second type of terms is responsible for instabilities in hydrodynamic problems such as the Bénard instability studied in Chapter XI.

From now, the source term $\sigma[\delta Z]$ corresponding to the excess entropy balance equation (7.56) will be called the *generalized excess entropy production*, and the corresponding flow term, the *generalized excess entropy flow*.

Alternative forms of the excess balance equations may also be obtained using suitable weighting functions, as shown in § 2 for the heat conduction problem (also § 12).

It should be emphasized that the excess entropy balance equation (7.56), used together with (7.57) is written in an inertial frame of reference. To obtain the appropriate expression in non-inertial frames of reference, one has to perform the following operations:

1°/Introduce into (7.57) the well known corresponding *inertial forces* as e.g. centrifugal and Coriolis forces, in addition to the external forces F_i.

2°/Interpret in (7.56) and (7.57): v_j, δv_j and operator ∂_t as respectively, the velocity, its perturbation and the local time derivative, with respect to the moving axis. Of course, the motion of the non-inertial frame of reference is assumed to be unperturbed.

† The so-called 'excess entropy production' should not be confused with the excess of the entropy production $\delta(\sum_{\alpha} J_{\alpha} X_{\alpha})$ (cf. § 2).

7. EXPLICIT STABILITY CRITERION FOR DISSIPATIVE PROCESSES

As in Chapter VI, we first consider separately the important special case of systems without convective effects. We have therefore: $v_j = \delta v_j = 0$, and also according to (6.17): $z = s$. Introducing these equalities into (7.56) and (7.57), we obtain a much simpler expression for the excess entropy balance equation:

$$\tfrac{1}{2}\,\partial_t \delta^2(\rho s) = \sigma[\delta S] - [\delta W_j \delta T^{-1} - \sum_{\gamma} \delta(\rho_\gamma \Delta_{\gamma j})\delta(\mu_\gamma T^{-1})]_{,j} \quad (7.60)$$

with

$$\sigma[\delta S] = \sum_{\alpha} \delta J_\alpha \delta X_\alpha \quad (7.61)$$

where the J_α and X_α are given by (2.21).

Let us observe that the generalized excess entropy flow and generalized excess entropy production including the effect of convection, simply reduce here, respectively to the excess entropy flow and to the excess entropy production. This may be checked term by term on the expressions (2.22) and (2.23) of the entropy flow and entropy production, taking into account the condition $v_j = 0$.

Once we have the excess entropy balance equation, we obtain directly using (6.12), the explicit stability conditions after integration over the volume. For fixed boundary conditions, we obtain in this way:

$$P[\delta S] = \int \sum_{\alpha} \delta J_\alpha \delta X_\alpha \mathrm{d}V \geqslant 0 \quad (>0) \qquad (t \geqslant t_0) \quad (7.62)$$

As the result, the explicit stability condition is determined by the sign of *the excess entropy production.*

Examples of fixed boundary conditions are:

$$(\delta T^{-1})_\Omega = 0 \quad \text{or} \quad (\delta W_n)_\Omega = 0 \quad (7.63)$$

$$[\delta(\mu_\gamma T^{-1})]_\Omega = 0 \quad \text{or} \quad [\delta(\rho_\gamma \Delta_\gamma)_n]_\Omega = 0 \quad (7.64)$$

However, as already pointed out in (7.23), the same procedure holds for more general boundary conditions. We have then a supplementary condition prescribed by the excess entropy flow:

$$\int_{\Omega_1} \Big[\sum_{\gamma} \delta(\rho_\gamma \Delta_{\gamma n})\delta(\mu_\gamma T^{-1}) - \delta W_n \delta T^{-1}\Big]\, \mathrm{d}\Omega \geqslant 0 \quad (>0) \quad (7.65)$$

where Ω_1 represents a part or the whole of the boundary. In the special case of a reference system in a steady state, with fixed boundary conditions, the stability condition (7.62) may be alternatively written as:

$$P_{st}[\delta S] = \int \sum_{\alpha} J_{\alpha} \delta X_{\alpha} \, dV \geqslant 0 \quad (>0) \tag{7.66}$$

Indeed, in this case:

$$\int \sum_{\alpha} (J_{\alpha})_{st} \delta X_{\alpha} \, dV = 0 \tag{7.67}$$

This relation may be proved easily, with the help of the mass and energy balance equations (1.26) and (1.42). For a steady state, without convection, these equations become:

$$0 = \sum_{\rho} \nu_{\gamma\rho} M_{\gamma} w_{\rho} - [\rho_{\gamma} \Delta_{\gamma j}]_{,j} \tag{7.68}$$

$$0 = \sum_{\gamma} F_{\gamma j} \rho_{\gamma} \Delta_{\gamma j} - W_{j,j} \tag{7.69}$$

We multiply (7.68) and (7.69) respectively by $-\delta(\mu_{\gamma} T^{-1})$ and δT^{-1}. We then add these equations and integrate by parts. The boundary term cancels and we obtain:

$$\int \{ W_j \delta T^{-1}_{,j} - \sum_{\gamma} (\rho_{\gamma} \Delta_{\gamma j}) \delta[(\mu_{\gamma} T^{-1})_{,j} - T^{-1} F_{\gamma j}] + \sum_{\rho} w_{\rho} \delta(A_{\rho} T^{-1}) \}_{st} \, dV = 0 \tag{7.70}$$

Obviously, this relation is identical to (7.67).

This derivation holds also in the case of the Helmholtz theorem (§ 3). Indeed, the momentum balance equation may then be used separately instead of (7.68) and (7.69). Multiplying by $T^{-1} \delta v_i$ and performing the same operations, we recover again equation (7.67). Some further comments on equation (7.67) are made in § 9.

8. STABILITY AND LINEAR THERMODYNAMICS

So far, the stability criterion (7.62) appears as a strictly thermodynamical condition, independent of the phenomenological laws.

Let us now restrict ourselves to the case of the linear phenomenological laws (3.2), with constant values of the coefficients $L_{\alpha\beta}$. The criterion (7.62) then becomes

$$P[\delta S] = \int \sum_{\alpha\beta} L_{\alpha\beta} \delta X_{\alpha} \delta X_{\beta} \, dV \geqslant 0 \quad (>0) \tag{7.71}$$

The r.h.s corresponds to the same type of quadratic form as the entropy production (3.8). Therefore, we may conclude that the stability condition (7.71) is identically fulfilled. We recover here a property we have already pointed out in the examples of heat conduction (7.8), (7.14) and chemical reactions (7.41).

Moreover, as a rule, the stability is in this case asymptotic. Indeed, for linear laws and given boundary conditions only one solution may exist. This means that the r.h.s. of (7.71) is then a definite function in the increments of the thermodynamic independent variables e, v, N_γ or ρe, ρ_γ (vanishing values of these increments imply vanishing of $\delta^2 S$ and *vice versa*).

The main conclusion to be drawn from condition (7.71) is that no instability may occur in the whole region of linear thermodynamics, when the boundary conditions are fixed, and in absence of inertial effects. An instability can only be expected beyond the region of linearity, i.e. far from equilibrium.

This is of course the main reason to extend the thermodynamic of irreversible processes to the non-linear domain.

9. STABILITY AND ENTROPY PRODUCTION

Let us now take into account the Onsager reciprocity relations (3.9). We obtain successively:

$$\begin{aligned}\sum_\alpha J_\alpha \delta X_\alpha &= \sum_{\alpha\beta} L_{\alpha\beta} X_\beta \delta X_\alpha \\ &= \sum_{\alpha\beta} X_\beta \delta(L_{\beta\alpha} X_\alpha) \\ &= \sum_\alpha X_\alpha \delta J_\alpha \end{aligned} \tag{7.72}$$

Therefore the variation of the entropy production takes the form:

$$\delta P[S] = \int \sum_\alpha \delta(J_\alpha X_\alpha)\, \mathrm{d}V = 2\int \sum_\alpha J_\alpha \delta X_\alpha\, \mathrm{d}V \tag{7.73}$$

For a reference system in a steady state with fixed boundary conditions, we have according to (7.67)

$$(\delta P)_{st} = 0 \tag{7.74}$$

Therefore we are left with:

$$(\Delta P)_{st} = \tfrac{1}{2}(\delta^2 P)_{st} = \int \sum_\alpha \delta J_\alpha \delta X_\alpha\, \mathrm{d}V \tag{7.75}$$

This result enables us to introduce the variation $(\Delta P)_{st}$ into our stability criterion (7.62). One obtains in this way an alternative and interesting proof of the *theorem of minimum entropy production* which has already been discussed in Chapter III, § 4. Indeed, we see that the entropy production in the perturbed state is always larger than in the reference steady state.

10. STABILITY AND EQUILIBRIUM

In this section we show that the stability theory of equilibrium developed in Chapters IV and V, may be derived as a particular case of the general approach outlined in this chapter. For an unperturbed state corresponding to thermodynamic equilibrium all the generalized forces in the entropy production (2.21) vanish. Hence:

$$A_\rho = 0; \quad T^{-1}_{,j} = 0; \quad F_\gamma T^{-1} \quad - (\mu_\gamma T^{-1})_{,j} = 0; \quad v_i = 0 \quad (7.76)$$

Introducing (7.76) into the excess balance equation (7.57) and considering the case of unperturbed boundary conditions, we obtain the global relation:

$$\tfrac{1}{2}\, \partial_t \delta^2 Z = P[\delta Z] = \int \sum_\alpha J_\alpha X_\alpha \, \mathrm{d}V = P[S] \geqslant 0 \qquad (7.77)$$

Observe that in contrast to the problem treated in § 7, the velocity perturbations are not excluded in the present case ($\delta v_j \neq 0$). According to (2.6), we see again that the inequality (7.77) is identically satisfied.

Integrating (7.77) over the time, we obtain:

$$-\tfrac{1}{2}\delta^2 Z = \int_i^e P \, \mathrm{d}t > 0 \quad (i = \text{initial}; \quad e = \text{equilibrium}) \qquad (7.78)$$

This implies (cf. 6.17):

$$\delta^2 s - T^{-1}(\delta \mathrm{v})^2 < 0 \qquad (7.79)$$

Comparison with (5.18) shows that we recover exactly the same stability conditions as before, since the coefficient T^{-1} of the additional term is a strictly positive quantity. Nevertheless (7.79) provides a supplementary piece of information. Indeed, we see that spontaneous excitation of internal convection cannot be generated from a state at rest in a system which is at thermodynamic equilibrium. This is of course a specific property of the equilibrium state. Indeed, as will be emphasized in Chapter XI, onset of free convection becomes possible starting from a steady non-equilibrium state, even in the linear region (Bénard problem).

In the general stability problem of equilibrium involving variable boundary conditions (cf. § 2), the condition (7.78) has to be replaced by (cf. 7.56); ($v_i = 0$ at equilibrium):

$$-\tfrac{1}{2}\delta^2 Z = \int_i^e P\,dt - \int_i^e dt \int_\Omega [W_j \delta T^{-1} - \sum_\gamma \rho_\gamma \Delta_{\gamma j} \delta(\mu_\gamma T^{-1}) - T^{-1} p_{ij} v_i - T^{-1} v_j \delta p] \alpha_j \, d\Omega > 0 \quad (7.80)$$

The α_j denote the components of the external normal along the x_j axis. As the first term on the r.h.s. is always a positive quantity, the stability condition for equilibrium will be certainly fulfilled if the additional inequality:

$$\int_\Omega [-W_j \delta T^{-1} + \sum_\gamma \rho_\gamma \Delta_{\gamma j} \delta(\mu_\gamma T^{-1}) + T^{-1} p_{ij} v_i + T^{-1} v_j \delta p] \alpha_j \, d\Omega > 0 \quad (7.81)$$

is separately satisfied. We observe that this supplementary condition involves the excess entropy condition flow, and terms linked to the pressure tensor. However, it no longer contains the excess entropy *convection flow* $v_\pi \delta(\rho s)$ as might have been expected from (2.22). A detailed analysis is given in the next paragraph. If (7.81) was not satisfied, external perturbations could induce instabilities in the system and e.g. convective effects could be generated.

11. COMPARISON WITH THE ENTROPY BALANCE EQUATION†

The stability criterion for equilibrium systems, derived in § 10 from the *excess entropy balance equation*, was already discussed in Chapter V, for purely dissipative systems starting from the *entropy balance equation* (5.1).

Let us now show that, using the excess balance equations for mass, momentum and energy (7.49)–(7.52), one can extend the latter method to situations involving inertial effects arising from perturbations of the barycentric velocity about the state at rest, ($v_i = 0$, $\delta v_i \neq 0$). In this way the agreement between the two different methods will be established. At the same time, the reason for the disappearance of the excess entropy convection flow in (7.81) will become clear. As we shall see, the excess entropy balance method used in § 10 is by far the simplest.

As already observed in Chapter V, § 1, we have to separate the second order terms in $\partial_t(\delta S)_e$ and $\Phi[S]_e$, before we may split the

† The results derived in this section are no longer used subsequently.

entropy balance equation (5.1) into two parts, containing respectively the first and second order terms.

Around the equilibrium state, one can write

$$W_n = \Delta W_n = \delta W_n + \tfrac{1}{2}\delta^2 W_n \tag{7.82}$$

and likewise for the flows $\rho_\gamma \Delta_{\gamma n}$ and v_n involved in (5.4). The contribution to the second order terms arising from $\Phi[S]_e$ is obtained in the form:

$$\tfrac{1}{2}\int [T^{-1}\delta^2 W_n - \sum_\gamma (\mu_\gamma T^{-1})\delta^2(\rho_\gamma \Delta_\gamma)_n + \rho s \delta^2 v_n] \, d\Omega \tag{7.83}$$

On the other hand, the contribution to the second order terms arising from $\partial_t(\delta S)_e$ may be calculated using the Gibbs equation in the form (2.60). This gives

$$\partial_t \delta(\rho s) = T^{-1} \, \partial_t \Delta(\rho e) - \sum_\gamma \mu_\gamma T^{-1} \, \partial_t \Delta \rho_\gamma \tag{7.84}$$

around an equilibrium state. The explicit form of $\partial_t \Delta(\rho e)$ and $\partial_t \Delta \rho_\gamma$ may be deduced from the *non-linearized* excess balance equations (7.49) and (7.52), that is by taking the δ as a finite increment ($\delta = \Delta$, § 5).

In this respect, let us emphasize that in the present case, the use of linearized equations would imply cancellation of the second order terms in the entropy balance equation (5.1). Therefore $\delta^2 S$ itself could then be neglected and the entropy balance equation would then reduce to the single first order equality (5.7). This equation is of no interest for stability theory.

The non-linearized excess balance equations (7.49) and (7.52), around equilibrium may be written as:

$$\partial_t \Delta \rho_\gamma = \sum_\rho \nu_{\gamma\rho} M_\gamma \Delta w_\rho - [\Delta(\rho_\gamma \Delta_{\gamma j}) + \rho_\gamma \Delta v_j + \Delta \rho_\gamma \Delta v_j]_{,j} \tag{7.85}$$

$$\begin{aligned} \partial_t \Delta(\rho e) = \sum_\gamma F_{\gamma j} \Delta(\rho_\gamma \Delta_{\gamma j}) - p \Delta v_{j,j} - \Delta P_{ij} \Delta v_{i,j} \\ - [\Delta W_j + \rho e \Delta v_j + \Delta(\rho e) \Delta v_j]_{,j} \end{aligned} \tag{7.86}$$

Let us introduce these relations into (7.84) and use the equilibrium conditions:

$$A_\rho = 0; \quad F_{\gamma j} = \mu_{\gamma,j}; \quad T_{,j} = 0 \tag{7.87}$$

The last condition implies also by virtue of the Gibbs–Duhem formula (2.46):

$$\sum_\gamma \rho_\gamma \mu_{\gamma,j} = p_{,j} \tag{7.88}$$

Using (2.60) and (2.40) in the form:

$$\sum_{\gamma} \rho_\gamma \mu_\gamma = \rho e + p - \rho s T \tag{7.89}$$

we obtain, after elementary manipulations:

$$T\, \partial_t \delta(\rho s) = -\Delta P_{ij} \Delta v_{i,j} - \sum_{\gamma} \mu_{\gamma,j} \Delta \rho_\gamma \Delta v_j - [\Delta W_j - \sum_{\gamma} \mu_\gamma \Delta(\rho_\gamma \Delta_{\gamma j}) + T \delta(\rho s) \Delta v_j + \rho s T \Delta v_j]_{,j} \tag{7.90}$$

Likewise, using again the second equality (7.87) and neglecting the third order terms such as $\Delta v_i (\rho_\gamma \Delta v_i \Delta v_j)_{,j}$, as well as $(\Delta v)^2 \partial_t \rho$, the non-linearized excess momentum balance (7.51) yields:

$$\tfrac{1}{2}\, \partial_t [\rho (\Delta v)^2] = \sum_{\gamma} \mu_{\gamma,i} \Delta v_i \Delta \rho_\gamma - \Delta v_i (\Delta P_{ij})_{,j} \tag{7.91}$$

Substituting (7.91) into (7.90), we obtain:

$$\partial_t \delta(\rho s) = -[T^{-1} \Delta W_j - \sum_{\gamma} \mu_\gamma T^{-1} \Delta(\rho_\gamma \Delta_{\gamma j}) + (\rho s + \delta(\rho s)) \Delta v_j + T^{-1} \Delta P_{ij} \Delta v_i]_{,j} - \tfrac{1}{2} T^{-1}\, \partial_t [\rho (\Delta v)^2] \tag{7.92}$$

Finally, one sees that the contribution to the second order terms arising from $\partial_t (\delta S)_e$ in the entropy balance equation (5.1) is given by:

$$-\tfrac{1}{2} \int [T^{-1} \delta^2 W_n - \sum_{\gamma} (\mu_\gamma T^{-1}) \delta^2 (\rho_\gamma \Delta_\gamma)_n + \rho s \delta^2 v_n]\, d\Omega - \int [\delta(\rho s) + T^{-1} \delta p] v_n\, d\Omega - \int T^{-1} p_{ij} v_i \alpha_j\, d\Omega - \partial_t \int \tfrac{1}{2} \rho T^{-1} v^2\, dV \tag{7.93}$$

This expression together with (7.83) represent the quantities we have to add to the l.h.s. of the second order entropy balance equation (5.8), used in Chapter V. We observe that in the absence of convective effects ($\delta v_i = v_i = 0$) expressions (7.83) and (7.93) cancel in the entropy balance equation. This remark confirms the validity of the method followed in Chapter V, when applied to purely dissipative systems. On the other hand, for more general situations involving convection, we recover the conditions obtained in the preceding paragraph. Indeed, due to the last term of (7.93) the quantity $\partial_t \delta^2 S$ is replaced by $\partial_t \delta^2 Z$, in agreement with the definitions (6.17) and (6.29). Moreover, the second term of (7.93) shows that the excess entropy convection flow $v_n \delta(\rho s)$ included in $\Delta\Phi[S]$ is also eliminated. The third term of (7.93) introduces the excess flow due to the pressure tensor, in agreement with our remarks at the end of § 10.

The reader will observe that the method used in § 10, which is based on the excess entropy balance equation, appears as more natural and more straightforward than the method used in this paragraph which starts from the entropy balance equation (5.1). For non-equilibrium systems, this second method is quite impracticable.

12. HYDRO-THERMODYNAMIC STABILITY

Going back to the complete expression of the excess entropy balance equation (7.56) and using (6.31), we deduce for fixed boundary conditions, the general stability criterion in the form:

$$P[\delta Z] = \int \sigma[\delta Z] \, \mathrm{d}V \geqslant 0 \quad (> 0) \qquad (t \geqslant t_0) \tag{7.94}$$

Therefore the explicit form of the general criterion concerns the sign of the generalized excess entropy production of the whole system. The explicit form of $\sigma[\delta Z]$ has been given in (7.57). We see that the stability condition for general macroscopic systems involving both dissipative and inertial effects, depends on three types of contributions:

(i) the excess entropy production which is the only contribution in the absence of convection;

(ii) terms containing as a factor gradients acting in the reference state.

(iii) terms containing as a factor the macroscopic velocity.

For variable boundary conditions, the additional inequality:

$$\int \{\sum_\gamma \delta(\rho_\gamma \Delta_{\gamma j})\delta(\mu_\gamma T^{-1}) - \delta W_j \delta T^{-1} + T^{-1}[\delta P_{ij}\delta \mathrm{v}_i + \tfrac{1}{2}\rho \mathrm{v}_j(\delta \mathbf{v})^2] - \mathrm{v}_j \delta^2(\rho s)\} \alpha_j \, \mathrm{d}\Omega \geqslant 0 \quad (> 0)(t \geqslant t_0) \tag{7.95}$$

has to be satisfied. The l.h.s. of (7.95) represents the generalized excess entropy flow. The general conditions (7.94) and (7.95) will be called the conditions of *hydrothermodynamic stability* as they involve both the thermodynamic and the hydrodynamic description of the system.

To understand the physical implications of these conditions it is essential to study various particular cases. This will be undertaken in subsequent chapters. In this respect, the separate thermodynamic and hydrodynamic conditions given in (6.42) and (6.43) are often useful (Chapter XI).

13. EXPLICIT FORM OF THE SEPARATE THERMODYNAMIC AND HYDRODYNAMIC STABILITY CRITERIA

To obtain the explicit form of the separate thermodynamic criterion (6.42), we introduce the linearized excess balance equations (7.49), (7.52) for mass and energy into (6.16), and we multiply the two sides by the weighting function ε^2 introduced in Chapter VI, § 9. Then, we proceed exactly as in § 6. Using (7.58) and (7.59) on the r.h.s. and recalling that the time derivative on the l.h.s. of (6.16) is taken for constant values of the coefficients, we get after elementary manipulations:

$$\begin{aligned}\tfrac{1}{2}\,\partial_t[\varepsilon^2\delta^2(\rho s)] &= \sum_\alpha \delta J_\alpha \delta X'_\alpha \\ &- \varepsilon^2 \delta \mathrm{v}_j\{T^{-1}_{,j}\delta(\rho e) + \sum_\gamma [F_{\gamma j}T^{-1} - (\mu_\gamma T^{-1})_{,j}]\delta\rho_\gamma\} \\ &+ \mathrm{v}_j\{\delta(\rho e)[\varepsilon^2\delta T^{-1}]_{,j} - \sum_\gamma \delta\rho_\gamma[\varepsilon^2\delta(\mu_\gamma T^{-1})]_{,j}\} \\ &+ \varepsilon^2\{F_j T^{-1}\delta\rho\delta \mathrm{v}_j + T^{-1}\delta P_{ij}\delta \mathrm{v}_{i,j} - \mathrm{v}_{i,j}\delta T^{-1}\delta P_{ij}\} \\ &- \{\varepsilon^2[\delta W_j \delta T^{-1} - \sum_\gamma \delta(\rho_\gamma\Delta_{\gamma j})\delta(\mu_\gamma T^{-1}) + \mathrm{v}_j\delta^2(\rho s)]\}_{,j}\end{aligned} \tag{7.96}$$

Here, the X'_α denote weighted generalized forces, namely:

$$\begin{aligned}\sum_\alpha \delta J_\alpha \delta X'_\alpha &= \delta W_j \delta(\varepsilon^2 T^{-1})_{,j} \\ &+ \sum_\gamma \delta(\rho_\gamma \Delta_{\gamma j})\delta[\varepsilon^2 F_{\gamma j}T^{-1} - (\varepsilon^2\mu_\gamma T^{-1})_{,j}] \\ &- \delta(p_{ij}T^{-1})\delta(\varepsilon^2 \mathrm{v}_{i,j}) - \sum_\rho \delta w_\rho \delta(\varepsilon^2 A_\rho T^{-1})\end{aligned} \tag{7.97}$$

The explicit expression of the thermodynamic stability criterion (6.42) arises directly from the balance equation (7.96) by integration over the whole volume. For fixed boundary conditions the last term of (7.96) vanishes by integration and the thermodynamic stability criterion takes the form:

$$P[\varepsilon^2\delta S] \geqslant 0 \quad (>0) \qquad (t \geqslant t_0) \tag{7.98}$$

For more general boundary conditions, one obtains the additional stability condition:

$$\int [\sum_\gamma \delta(\rho_\gamma \Delta_{\gamma n})\delta(\mu_\gamma T^{-1}) - \delta W_n \delta T^{-1} - \mathrm{v}_n \delta^2(\rho s)]\varepsilon^2 \, \mathrm{d}\Omega \geqslant 0 \quad (>0) \qquad (t \geqslant t_0) \tag{7.99}$$

As expected, the balance equation (7.96) shows that the conditions (7.98) and (7.99) reduce respectively to (7.62) and (7.65) for dissipative systems ($v_j = \delta v_j = 0$).

Likewise, to calculate the explicit expression of the separate hydrodynamic criterion (6.43), we use the weighting function $\tau^2 T_0^{-1}$ considered in Chapter VI, § 9. Then, taking into account (6.23), we obtain for the time derivative (again for constant values of the coefficients):

$$\tfrac{1}{2}\, \partial_t \left[\frac{\tau^2}{2}\, \delta^2(\rho \mathrm{v}^2)\right] = \tau^2 \delta \mathrm{v}_i\, \partial_t \delta(\rho \mathrm{v}_i) - \tau^2 \delta \left(\frac{\mathrm{v}^2}{2}\right) \quad \partial_t \delta\rho \quad (7.100)$$

Let us now introduce in the r.h.s., the excess balance equations for the total mass (7.50) and momentum (7.51). We find after manipulations similar to those above (cf. 6.24):

$$\begin{aligned}\tfrac{1}{2}\, \partial_t \left[\frac{\tau^2}{2}\, \delta^2(\rho \mathrm{v}^2)\right] &= \tfrac{1}{2}\, \partial_t[\rho\tau^2(\delta \mathrm{v})^2] \\ &= \tau^2 F_i \delta\rho \delta \mathrm{v}_i + \delta P_{ij}(\tau^2 \delta \mathrm{v}_i)_{,j} - \tau^2 \mathrm{v}_{i,j} \delta \mathrm{v}_i \delta(\rho \mathrm{v}_j) \\ &\quad - \tfrac{1}{2}(\delta \mathrm{v})^2 (\tau^2 \rho \mathrm{v}_j)_{,j} - \left[\tau^2 \delta \mathrm{v}_i \delta P_{ij} + \frac{\tau^2}{2} \rho \mathrm{v}_j (\delta \mathrm{v})^2\right]_{,j}\end{aligned} \quad (7.101)$$

For $\tau = 1$, this equality reduces to the excess balance equation for kinetic energy. In general, it provides us with the explicit expression of the hydrodynamic stability criterion (6.43). For fixed boundary conditions the last term of (7.101) vanishes after integration over the volume and the hydrodynamic stability criterion takes the form:

$$P[\tau^2 \delta E_{kin}] \leqslant 0 \qquad (t \geqslant t_0) \quad (7.102)$$

Again, for more general boundary conditions, one obtains the additional stability condition over the surface Ω:

$$\int [\delta \mathrm{v}_i \delta P_{ij} + \tfrac{1}{2} \rho \mathrm{v}_j (\delta \mathrm{v})^2] \tau^2 \alpha_j \, d\Omega \geqslant 0 \qquad (t \geqslant t_0) \quad (7.103)$$

as in (7.99). In particular, for $\varepsilon^2 = 1$ and $\tau^2 = T^{-1}$, the sum of (7.99) and (7.103) restores (7.95).

The physical meaning of the two separate thermodynamic and hydrodynamic stability criteria is discussed in Chapter XI in connection with the Bénard stability problem.

CHAPTER VIII

Stability and Fluctuations

1. EINSTEIN'S FLUCTUATION FORMULA

The thermodynamic stability theory as developed in Chapters VI and VII is essentially based on the inequalities (6.12), (6.13) (we neglect here macroscopic motion therefore: $Z = S$; cf. (6.17)):

$$\delta^2 S < 0 \tag{8.1}$$

and

$$\partial_t \delta^2 S > 0 \tag{8.2}$$

We want now to present a discussion of these inequalities from the point of view of fluctuation theory. This will give us a deeper insight into the meaning of stability theory from the molecular point of view.

Let us first consider equilibrium situations. The probability of fluctuations in an isolated system may be expressed through the basic Einstein formula (cf. H. Callen 1965, for an excellent summary of fluctuation theory):

$$Pr \sim \exp \frac{\Delta S}{k} \tag{8.3}$$

where ΔS is the change of entropy starting from equilibrium ($\Delta S < 0$) associated with the fluctuations and k the Boltzmann constant. Let us expand as in (5.2) the entropy around its value at equilibrium

$$S = S_e + (\delta S)_e + \tfrac{1}{2}(\delta^2 S)_e \tag{8.4}$$

For an isolated system

$$(\delta S)_e = 0 \tag{8.5}$$

Therefore (8.3) may also be written

$$Pr \sim \exp\left[\tfrac{1}{2}\frac{(\delta^2 S)_e}{k}\right] \tag{8.6}$$

The work of Greene and Callen (1951), Tisza and Quay (1963) has definitively established the validity of (8.6) when the fluctuations are small. Does (8.6) remain valid for non-equilibrium conditions? As inequality (8.1) is a direct consequence of the basic local equilibrium assumption adopted in this monograph, it is very tempting to also assume the validity of (8.6) over the same range.

The validity of Einstein's formula for non-equilibrium fluctuations was indeed postulated by one of us some years ago. Unfortunately, very little work has been done in this direction (Prigogine, 1954, Prigogine and Mayer, 1955, Lax, 1960). Recently however, Nicolis and Babloyantz (1969) have studied in detail various simple situations and have confirmed the validity of Einstein's formula for non-equilibrium patterns, at least when well defined conditions on the relaxation times are satisfied. These conditions refer to the *separation of the time scales* between the fluctuating system and the *outside world.* The time scales associated with the fluctuating system have to be much shorter than the time scales associated with the outside world, so that the state of the outside world may be considered as independent of the instantaneous state of the fluctuating system. This condition is in agreement with the idea that *given* boundary conditions maintain the non equilibrium state of the fluctuating system. In § 2, we consider a simple example.

2. CHEMICAL REACTIONS

The simplest case one may test Einstein's formula for non-equilibrium conditions refers to fluctuations in homogeneous non-equilibrium steady states which result from chemical reactions in open systems as considered in Chapter III, § 4. The simplifying features are then the following:

(*a*) The fluctuations involve a discrete finite set of variables instead of an infinite set, when transport phenomena occur.

(*b*) By varying the ratio of the initial and final products we can realize situations arbitrarily far from thermodynamic equilibrium.

In this simple case Einstein's formula (8.6) becomes (see 2.50):

$$Pr \sim e^{-\frac{1}{2kT}\sum_{\gamma\gamma'} \mu_{\gamma\gamma'}\delta N_\gamma \delta N_{\gamma'}} \tag{8.7}$$

where N_γ now denotes the *number of particles* of species γ. Thermodynamic quantities such as the chemical potentials are functions of the average number of particles $\overline{N}_\gamma$. Therefore for an *ideal* system (see 4.20) we have:

$$\mu_{\gamma\gamma'} = \frac{kT}{\overline{N}_\gamma}\delta_{\gamma\gamma'} \qquad \delta_{\gamma\gamma'} = \begin{cases} 0 & \gamma \neq \gamma' \\ 1 & \gamma = \gamma' \end{cases} \tag{8.8}$$

Then (8.7) reduces to

$$Pr \sim e^{-\sum_\gamma \frac{(\delta N_\gamma)^2}{2\overline{N}_\gamma}} \tag{8.9}$$

This is a gaussian distribution, predicting independent fluctuations for each component in an ideal system.

To test this formula let us consider the two monomolecular reactions:

$$A \underset{k_{21}}{\overset{k_{12}}{\rightleftharpoons}} X \underset{k_{32}}{\overset{k_{23}}{\rightleftharpoons}} F \tag{8.10}$$

The affinity of the overall reaction $A \rightarrow F$ is (cf. 3.43)†

$$\frac{\mathscr{A}}{RT} = \log \frac{K}{FA^{-1}} \tag{8.11}$$

where K is the equilibrium constant.

The affinity may take an arbitrary value depending on the given values of the concentrations of F and A. The chemical kinetic equations lead to the rate of change of X in the form of the difference between two partial reaction rates as (see 3.45):

$$\frac{dX}{dt} = w = (k_{12}A + k_{32}F) - (k_{21} + k_{23})X \tag{8.12}$$

The steady-state value of X is therefore

$$X_{st} = \frac{k_{12}A + k_{32}F}{k_{21} + k_{23}} \tag{8.13}$$

The description in terms of chemical kinetic equations such as (8.12) neglects fluctuations. In (8.12), X is the *average concentration* of component X. We want now to go to a more refined description

† In this chapter we use the symbol $\mathscr{A}$ instead of A for the chemical affinity, to avoid confusion with the initial product of the chemical reaction (8.10).

of the chemical processes which includes fluctuations. To this end we introduce the function $P(A, X, F, t)$ which gives the probability for finding prescribed values of the concentrations A, X, F at time t. There exists classical methods which allow an equation for the time change of P based on the so-called *stochastic* approach to be derived (some basic papers on this subject are collected in a special volume by Wax, 1954). A discussion of the molecular derivation of the stochastic equations method is beyond the scope of the present monograph. Let us only notice that for reaction rates not too-fast this approach is certainly justified (see Chapter II).

The specific applications of stochastic methods to chemical reactions has been very well reviewed by McQuarrie, 1967. For more details as well as many examples the reader is referred to this paper.

Consider the first reaction $A \rightarrow X$. In order to have, at time $t + \Delta t$, the state (A, X, F) it is necessary to start, at time t, with a state $(A + 1, X - 1, F)$ if Δt is sufficiently small. Therefore

$$\begin{aligned} P(A, X, F, t + \Delta t) = {} & (\text{Probability of reaction}) \\ & \times P(A + 1, X - 1, F, t) \\ & + (\text{Probability of no reaction}) \times P(A, X, F, t) \end{aligned}$$

Now the probability for a reaction to take place in the state (A, X, F) is clearly

$$\text{Probability} = k_{12} A \Delta t + 0(\Delta t)^2$$

where k_{12} is the rate of the reaction $A \rightarrow F$. It follows that

$$\begin{aligned} P(A, X, F, t + \Delta t) = {} & k_{12}(A + 1)\Delta t P(A + 1, X - 1, F, t) \\ & + (1 - k_{12} A \Delta t) P(A, X, F, t) \end{aligned}$$

The equation of evolution of $P(A, X, F, t)$ is obtained by adding the contributions of all four partial reactions (8.10). In addition we take the limit $\Delta t \rightarrow 0$ as we deal with a process which is continuous in time. In this way we finally obtain the following differential equation for P:

$$\begin{aligned} \partial_t P = {} & k_{12}(A + 1)P(A + 1, X - 1, F) - k_{12} A P \\ & + k_{21}(X + 1)P(A - 1, X + 1, F) - k_{21} X P \\ & + k_{23}(X + 1)P(A, X + 1, F - 1) - k_{23} X P \\ & + k_{32}(F + 1)P(A, X - 1, F + 1) - k_{32} F P \end{aligned} \tag{8.14}$$

This equation is difficult to handle as it contains finite differences. For this reason it is convenient to switch from A, X, F to new variables $\mathscr{S}_A$, $\mathscr{S}_X$, $\mathscr{S}_F$, corresponding to the moment generating function $\mathscr{F}(\mathscr{S}_A, \mathscr{S}_X, \mathscr{S}_F, t)$ through the relation:

$$\underset{|\mathscr{S}_i| \leqslant 1}{\mathscr{F}(\mathscr{S}_A, \mathscr{S}_X, \mathscr{S}_F, t)} = \sum_{A,X,F} \underset{(i=A,X,F)}{\mathscr{S}_A^A \mathscr{S}_X^X \mathscr{S}_F^F P(A, X, F, t)} \tag{8.15}$$

The condition $|\mathscr{S}_i| \leqslant 1$ is necessary for the sum over A, X, F to be convergent.

For convenience we compile here some of the fundamental properties of the moment generating function. Taking $\mathscr{S}_A, \mathscr{S}_X, \mathscr{S}_F \to 1$ in (8.15) we obtain because of the normalization of the distribution function P:

$$\lim_{\mathscr{S}_A, \mathscr{S}_X, \mathscr{S}_F \to 1} \mathscr{F}(\mathscr{S}_A, \mathscr{S}_X, \mathscr{S}_F, t) \equiv (\mathscr{F})_{\mathscr{S}=1} = \sum_{A,X,F} P(A, X, F, t) = 1$$

Moreover

$$\left(\frac{\partial \mathscr{F}}{\partial \mathscr{S}_X}\right)_{\mathscr{S}=1} = \sum_{A,X,F} XP(A, X, F, t) = \text{Average value of } X = \bar{X} \tag{8.16}$$

and

$$\frac{\partial}{\partial \mathscr{S}_X}\left(\mathscr{S}_X \frac{\partial \mathscr{F}}{\partial \mathscr{S}_X}\right)_{\mathscr{S}=1} = \sum_{A,X,F} X^2 P(A, X, F, t) = \overline{X^2} \tag{8.17}$$

There exist therefore simple relations between the partial derivatives of $\mathscr{F}$ taken at $\mathscr{S} = 1$ and the *moments* of P such as $\bar{X}^n$. For instance:

Mean square fluctuation of $X = \overline{\Delta X^2} = \overline{(X - \bar{X})^2} = \overline{X^2} - \bar{X}^2$

becomes:

$$\overline{\Delta X^2} = \frac{\partial}{\partial \mathscr{S}_X}\left(\mathscr{S}_X \frac{\partial \mathscr{F}}{\partial \mathscr{S}_X}\right)_{\mathscr{S}=1} - \left(\frac{\partial \mathscr{F}}{\partial \mathscr{S}_X}\right)^2_{\mathscr{S}=1}$$

We now take the time derivative of $\mathscr{F}$ and replace in (8.15) the time derivative of P by its value (8.14). Simple manipulations lead directly to the partial differential equation:

$$\partial_t \mathscr{F} = k_{12}(\mathscr{S}_X - \mathscr{S}_A) \frac{\partial \mathscr{F}}{\partial \mathscr{S}_A} + k_{21}(\mathscr{S}_A - \mathscr{S}_X) \frac{\partial \mathscr{F}}{\partial \mathscr{S}_X}$$
$$+ k_{23}(\mathscr{S}_F - \mathscr{S}_X) \frac{\partial \mathscr{F}}{\partial \mathscr{S}_X} + k_{32}(\mathscr{S}_X - \mathscr{S}_F) \frac{\partial \mathscr{F}}{\partial \mathscr{S}_F} \tag{8.18}$$

which is simpler to study than the finite difference equation (8.14).

This equation will in general admit no other time-independent solution than that corresponding to complete thermodynamic equilibrium between the products A, X, F in (8.10). Here we are however interested mainly in non-equilibrium steady states which are obtained by imposing the values of the concentrations of the initial and final products. We have therefore to introduce these conditions into (8.18). To do so we follow a method due to Nicolis and Babloyantz, (1969). We first define the reduced generating function

$$f(\mathscr{S}_X, t) = \mathscr{F}(\mathscr{S}_A = 1, \mathscr{S}_X, \mathscr{S}_F = 1, t) = \sum_X \mathscr{S}_X^X \{\sum_{A,F} P(A, X, F, t)\} \tag{8.19}$$

On the other hand, taking in (8.18) the limit $\mathscr{S}_A = \mathscr{S}_F = 1$, we obtain the equation:

$$\partial_t f = k_{12}(\mathscr{S} - 1)\sum_X \mathscr{S}^X\{\sum_{A,F} AP\} + (k_{21} + k_{23})(1 - \mathscr{S})\frac{\partial f}{\partial \mathscr{S}} + k_{32}(\mathscr{S} - 1)\sum_X \mathscr{S}^X\{\sum_{A,F} FP\} \tag{8.20}$$

where we have written $\mathscr{S}$ for $\mathscr{S}_X$.

We now assume that the *conditional* averages

$$\sum_{A,F} AP(A, X, F) \quad \text{and} \quad \sum_{A,F} FP(A, X, F) \tag{8.21}$$

do not depend on the concentration of X. This corresponds to the physical idea that the concentrations of the initial and final components play the role of *boundary conditions* and do not depend on the 'internal' state of the system. This assumption leads then to the equation:

$$\partial_t f = (1 - \mathscr{S})\left[(k_{23} + k_{21})\frac{\partial f}{\partial \mathscr{S}} - (k_{12}A + k_{32}F)f\right] \tag{8.22}$$

A, F are now the given concentrations of A and F.

The important feature is that equation (8.22) admits a time independent solution which is, when properly normalized [$f(\mathscr{S}) = 1$ for $\mathscr{S} = 1$, see (8.19)]:

$$f = \exp X_{st}(\mathscr{S} - 1) \tag{8.23}$$

Here X_{st} is the average concentration of X at the steady state as given by (8.13). Clearly, (8.23) is the generating function corresponding to the Poisson distribution

$$\rho(X) = e^{-X_{st}} \cdot \frac{(X_{st})^X}{X!} \tag{8.24}$$

Indeed [see (8.19)]:

$$e^{-X_{st}} \cdot \sum_X \frac{(X_{st})^X}{X!} \mathscr{S}^X = e^{-X_{st}} e^{X_{st}\mathscr{S}} \tag{8.25}$$

For small fluctuations around X_{st} (8.24) reduces to the Gaussian distribution

$$\rho(X) \sim e^{-\frac{(\delta X)^2}{2X_{st}}} \tag{8.26}$$

This relation is in agreement with Einstein's formula (8.9). The only change with respect to equilibrium fluctuations is in the value of X_{st} which is now given by (8.13) while for equilibrium fluctuations we would have [put $\mathscr{A} = 0$ in (8.11)]:

$$X_{st} = \frac{k_{12}}{k_{21}} A \quad \text{at thermodynamic equilibrium} \tag{8.27}$$

It may also be verified that if we start at $t = 0$ with an arbitrary function f it will approach in time the solution (8.23). Similar calculations apply to more general systems of chemical reactions (Nicolis and Babloyantz, 1969). Once it is assumed that the distribution of the initial and final components do not depend on the internal state of the system, Einstein's formula follows directly for small fluctuations.

3. FLUCTUATIONS OF TEMPERATURE

These considerations can also be extended to other irreversible processes such as diffusion or heat conduction. In the special case of fluctuations near *equilibrium* Einstein's formula (8.6) gives for temperature fluctuations (e.g. Landau and Lifshitz, 1958)

$$Pr \sim e^{-\frac{1}{2k}\int \frac{c_v^e}{T_e^2}(\delta T)^2 \, dV} \tag{8.28}$$

This formula permits us to calculate the probability of a temperature distribution $T(x, t)$, knowing the equilibrium temperature T_e. Let us notice once more the essential role of the thermodynamic stability condition (8.1) which implies $c_v^e > 0$.

We may now introduce a stochastic model for the exchange of energy similar to that used in § 2 for chemical reactions. The most expedient is to subdivide first the system into a finite number of boxes between which the exchange of energy proceeds.

The system as a whole is placed inside heat reservoirs. It is again assumed that the state of the heat reservoirs does *not* depend on the distribution of temperature inside the system. This decoupling procedure is similar to that used in § 2. Once this assumption is admitted the generalized Einstein formula follows directly (see for more detail Nicolis and Babloyantz, 1969) and the probability distribution takes the form

$$Pr \sim e^{-\frac{1}{2k}\int \frac{C_v^0}{T_0^2}(\delta T)^2 dV} \tag{8.29}$$

The steady state distribution T_0 is now space dependent. Also δT is now the deviation from the local steady state distribution. Formula (8.29) remains even valid for fluctuations around a time dependent macroscopic temperature distribution $T_0(x_i, t)$.

All these conclusions are based on the *decoupling assumption.* the internal state of the system does not influence directly the state of the reservoirs. One obtains then as in § 2 a closed master equation in which the state variables of the reservoirs appear as parameters, and which leads to an Einstein formula for small fluctuations.

4. REGRESSION OF FLUCTUATIONS

The validity of Einstein's formula for the problem treated in this Chapter permits to interpret the stability theory developed in Chapter V–VII in very simple physical terms. Inequality (8.1) expresses that the steady state we consider (or the macroscopic evolution) is more probable than any of the neighbouring states reached through a small fluctuation. However this is not enough to insure the stability. Let us indeed consider the probability Pr as a function of some fluctuating variable ξ. In Fig. (8.1(b)) we consider the case where there exists a second maximum of Pr which is even higher. We may then expect that the fluctuations will grow till the system will reach the second maximum. A sufficient condition that this will *not* happen and that *the fluctuations will regress* is given by inequality (8.2).

As we have seen the transition between stable and unstable states is due to the breakdown of inequality (8.2) for the critical normal mode (Chapter VII).

On the other hand $\delta^2 S$ will in general not vanish. From this point of view there is a parallelism between instabilities and phase transitions. As well known from equilibrium thermodynamics when we approach the boundary separating one phase from another in the

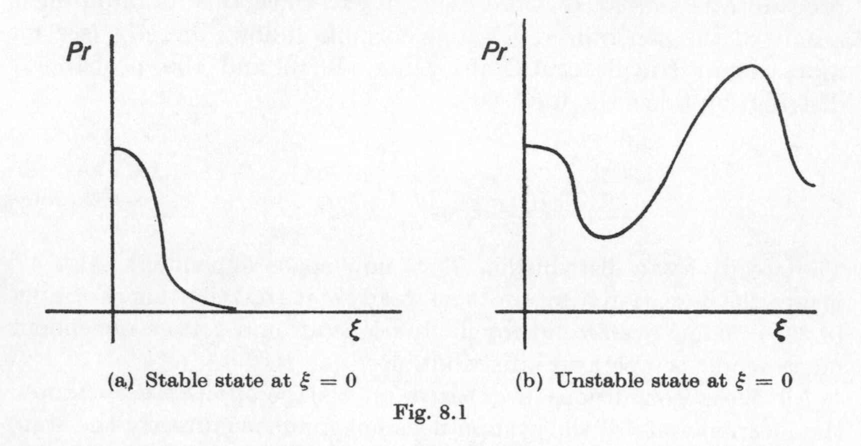

(a) Stable state at $\xi = 0$ (b) Unstable state at $\xi = 0$

Fig. 8.1

phase diagram, $\delta^2 S$ does not vanish. The situation is quite different near a critical point where $\delta^2 S \rightarrow 0$.

Beyond the critical normal mode, inequality (8.2) is no longer satisfied; the fluctuations will then grow. Within the framework of a linear theory we have to expect infinite fluctuations. In fact the latter will be damped due to the increasing influence of the non-linear terms neglected (Landau and Lifshitz, 1958).

5. CAUSAL DESCRIPTION AND FLUCTUATIONS

One of the most fascinating aspects of stability theory is its position at the border line between the deterministic description of matter in terms of macroscopic equations (such as the Stokes–Navier equation of motion . . .) and the theory of random processes. The very existence of spontaneous fluctuations is of course a manifestation of the many body aspect (or the atomistic aspect) of the systems considered. However when the system is stable, the fluctuations are of no importance as they regress. Their only effect is to add to the average evolution a kind of statistical noise. The situation changes radically when instabilities occur. Then the fluctuations are amplified and reach a macroscopic level. Once a new stable state is reached (steady or unsteady) the macroscopic description becomes again valid. However even so the statistical aspect of the time evolution remains essential because the type of new stable state which occurs may depend of the initial random fluctuation. As we shall see later, instabilities may give rise to a variety of new situations (see specially the case of symmetry breaking instabilities studied in Chapter XIV). It is the initial fluctuation which will determine the future situation.

Therefore the time evolution of such systems can only be understood using simultaneously deterministic and stochastic methods. It would of course be very gratifying to use a more unified approach based only on a stochastic equation of the type (8.14). It would then be no more necessary to add arbitrary perturbations to test the stability and the time evolution would be determined once the distribution function would be known at the initial time.

This approach would also give us the time delay involved in the formation of the new state, once the unstable region is reached.

This direction of research is being actively persued, but it is premature to present it now. The difficulty is to solve master equations such as those studied in § 2 of this chapter, when instabilities are involved. Also we are fully aware that it should be possible to establish the relation between $\delta^2 S$ and fluctuation theory in a more general way.

It should be noticed that Einstein's formula is not necessary for the validity of our thermodynamic approach. Even if for some classes of non linear patterns of evolution, it would appear that the stochastic analysis does not lead to Einstein's formula, still $\delta^2 S$ would keep its meaning as a *measure* of the fluctuation. Indeed, $\delta^2 S$ has a definite sign for stable systems and vanishes only at the steady state.

CHAPTER IX

The General Evolution Criterion

1. INTRODUCTION

Starting again from the local equilibrium assumption, we establish in this chapter a general inequality valid in the whole range of macroscopic physics for fixed boundary conditions. Owing to its high degree of generality, this inequality represents really a 'universal' evolution criterion.

In this respect, the linear stability theory developed in the previous chapters appears here as a simple particular case corresponding to evolution in the neighbourhood of a given reference state. In fact, this was the way we followed in our first papers on this subject (Glansdorff and Prigogine, 1964, 1965), before we adopted the separate treatment of stability as outlined in the preceding chapters. Accordingly, we shall no longer return to the infinitesimal stability problem in the present chapter, except to derive some supplementary results about the behaviour of a system around a steady state.

We shall rather focus our attention on some general properties related to this evolution criterion. In this respect, it is useful to distinguish two different types of problems: those which can be reduced to a variational formulation due to the existence of a so-called 'kinetic potential' and the others for which a potential does not exist.

Let us first consider the case of boundary conditions compatible with the maintenance of an equilibrium state. Then, the second law of thermodynamics immediately provides a criterion describing the evolution towards equilibrium. Indeed, as we have seen, the entropy production is never negative (cf. 2.2),

$$\mathrm{d}_i S \geqslant 0 \tag{9.1}$$

The equality sign corresponds to equilibrium (or reversible processes). As the entropy production represents only the part of the entropy increase due to changes inside the system, the criterion (9.1) appears in the form of a *non-exact differential.* However, when the evolution proceeds under conditions such that a thermodynamic potential exists, inequality (9.1) may be transformed into an exact differential. As an example, according to (5.20), for systems at constant temperature and volume, the entropy production becomes the exact differential

$$d_iS = -T^{-1}\,(\mathrm{d}F)_{TV} \geqslant 0 \tag{9.2}$$

This quantity is then directly related to the change of Helmholtz's free energy F. The evolution proceeds in the direction of decreasing F, until for stable equilibrium,the minimum of free energy is reached.

Now let us consider fixed boundary conditions which are incompatible with equilibrium. Then the system may eventually reach a steady non-equilibrium state, as e.g. in the heat conduction problem with temperature prescribed on the surface of the system. The general evolution criterion we shall establish supplements inequality (9.1), by introducing a new non-exact differential expression, say $\mathrm{d}\mathscr{D}$ satisfying the inequality

$$\mathrm{d}\mathscr{D} \leqslant 0 \tag{9.3}$$

with

$$\mathrm{d}\mathscr{D} < 0 \quad \text{for time-dependent processes} \tag{9.4}$$

$$\mathrm{d}\mathscr{D} = 0 \quad \text{for steady states} \tag{9.5}$$

The difference between these relationships and equation (9.1) which of course remains valid, is that condition (9.5) is now required for all time independent situations compatible with the boundary conditions, and not only for equilibrium states as in the case of inequality (9.1). The introduction in (9.3) of the sign $<$, as opposed to $>$ in (9.1), is a simple matter of definition, adopted for convenience.

On the other hand, the generalization of inequality (9.2) to non-equilibrium situations leads to the following question: Does there exist a potential such as F which determines by its sign the direction of evolution? In other words, is it possible to find a positive integrating factor ε such that

$$\varepsilon \mathrm{d}\mathscr{D} = \mathrm{d}\Phi \tag{9.6}$$

where $\mathrm{d}\Phi$ denotes the exact differential of some function Φ? When such a function exists in non equilibrium conditions, it will be called a kinetic potential, to avoid confusion with the usual thermodynamic potentials, as well as with the velocity potentials in hydrodynamics. It is in fact this question which was studied first (Prigogine, 1945), even before a general evolution criterion was considered. We have shown in Chapter III, § 4, that steady states which occur sufficiently near equilibrium may be characterized by the minimum of the entropy production $P[S]$. Therefore, a kinetic potential always exists for this class of evolutions, which then proceed in the direction of minimum entropy production. In several special cases, it is also possible to establish other kinetic potentials (see Chapter III, §§ 4 and 6).

In the general case however, the search for a 'universal' kinetic potential has proved to be unsuccessful (also Gage, Schiffer, Kline and Reynolds, 1965).

This is essentially due to the wide variety of macroscopic behaviours which may occur in systems far from equilibrium. To illustrate this point, let us consider the chemical non-equilibrium system (Chapter III, § 5; Chapter VII, § 4):

$$\{A\} \rightleftharpoons \boxed{\{X\}} \rightleftharpoons \{B\}$$

$\{A\}$ is the set of initial components, $\{B\}$ of the final ones. The concentrations of both $\{A\}$ and $\{B\}$ are maintained constant in time (fixed boundary conditions). On the contrary, the $\{X\}$ are intermediate components and their concentrations may vary in time. Suppose the $\{X_0\}$ correspond to a steady state (stable or not).

The behaviour of such a system is quite different according to the position of the steady state in respect to equilibrium. If the steady state is close to equilibrium, it is necessarily stable (Chapter VII, § 8). The system, if slightly perturbed, will go back directly to the steady state (Figure 9.1). Steady states far from equilibrium are stable or unstable as the case may be. Even for the stable cases, we may either have the behaviour represented in Figure 9.1, or that corresponding to Figure 9.2 or 9.3.

Examples of both types will be studied later on (§ 5 and Chapter XIV). Finally in the case of unstable steady states, an even larger variety of possibilities may arise.

Now it is clear that a rotation around a steady state, in the space of the concentrations X, as represented in Figure 9.3, with an amplitude which depends on the initial perturbation is in general incompatible with the existence of a kinetic potential in the form of a one-valued function. Indeed on each closed line, such as drawn on

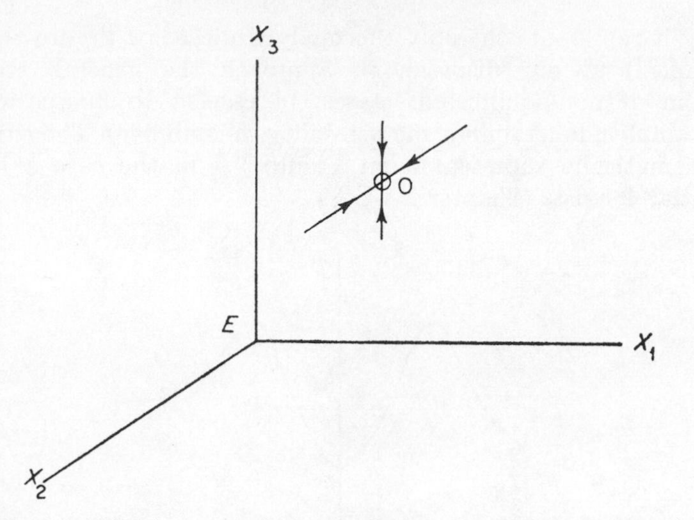

Fig. 9.1. Behaviour around the steady-state 0, near thermodynamic equilibrium E.

Figure 9.3, the kinetic potential would have to decrease continuously as a result of (9.6) and still to recover its initial value after each rotation. This is of course impossible (we shall give a simple example in § 5).

On the contrary, the construction of the non exact differential (9.3) first obtained for systems in mechanical equilibrium (Glansdorff and Prigogine, 1954), can be achieved in full generality to include mechanical convection processes as well (Glansdorff, 1960, Glansdorff and Prigogine, 1964). This 'universal' criterion is closely related to the properties of the entropy production. In the present states of the

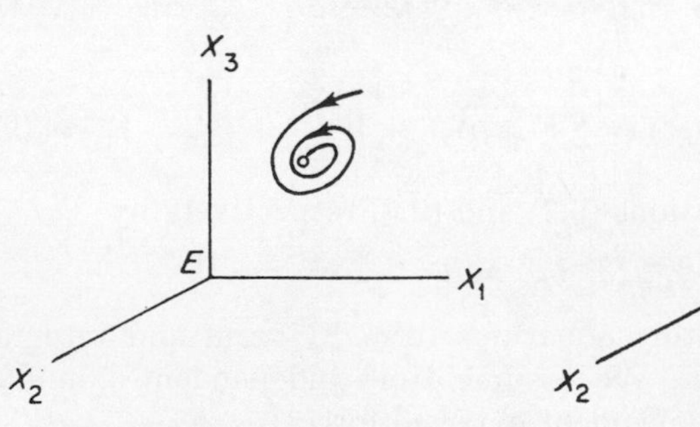

Fig. 9.2. Spiral approach to the steady state.

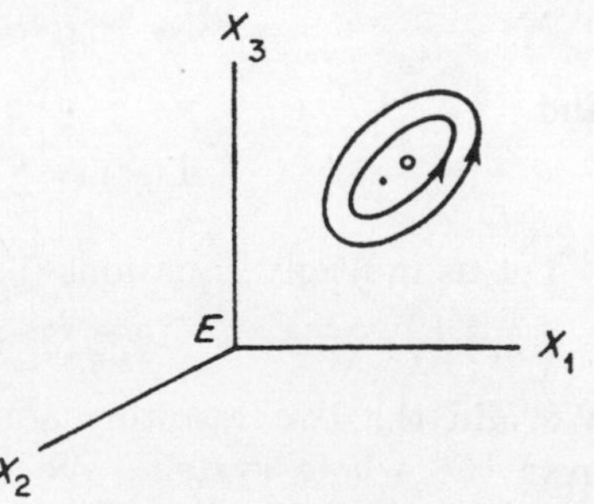

Fig. 9.3. Rotation around the steady state.

theory it represents the only thermodynamical law (hydro-thermodynamical) at our disposal, to approach the general stability problem of non-equilibrium states, in respect to fluctuations of finite amplitude (stability, metastability, unstability). The situation is schematically represented on Figure 9.4 in the case a kinetic potential Φ exists (Chapter XVI, §3).

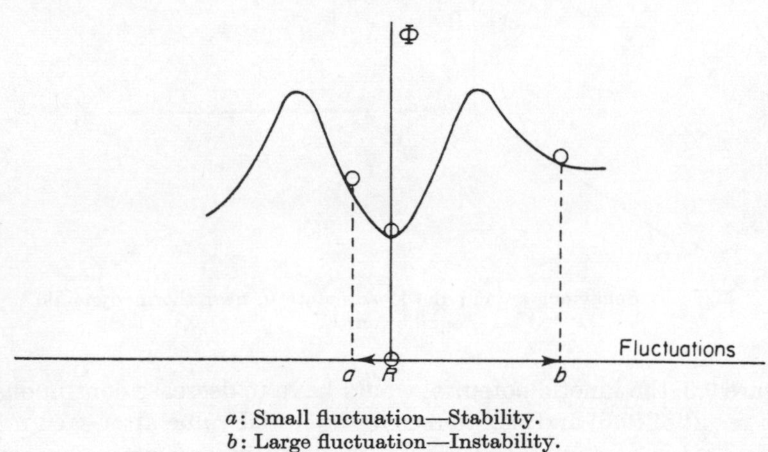

a: Small fluctuation—Stability.
b: Large fluctuation—Instability.
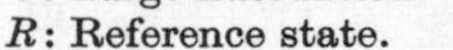
R: Reference state.

Fig. 9.4. Metastability in the case of arbitrary fluctuations.

2. EVOLUTION CRITERION FOR DISSIPATIVE PROCESSES

We first consider systems in the absence of any convective motion. The balance equations for mass and energy (1.28) and (1.42), take then the simple form [$\mathrm{d}_t = \partial_t$, see (1.16)]:

$$\mathrm{d}_t\rho_\gamma = \sum_\rho \nu_{\gamma\rho} M_\gamma w_\rho - [\rho_\gamma \Delta_{\gamma j}]_{,j} \tag{9.7}$$

and

$$\mathrm{d}_t(\rho e) = \sum_\gamma F_{\gamma j}\rho_\gamma \Delta_{\gamma j} - W_{j,j} \tag{9.8}$$

Let us multiply equations (9.7) and (9.8) respectively by

$$-\mathrm{d}_t(\mu_\gamma T^{-1}) \quad \text{and} \quad \mathrm{d}_t T^{-1}$$

We add the two resulting equations term by term and integrate over the whole system. We assume time independent boundary conditions and time independent external forces:

$$(\mathrm{d}_t\mu_\gamma)_\Omega = (\mathrm{d}_t T)_\Omega = 0; \quad \mathrm{d}_t F_{\gamma j} = 0 \tag{9.9}$$

Using equation (2.64) and integrating by parts in the r.h.s., we obtain the relation:

$$-\int \frac{\rho}{T}\left[\frac{c_v}{T}(\mathrm{d}_t T^{-1})^2 + \frac{\rho}{\chi}(\mathrm{d}_t v)^2_{N_\gamma} + \sum_{\gamma\gamma'} \mu_{\gamma\gamma'}\, \mathrm{d}_t N_\gamma\, \mathrm{d}_t N_{\gamma'}\right] \mathrm{d}V$$

$$= \int \{W_j\, \mathrm{d}_t T^{-1}_{,j} - \sum_\gamma \rho_\gamma \Delta_{\gamma j}\, \mathrm{d}_t[(\mu_\gamma T^{-1})_{,j} - T^{-1} F_{\gamma j}]$$

$$+ \sum_\rho w_\rho\, \mathrm{d}_t(A_\rho T^{-1})\}\, \mathrm{d}V \leqslant 0 \qquad (9.10)$$

The inequality sign is prescribed by the local equilibrium stability conditions (4.13–4.15).

It is only for a time independent state which may be either an equilibrium state or a non-equilibrium steady state, that (9.10) vanishes.

Let us compare inequality (9.10) to the expression of the entropy production written as (cf. 2.21, 2.24)

$$P[S] = \int \sum_\alpha J_\alpha X_\alpha\, \mathrm{d}V = \int \{W_j T^{-1}_{,j} - \sum_\gamma \rho_\gamma \Delta_{\gamma j}[(\mu_\gamma T^{-1})_{,j} - T^{-1} F_{\gamma j}]$$

$$+ \sum w_\rho A_\rho T^{-1}\}\, \mathrm{d}V \geqslant 0 \qquad (9.11)$$

It is convenient to split the time change of the entropy production P into two parts

$$\frac{\mathrm{d}P}{\mathrm{d}t} = \frac{\mathrm{d}_X P}{\mathrm{d}t} + \frac{\mathrm{d}_J P}{\mathrm{d}t} \qquad (9.12)$$

By definition:

$$\frac{\mathrm{d}_X P}{\mathrm{d}t} = \int \sum_\alpha J_\alpha\, \mathrm{d}_t X_\alpha\, \mathrm{d}V; \quad \frac{\mathrm{d}_J P}{\mathrm{d}t} = \int \sum_\alpha X_\alpha\, \mathrm{d}_t J_\alpha\, \mathrm{d}V \qquad (9.13)$$

Inequality (9.10) then becomes

$$\frac{\mathrm{d}_X P}{\mathrm{d}t} \leqslant 0 \qquad (9.14)$$

This is the compact form of our evolution criterion for dissipative systems. The change of the forces X_α proceeds always in a way as to lower the value of the entropy production. This criterion is independent of any assumption about the phenomenological relations between the rates and the forces.

On the other hand, it must be stressed that this criterion does not yield any information about the sign and magnitude of $\mathrm{d}_J P$. Therefore, as a rule, the sign of the exact differential (9.12) which represents the total change of the entropy production, is by no means prescribed by that of (9.14).

As shown by the l.h.s. of (9.10), the existence of such an evolution criterion is a direct consequence of the local equilibrium stability conditions and thus an indirect consequence of the second law of thermodynamics.

Let us also observe that the equality sign in (9.14) corresponds to the steady state if its exists. Using the mass and energy balances (9.7) and (9.8), it is easy to verify by the same line of reasoning as for (7.67), that at the steady state:

$$\left(\frac{\mathrm{d}_X P}{\mathrm{d}t}\right)_{st} = \int \sum_\alpha J_{\alpha st}\, \mathrm{d}_t X_\alpha\, \mathrm{d}V = 0 \tag{9.15}$$

This allows us to rewrite the evolution criterion in terms of the excess flows and excess forces, in respect to a steady state, as:

$$\frac{\mathrm{d}_X \Delta P}{\mathrm{d}t} = -\int \frac{\rho}{T}\left[\frac{c_v}{T}(\mathrm{d}_t \Delta T^{-1})^2 + \frac{\rho}{\chi}(\mathrm{d}_t \Delta v)^2_{N_\gamma} + \sum_{\gamma\gamma'} \mu_{\gamma\gamma'}\, \mathrm{d}_t \Delta N_\gamma\, \mathrm{d}_t \Delta N_{\gamma'}\right] \mathrm{d}V = \int \sum_\alpha \Delta J_\alpha\, \mathrm{d}_t \Delta X_\alpha\, \mathrm{d}V \leqslant 0$$

$$(\Delta J_\alpha = J_\alpha - J_{\alpha st};\quad \Delta X_\alpha = X_\alpha - X_{\alpha st}) \tag{9.16}$$

Therefore, in the neighbourhood of a steady state, the evolution criterion is independent of the first order terms, exactly as the entropy production around equilibrium (Chapter IV § 1).

3. EVOLUTION CRITERION AND THEOREM OF MINIMUM ENTROPY PRODUCTION

In the linear range of thermodynamics of irreversible processes (Chapter III) we have, using the Onsager reciprocity relations (3.9):

$$\begin{aligned}\sum_\alpha J_\alpha\, \mathrm{d}_t X_\alpha &= \sum_{\alpha\beta} L_{\alpha\beta} X_\beta\, \mathrm{d}_t X_\alpha \\ &= \sum_{\alpha\beta} X_\beta\, \mathrm{d}_t (L_{\beta\alpha} X_\alpha) \\ &= \sum_\beta X_\beta\, \mathrm{d}_t J_\beta\end{aligned} \tag{9.17}$$

There is then a complete symmetry between forces and rates, and the two terms in (9.12) are equal. Therefore

$$\mathrm{d}_X P = \mathrm{d}_J P = \tfrac{1}{2}\, \mathrm{d}P \leqslant 0 \tag{9.18}$$

As could have been expected we simply recover in this case the theorem of minimum entropy production (Chapter III, § 4). We see that this theorem is a very special case of the more general evolution criterion (9.14).

4. EVOLUTION CRITERION AND STEADY-STATE CONDITIONS

We have already stressed that $\mathrm{d}_X P$ exactly as $\mathrm{d}_i S$ is a non total differential. In spite of this, we may derive directly from $\mathrm{d}_i S$ both the classical equilibrium conditions and the corresponding stability conditions. Likewise, we may expect to derive from $\mathrm{d}_X P$ both the steady state conditions and the corresponding stability conditions.

To this end, let us consider first the case of chemical reactions in a homogeneous medium. The only forces X_α are the affinities A_ρ. We obtain (see 2.21 and 4.24):

$$T\,\mathrm{d}_i S = \sum_\rho A_\rho\,\mathrm{d}\xi_\rho \geqslant 0 \tag{9.19}$$

where ξ_ρ is the chemical variable corresponding to the reaction ρ. Let us consider arbitrary variations of the ξ_ρ. The system will be in stable equilibrium if, for all possible variations of the independent variables ξ_ρ, we have:

$$T\delta_i S = \sum_\rho A_\rho \delta\xi_\rho \leqslant 0 \tag{9.20}$$

This implies the equilibrium conditions

$$(A_\rho)_{eq} = 0 \tag{9.21}$$

and the stability conditions:

$$\sum_{\rho\rho'} \left(\frac{\partial A}{\partial \xi_{\rho'}}\right)_{eq} \delta\xi_\rho \delta\xi_{\rho'} < 0 \tag{9.22}$$

according to (3.1) and (4.29).

Let us now proceed similarly with $\mathrm{d}_X P$ for a steady state.

Considering again chemical reactions in a homogeneous medium where the only forces X_α are the affinities A_ρ, we have:

$$P = \sum_\rho w_\rho A_\rho T^{-1} \geqslant 0 \tag{9.23}$$

and (T = const.):

$$\mathrm{d}_X P = \sum_\rho w_\rho T^{-1}\,\mathrm{d}A_\rho \leqslant 0 \tag{9.24}$$

The geometrical interpretation of (9.23–9.24) is given on Figure 9.5. According to (9.23) the scalar product of the 'vector' **A** (of components $A_1, \ldots, A_r$) with 'vector' $\boldsymbol{w}$ is a positive quantity, while according to (9.24), the scalar product of 'vector' d**A** with 'vector'

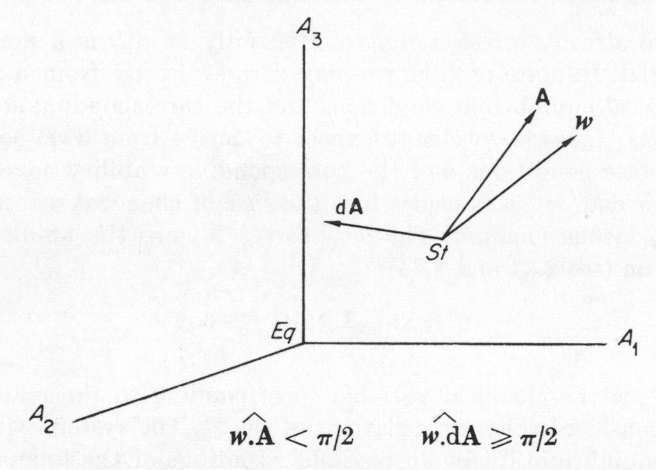

$\widehat{\boldsymbol{w}.\mathbf{A}} < \pi/2$ $\quad \widehat{\boldsymbol{w}.\mathrm{d}\mathbf{A}} \geqslant \pi/2$

Fig. 9.5. Geometrical interpretation of (9.23–9.24).

$\boldsymbol{w}$ is a negative quantity. As a consequence of (9.24) a sufficient condition for the stability of the steady state is given by inequality

$$\delta_X P = \sum_\rho w_\rho T^{-1} \delta A_\rho \geqslant 0 \tag{9.25}$$

since the temperature is here considered as a constant ($\delta T = 0$). The system will be at the steady state if the equality sign is fulfilled in (9.25). This implies that the coefficients of the *independent* affinities vanish [e.g. (9.32) below], namely

$$\mathscr{W}_\rho = 0 \tag{9.26}$$

where $\mathscr{W}_\rho$, denotes a *relative* velocity [e.g. (9.33) below]. Moreover (9.25) gives rise to the explicit stability condition for small disturbances around the steady state in the form:

$$\sum_\rho \delta w_\rho \delta A_\rho > 0 \tag{9.27}$$

This is again in agreement with our former results as we recover our condition (7.40) on the excess entropy production.

Likewise the whole stability theory of non-equilibrium steady states with respect to small disturbances, could be deduced from the evolution criterion (9.14). The method leads to the same results as those already obtained in Chapter VII. Therefore, we shall not go into more detail.

Let us indicate an alternative form of the steady state condition (9.26). Inequality (9.25) can also be written as

$$\delta_X P = \delta P - \sum_\rho A_\rho T^{-1} \delta w_\rho \geqslant 0 \tag{9.28}$$

If the steady state depends on some parameter which may be for instance, the concentration Y of some intermediate component, inequality (9.28) leads to the steady state condition:

$$\frac{\partial P}{\partial Y} - \sum_\rho A_\rho T^{-1} \frac{\partial w_\rho}{\partial Y} = 0 \tag{9.29}$$

Whereas (9.26) represents the kinetic form of the steady state condition, the relation (9.29) appears as its thermodynamic form. In this respect, it is easy to verify that close to equilibrium, the second term of (9.29) may be combined with the first such as to recover the theorem (9.18) of minimum entropy production in the form (for an explicit example, see Prigogine, 1965):

$$\frac{\partial P}{\partial Y} = 0 \tag{9.30}$$

Equation (9.29) provides therefore a generalization of this theorem for steady states far from equilibrium.

Let us finally go back to the interpretation of the kinetic condition (9.26). We consider the sequence of chemical reactions:

$$A \rightleftharpoons X \rightleftharpoons Y \rightleftharpoons B \tag{9.31}$$

already discussed in § 1. The total affinity:

$$A_{\text{tot}} = A_1 + A_2 + A_3 = \text{constant} \tag{9.32}$$

is fixed, since the concentrations A and B are prescribed. In terms of the *independent* affinities, (9.25) may be rewritten as:

$$\delta_X P = (w_1 - w_3) T^{-1} \delta A_1 + (w_2 - w_3) T^{-1} \delta A_2 \geqslant 0 \tag{9.33}$$

This leads to steady state conditions:

$$w_1 - w_2 = w_2 - w_3 = 0 \tag{9.34}$$

which represent the explicit form of (9.26). As a result:

$$w_1 = w_2 = w_3 \tag{9.35}$$

Clearly at the steady state, the rates of the three processes have to be equal.

5. ROTATION AROUND STEADY STATES—KINETIC POTENTIAL

We start again with the evolution criterion (9.24) related to the chemical problem discussed in § 4. We consider more especially evolution in the neighbourhood of a steady state. Using the excess form (9.16) we may write:

$$T\,\mathrm{d}_X P = \sum_\rho \Delta w_\rho \,\mathrm{d}\,\Delta A_\rho \leqslant 0 \tag{9.36}$$

As we limit ourselves to the neighbourhood of the steady state, we may expand the Δw_ρ in terms of the deviations $\Delta A_{\rho'}$ of the affinities and retain only the first order terms. Such an expansion implies that the deviations of w_ρ depend only on those of the affinities. In this particular case we obtain:

$$\Delta w_\rho = \delta w_\rho = \sum_{\rho'} l_{\rho\rho'} \delta A_\rho \tag{9.37}$$

where the coefficients

$$l_{\rho\rho'} = \frac{\partial w_\rho}{\partial A_{\rho'}} \tag{9.38}$$

refer to the steady state of reference. In Chapter XIV we shall consider some specific examples of chemical reactions and give explicit expressions of these coefficients. Here, we wish only to present a few theoretical remarks (Prigogine and Balescu, 1955, 1956; Prigogine, 1967). Exactly as in (3.7) we may split $l_{\rho\rho'}$, into a symmetric and an antisymmetric part, as

$$l_{\rho\rho'} = l_{(\rho\rho')} + l_{[\rho\rho']} \tag{9.39}$$

with

$$l_{(\rho\rho')} = l_{(\rho'\rho)}; \quad l_{[\rho\rho']} = -l_{[\rho'\rho]} \tag{9.40}$$

We then obtain for (9.36)

$$T\,\mathrm{d}_X P = \sum_{\rho\rho'} [l_{(\rho\rho')} \delta A_{\rho'}\, \mathrm{d}(\delta A_\rho) + l_{[\rho\rho']} \delta A_{\rho'}\, \mathrm{d}(\delta A_\rho)] \leqslant 0 \tag{9.41}$$

Let us now consider the following special cases:

(*a*) the steady state is sufficiently close to equilibrium to apply linear thermodynamics of irreversible processes. Then the coefficients $l_{\rho\rho'}$, become identical to the phenomenological coefficients $L_{\rho\rho'}$ introduced in Chapter III. Because of the Onsager reciprocal relations (3.9), we then have simply:

$$T\,\mathrm{d}_X P = \sum_{\rho\rho'} L_{(\rho\rho')}\delta A_{\rho'}\,\mathrm{d}(\delta A_\rho) = T\,\mathrm{d}_J P = \tfrac{1}{2}T\,\mathrm{d}P \leqslant 0 \tag{9.42}$$

According to (9.18) we recover again the theorem of minimum entropy production.

(*b*) the steady state is too far from equilibrium to apply the linear theory of Chapter III. Still we can consider situations, where the antisymmetric coefficients vanish. Thus

$$l_{[\rho\rho']} = 0 \tag{9.43}$$

This is a *sufficient* condition for the existence of a kinetic potential, as we have then according to (9.41):

$$T\,\mathrm{d}_X P = \mathrm{d}\Phi \leqslant 0 \tag{9.44}$$

where the homogeneous function of the second degree with respect to this δA_ρ:

$$\Phi = \tfrac{1}{2}\sum_{\rho\rho'} l_{(\rho\rho')}\delta A_\rho \delta A_{\rho'} \tag{9.45}$$

denotes this kinetic potential.

(*c*) The antisymmetric part does not vanish. The approach to the steady state will then take the form represented on Figure 9.2. Still in some cases it is possible to find a kinetic potential through the introduction of a suitable integrating factor [see (9.6)]. However as we already mentioned in § 1, in general no kinetic potential exists.

To illustrate this point let us consider the extreme case of two chemical processes described by a purely antisymmetric matrix $l_{\rho\rho'}$, that is:

$$l_{(\rho\rho')} = 0; \quad l_{[12]} = -l_{[21]} = l \tag{9.46}$$

Then (9.41) becomes:

$$T\,\mathrm{d}_X P = l(\delta A_2\,\mathrm{d}\delta A_1 - \delta A_1\,\mathrm{d}\delta A_2) \leqslant 0 \tag{9.47}$$

Introducing polar coordinates r, θ around the steady state, (9.47) takes the simpler form:

$$T\,\mathrm{d}_X P = -lr^2\,\mathrm{d}\theta \leqslant 0 \tag{9.48}$$

This inequality prescribes an irreversible *direction of rotation* around the steady state. However the function:

$$\Phi = lr^2\theta \tag{9.49}$$

is not a suitable kinetic potential as it corresponds to a multi-valued function which increases by the amount $-2\pi lr^2$, every time the system performs a rotation around the steady state. As a matter of fact, one could obtain a uniform function by using in (9.48) as integrating factor:

$$\lambda = \sin\theta$$

This would give the single-valued potential:

$$\Phi = lr^2\cos\theta \tag{9.50}$$

Unfortunately, the inequality $\mathrm{d}\Phi \leqslant 0$ is no longer fulfilled for this kinetic potential. It is clearly impossible to obtain a one-valued function which still satisfies inequality (9.48) by means of a non-vanishing integrating factor. Therefore one may conclude that in such situations it is impossible to construct non-singular kinetic potentials.

6. BEHAVIOUR OF NORMAL MODES AROUND A STEADY STATE IN DISSIPATIVE SYSTEMS

Let us now investigate the behaviour of a single normal mode corresponding to a small disturbance around the steady state in a dissipative system (i.e. a system without convection). To this end we start from expression (9.16) of the evolution criterion limited to the second-order terms. Using (2.76), we write (9.16) in complex variables as:

$$-\int \frac{\rho}{T}\left[\frac{c_v}{T}\,\mathrm{d}_t\delta T^{-1}\,\mathrm{d}_t\delta T^{-1*} + \frac{\rho}{\chi}(\mathrm{d}_t\delta v)_{N_\gamma}(\mathrm{d}_t\delta v)^*_{N_\gamma}\right.$$
$$\left. + \tfrac{1}{2}\sum_{\gamma\gamma'}\mu_{\gamma\gamma'}(\mathrm{d}_t\delta N_\gamma\,\mathrm{d}_t\delta N^*_{\gamma'} + \mathrm{d}_t\delta N^*_\gamma\,\mathrm{d}_t\delta N_{\gamma'})\right]\mathrm{d}V$$
$$= \tfrac{1}{2}\int\sum_\alpha(\delta J_\alpha\,\mathrm{d}_t\delta X^*_\alpha + \delta J^*_\alpha\,\mathrm{d}_t\delta X_\alpha)\,\mathrm{d}V \leqslant 0 \tag{9.51}$$

As observed previously the l.h.s. represents a real negative definite quantity due to the local equilibrium stability conditions (4.13–4.15). For a single normal mode (9.51) becomes:†

$$(\omega_r^2 + \omega_i^2)\delta^2 S = \omega_r \delta P + \omega_i \delta \Pi \leqslant 0 \tag{9.52}$$

where

$$\delta P = \tfrac{1}{2}\int \sum_\alpha (\delta J_\alpha \delta X_\alpha^* + \delta J_\alpha^* \delta X_\alpha)\, \mathrm{d}V \tag{9.53}$$

is the excess entropy production. Likewise

$$\delta \Pi = -\frac{i}{2}\int \sum_\alpha (\delta J_\alpha \delta X_\alpha^* - \delta J_\alpha^* \delta X_\alpha)\, \mathrm{d}V \tag{9.54}$$

is a real quantity, the sign of which determines the direction of the irreversible rotations around the steady state in the X_j space (Fig. 9.2–9.3). Indeed, equality:

$$\tfrac{1}{2}\, \mathrm{d}_t \delta^2 S = \omega_r \delta^2 S = \delta P = P[\delta S] \tag{9.55}$$

arising from (6.34), (6.39) and (7.62) allows us to split inequality (9.52) into two separate relations, namely:

$$\omega_r^2 \delta^2 S = \omega_r \delta P \leqslant 0 \tag{9.56}$$

and

$$\omega_i^2 \delta^2 S = \omega_i \delta \Pi \leqslant 0 \tag{9.57}$$

The first inequality again reduces to the stability criterion for a single normal mode ($\delta P > 0$ or $\omega_r < 0$), while the second links the sign of $\delta\Pi$ to that of the angular frequency ω_i, that is to the rotation in the X_j space.

Condition $\delta P = 0$, corresponds to the marginal state of critical stability, when satisfied for non trivial zero values of the disturbances ($\omega_r = 0$). Similarly, condition $\delta \Pi = 0$, corresponds to the critical state for the appearance of an aperiodical motion ($\omega_i = 0$).

On the other hand, taking once more the time derivative of (9.55), we obtain:

$$\mathrm{d}_t \delta P = 2\omega_r \delta P = 2\omega_r^2 \delta^2 S \leqslant 0 \tag{9.58}$$

This result allows us to interpret the excess entropy production as a potential which decreases on time and corresponds to a minimum

† For the sake of simplicity, subscript m is here everywhere implied in $\delta_m^2 S$, $\delta_m P$, $\delta_m \Pi$.

for a stable steady state ($\delta P \geqslant 0$), and to a maximum in the unstable case ($\delta P < 0$). In both cases the extremum corresponds to a vanishing value of δP.

As in (9.12) we may split the time change of δP into two parts as:

$$\mathrm{d}\delta P = \mathrm{d}_x\delta P + \mathrm{d}_j\delta P \quad (x \equiv \delta X; \;\; j \equiv \delta J) \tag{9.59}$$

Observe that, according to (9.53), the term $\mathrm{d}_x\, \delta P/\mathrm{d}t$ represents the r.h.s. of (9.51). Therefore, by comparison with (9.52), we get

$$\frac{\mathrm{d}_x\delta P}{\mathrm{d}t} = (\omega_r^2 + \omega_i^2)\delta^2 S \leqslant 0 \tag{9.60}$$

We observe $\mathrm{d}_x\delta P$ as well as $\mathrm{d}\delta P$ deduced from (9.58), are both negative quantities, while the sign of $\mathrm{d}_j\delta P$ is not prescribed. One obtains by difference:

$$\frac{\mathrm{d}_j\delta P}{\mathrm{d}t} = (\omega_r^2 - \omega_i^2)\delta^2 S = \omega_r\delta P - \omega_i\delta\Pi \tag{9.61}$$

The sign depends therefore on the relative values of $|\omega_r|$ and $|\omega_i|$, that is finally on the logarithmic decrement of the normal mode considered.

Let us now consider the case for which the excess flows may be expanded in terms of the excess forces as:

$$\delta J_\alpha = \sum_\beta l_{\alpha\beta}\delta X_\beta \tag{9.62}$$

with

$$l_{\alpha\beta} = \frac{\partial J_\alpha}{\partial X_\beta} \tag{9.63}$$

This is the straightforward generalization of the chemical problem treated in § 5 (compare with 9.37 and 9.38). In this case, (9.53) and (9.54) become respectively:

$$\delta P = \tfrac{1}{2}\int \sum_{\alpha\beta} l_{(\alpha\beta)}\,(\delta X_\alpha^*\delta X_\beta + \delta X_\alpha\delta X_\beta^*)\,\mathrm{d}V$$
$$= \int \sum_{\alpha\beta} l_{(\alpha\beta)}\,\delta X_\alpha^*\delta X_\beta\,\mathrm{d}V \tag{9.64}$$

and

$$\delta\pi = \frac{i}{2}\int \sum_{\alpha\beta} l_{[\alpha\beta]}\,(\delta X_\alpha^*\delta X_\beta - \delta X_\alpha\delta X_\beta^*)\,\mathrm{d}V$$
$$= i\int \sum_{\alpha\beta} l_{[\alpha\beta]}\,\delta X_\alpha^*\delta X_\beta\,\mathrm{d}V \tag{9.65}$$

Relation (9.64) introduced into (9.56) allows to relate the sign of ω_r to the symmetric matrix $l_{(\alpha\beta)}$.

Likewise, (9.65) introduced in (9.57) allows to connect the direction of the rotation ω_i to the antisymmetric matrix elements $l_{[\alpha\beta]}$.

However as a rule, the expansion (9.62) is not complete. Therefore equation (9.64) and (9.65) are of a rather limited interest. Indeed, the excess flows δJ_α depend not only on the excess forces δX_β but also on the increments of other state functions. For instance, according to Fourier's law (3.13) the excess heat flow $\delta \mathbf{W}$ has to be written as

$$\delta \mathbf{W} = -\lambda \delta(\nabla T) - (\delta \lambda) \nabla T$$

One sees that it is only in the special case $\lambda =$ constant ($\delta\lambda = 0$), that this relation takes the form (9.62).

7. CONVECTION PROCESSES†

We now wish to extend the evolution criterion (9.14) to include convection processes. The only new feature will be the introduction of new generalized forces and flows which will include the convection effects.

We shall first treat the case of systems which are in a *mechanical steady state*. More precisely we assume:

$$\partial_t \mathrm{v}_i = \partial_t p = \partial_t F_\gamma = 0 \tag{9.66}$$

and on the boundaries (see 9.9)

$$(\partial_t \mu_\gamma)_\Omega = (\partial_t T)_\Omega = 0 \tag{9.67}$$

As already pointed out in (9.13) and (9.14), the formulation of a universal evolution criterion is mainly related to the bilinear character of the entropy source in terms of flows and generalized forces.The invariance condition (2.26) allows us to choose the set of flows and forces which is the most appropriate for this purpose, independently of all reference to phenomenological relations. Therefore, we may now introduce flows J'_α containing both the *conduction* flow J_α used so far, and an additional *convection* flow which remains to be

† The results derived in §§ 7 and 8 will no longer be used subsequently.

determined. Also the new generalized forces X'_α have to be selected in such a way that the condition:

$$\sigma[S] = \sum_\alpha J'_\alpha X'_\alpha \geqslant 0 \tag{9.68}$$

is satisfied.

We have here an example of a *non-equivalent* description in the sense of Chapter II, § 3, as the new forces and flows have now a quite different physical meaning.

The convection flows may be introduced into the entropy source (9.68) by means of the Gibbs–Duhem relation (2.47) written as

$$\rho h \mathrm{v}_j T^{-1}_{,j} + T^{-1}\mathrm{v}_j p_{,j} - \sum_\gamma \rho_\gamma \mathrm{v}_j (\mu_\gamma T^{-1})_{,j} = 0 \tag{9.69}$$

where operator δ is replaced by the gradient operator. Let us emphasize that as previously, the velocity v_j corresponds to the frame of reference for which the boundary conditions have been prescribed as locally time independent.

We now add (9.69) to the expression (2.21) of the entropy source. This then yields:

$$\begin{aligned}\sigma[S] = [W_j + \rho h \mathrm{v}_j] T^{-1}_{,j} - p_{ij} T^{-1} \mathrm{v}_{i,j} + \mathrm{v}_j T^{-1} p_{,j} \\ + \sum_\gamma \rho_\gamma \Delta_{\gamma_j} (T^{-1} F_{\gamma j}) - \sum_\gamma \rho_\gamma \mathrm{v}_{\gamma j} (\rho_\gamma T^{-1})_{,j} \\ + \sum_\rho w_\rho T^{-1} A_\rho \geqslant 0\end{aligned} \tag{9.70}$$

The definitions of the J'_α and X'_α are given in Table 9.1. In the absence of convection we recover the usual description.

Table 9.1. Generalized flows J'_α and forces X'_α for systems in mechanical steady state.

J'_α	X'_α
$W_j + \rho h \mathrm{v}_j$	$T^{-1}_{,j}$
p_{ij}	$-T^{-1}\mathrm{v}_{i,j}$
v_j	$T_0^{-1} p_{,j}$
$\rho_\gamma \Delta_{\gamma j}$	$T^{-1} F_{\gamma j}$
$\rho_\gamma \mathrm{v}_{\gamma j}$	$-(\mu_\gamma T^{-1})_{,j}$
w_ρ	$T^{-1} A_\rho$

We now proceed exactly as in § 2 to construct the evolution criterion. Using again (1.28), (1.42) and (2.64), one deduces now:

$$-\partial_t T^{-1}[W_j + \rho h \mathrm{v}_j]_{,j} - \partial_t T^{-1} p_{ij} \mathrm{v}_{i,j} + \partial_t T^{-1} \mathrm{v}_j p_{,j} + \partial_t T^{-1} \sum_\gamma \rho_\gamma \Delta_{\gamma j} F_{\gamma j} + \sum_\gamma \partial_t(\mu_\gamma T^{-1})[\rho_\gamma \mathrm{v}_{\gamma j}]_{,j} + \sum_\rho w_\rho \, \partial_t(T^{-1} A_\rho) \leqslant 0 \tag{9.71}$$

Let us integrate by parts the first and the fifth terms. Then using (9.66), (9.67) as well as the definitions listed in Table 9.1, we obtain the generalized form of the evolution criterion

$$\frac{\partial_{X'} P}{\partial t} = \int \sum_\alpha J'_\alpha \, \partial_t X'_\alpha \, \mathrm{d}V \leqslant 0 \tag{9.72}$$

The analogy with (9.12) and (9.14) is obvious. Again the time change of the entropy production:

$$P = \int \sum_\alpha J'_\alpha X'_\alpha \, \mathrm{d}V \tag{9.73}$$

may be split into two contributions:

$$\frac{\partial P}{\partial t} = \frac{\partial_{X'} P}{\partial t} + \frac{\partial_{J'} P}{\partial t} \tag{9.74}$$

The contribution (9.72) due to the change of the forces X'_α at constant flows J'_α has as before a well defined sign. Again as in the case of purely dissipative systems the second contribution

$$\frac{\partial_{J'} P}{\partial t} = \int \sum_\alpha X'_\alpha \, \partial_t J'_\alpha \, \mathrm{d}V \tag{9.75}$$

has in general no definite sign.

However in the present description we have no longer simple phenomenological relations between the flows J'_α and the forces X'_α. Moreover, equality (9.15) written with the flows J'_α at the steady state instead of J_α, is no longer valid. In spite of this difference the equality sign in the evolution criterion (9.72) refers as in (9.14) to steady states, because then the time variation of the forces vanish.

But in the neighbourhood of a steady state, $\partial_{X'} P/\partial t$ will now in general contain *both* first and second order terms. This is the reason why stability theory, when convection processes occur, is simpler to study by the methods of Chapter VII, than starting from the evolution criterion.

8. TIME-DEPENDENT CONVECTION PROCESSES

To derive the evolution criterion in its most general form we now relax the conditions (9.66). The boundary conditions will be given as previously by (9.67).

We once more proceed as in § 2, except that in addition we also take into account the balance equation (1.29) for the momentum. We multiply (1.29) by $-T^{-1}\,\partial_t \mathrm{v}_i$, and add this term to (9.10). Instead of (2.64) we now obtain in the left hand side:

$$-\frac{\rho}{T}\left[\frac{c_v}{T}(\partial_t T)^2 + \frac{\rho}{\chi}(\partial_t v)^2_{N\gamma} + \sum_{\gamma\gamma'} \rho_{\gamma\gamma'}\,\partial_t N_\gamma\,\partial_t N_{\gamma'} + (\partial_t \mathrm{v}_i)^2\right] \leqslant 0 \tag{9.76}$$

For the right hand side we obtain, after elementary manipulations. the inequality:

$$\begin{aligned}
&[-(W_j + \rho e \mathrm{v}_j)\,\partial_t T^{-1} + P_{ij}T^{-1}\,\partial_t \mathrm{v}_i + \sum_\gamma \rho_\gamma \mathrm{v}_{\gamma j}\,\partial_t(\mu_\gamma T^{-1})]_{,j} \\
&+ \left[W_j + \rho h \mathrm{v}_j + p_{ij}\mathrm{v}_i + \rho \mathrm{v}_j\frac{\mathrm{v}^2}{2}\right]\partial_t T^{-1}_{;j} \\
&- \sum_\gamma \rho_\gamma \mathrm{v}_{\gamma j}\,\partial_t[(\mu_\gamma T^{-1})_{,j} - T^{-1}F_{\gamma j}] - (P_{ij} + \rho \mathrm{v}_i \mathrm{v}_j)\,\partial_t(T^{-1}\mathrm{v}_i)_{,j} \\
&+ \rho \mathrm{v}_j\,\partial_t\left(T^{-1}\frac{\mathrm{v}^2}{2}\right)_{,j} - \rho F_j\,\partial_t(\mathrm{v}_j T^{-1}) - \sum_\gamma \rho_\gamma \mathrm{v}_{\gamma j} T^{-1}\,\partial_t F_{\gamma j} \\
&+ \sum_\rho w_\rho\,\partial_t(T^{-1}A_\rho) \leqslant 0
\end{aligned} \tag{9.77}$$

with the additional boundary condition (9.9):

$$[\partial_t(\mathrm{v}_i)]_\Omega = 0$$

The first contribution to (9.77) is a flow term.†

Using the boundary conditions (9.67) we obtain after integration over the volume

$$\int \sum_\alpha J''_\alpha\,\partial_t X''_\alpha\,\mathrm{d}V \leqslant 0 \tag{9.78}$$

This is the most general form of the evolution criterion. The definition

† The Gibbs–Duhem formula (2.47) may be written as

$$\rho h\,\partial_t T^{-1} + T^{-1}\,\partial_t p = \sum_\gamma \rho_\gamma\,\partial_t(\mu_\gamma T^{-1})$$

In this way the flow term in (9.77) becomes alternatively:

$$[-W_j\,\partial_t T^{-1} + \partial_t(pT^{-1}\mathrm{v}_j) + p_{ij}T^{-1}\,\partial_t \mathrm{v}_i + \sum_\gamma \rho_\gamma \Delta_{\gamma j}\,\partial_t(\mu_\gamma T^{-1})]_{,j}$$

of the flows and forces used here appear directly in (9.77). In the special case of conservative systems:

$$\partial_t F_{\gamma i} = 0 \tag{9.79}$$

It is then easy to verify that:

$$P = \int \sum_{\alpha} J'_{\alpha} X'_{\alpha} \, dV = \int \sum_{\alpha} J''_{\alpha} X''_{\alpha} \, dV + \oint_{\Omega} p T^{-1} v_n \, d\Omega \tag{9.80}$$

Therefore, for fixed boundary conditions, we have still

$$\frac{\partial P}{\partial t} = \frac{\partial}{\partial t} \int \sum_{\alpha} J''_{\alpha} X''_{\alpha} \, dV \quad (P > 0) \tag{9.81}$$

together with

$$\frac{\partial_{X''} P}{\partial t} = \int \sum_{\alpha} J''_{\alpha} \, \partial_t X''_{\alpha} \, dV \leqslant 0 \tag{9.82}$$

We may therefore conclude that, with the exception of non-conservative systems, the evolution criterion is always associated to the entropy production.

Various different forms of the evolution criterion (9.82) may be obtained by the same method. Instead of multiplying the mass, momentum and energy balances, respectively by the quantities

$$-\partial_t(\mu_\gamma T^{-1}); \quad -T^{-1} \, \partial_t v_i; \quad \partial_t T^{-1}$$

as performed above, we may use as well:

$$-\varepsilon_m^2 \, \partial_t(\mu_\gamma T^{-1}); \quad -\varepsilon_q^2 T^{-1} \, \partial_t v_i; \quad \varepsilon_e^2 \, \partial_t T^{-1}$$

where ε_m^2, ε_q^2 and ε_e^2 are suitable multipliers. We shall see later the practical interest of this possibility on some examples (Chapter X, § 2, 10.19). However, in order to preserve the fundamental negative definite form (2.64) we have to require

$$\varepsilon_m^2 = \varepsilon_e^2 = \varepsilon^2 \tag{9.83}$$

Therefore, it remains at our disposal only two distinct multiplicators, both taken as non negative quantities (Chapter VI, § 9):

$$\varepsilon^2 \geqslant 0 \quad \varepsilon_q^2 = \tau^2 \geqslant 0 \tag{9.84}$$

having the same dimensions, and no singularities in the space of the generalized forces X_α.

PART II

Variational Techniques and Hydrodynamic Applications

CHAPTER X

The Local Potential

1. CONSERVATION EQUATIONS AND VARIATIONAL CALCULUS

Let us consider the set of conservation equations (Chapter I), together with the phenomenological laws, which express the flows in terms of the generalized forces (Chapter III). We obtain in this way a set of partial differential equations which, in the special case of intensive variables independent of the space coordinates, reduce to ordinary differential equations. The solution of these equations corresponds, either to a boundary value problem (steady state), to an initial value problem (time-dependent homogeneous process) or to both a boundary and an initial value problem (time dependent non-homogeneous process).

As a rule, we are dealing here with a complicated problem and, except for a few simple cases, an exact solution cannot be expected. It is then necessary to introduce approximate analytic or numerical methods.

For linear equations many methods are available (finite—difference methods, variational techniques, . . .). They are carefully explained in excellent textbooks entirely devoted to such questions to which the reader may refer (e.g. Kantorovich and Krylov, 1958). But in the case of non-linear equations, the situation is far from being so favourable. An additional difficulty is that most problems involving irreversible processes lead to differential equations which are not self-adjoint (see Chapter XII) and do not derive therefore from some extremum (maximum or minimum) principle. For this reason such problems cannot be treated by the variational calculus in its classical form. Powerful methods such as the well known Rayleigh–Ritz method (cf. Kantorovich and Krylov, 1958) are then no longer applicable. However, let us notice that an extension of the variational calculus to some non-linear problems has been recently developed by Tonti (1969).

It is the purpose of this chapter, to show that our macroscopic theory provides an additional information leading to the concept of *local potential*, which allows us to extend the use of variational techniques to non self-adjoint problems. As a matter of fact, the minimum property involved has a simple physical meaning. It expresses that the solution of the problem corresponds to the most probable state with respect to small fluctuations, according to the Einstein formula discussed in Chapter VIII.

For the sake of simplicity, we shall first of all focus our attention on the non-linear heat conduction problem in an isotropic body. In this example it is easy to show how the concept of local potential may be introduced and also how it may be used as the basis of a variational formulation.

In some particular problems, it remains possible to construct *ad hoc* true potentials, which may then be treated by the usual variational techniques. For example, in the non-linear heat conduction problem corresponding to a steady state, one may use the quantity $(\Theta,_j)^2$ as a Lagrangian (see 7.15). But in the case of an anisotropic medium, this possibility no longer exists. For this reason, we shall not deal here with such Lagrangians.

The general expression of the local potentials will be given in § 8 and illustrated by a few specific examples in Chapter XII. In this respect, it must be emphasized that the local potential may also be used as a simple variational technique for practical purposes, independently of any physical interpretation such as e.g. in terms of the Einstein formula. In this case, the local equilibrium assumption need not be invoked, and the method of calculation may be extended to more general situations, involving e.g. rheological problems for which the local entropy may depend on supplementary variables. Some applications of this type have been treated by Schechter (1967).

2. LOCAL POTENTIAL FOR THE HEAT CONDUCTION PROBLEM

Let us start with the energy balance equation (1.44) for a solid:

$$\rho\, \partial_t e = -W_{j,j} \tag{10.1}$$

We multiply the two sides by the increment δT^{-1} and integrate over the whole volume. After integration by parts on the r.h.s. we obtain for fixed boundary conditions:

$$\int \rho \delta T^{-1}\, \partial_t e \,\mathrm{d}V = \int W_j \delta T^{-1}_{,j} \,\mathrm{d}V \tag{10.2}$$

or using Fourier's law (3.11):

$$\int \rho \delta T^{-1} \, \partial_t e \, \mathrm{d}V = \tfrac{1}{2}\int \lambda T^2 \delta (T_{,j}^{-1})^2 \, \mathrm{d}V \tag{10.3}$$

We consider in this paragraph the steady-state problem. Time dependent problems will be studied in § 3. In the vicinity of a steady-state characterized by the temperature $T_0(x_j)$ one has:

$$\partial_t e = \partial_t \delta e \tag{10.4}$$

and, neglecting higher order terms:

$$\lambda T^2 = \lambda_0 T_0^2 + \delta(\lambda T^2) \tag{10.5}$$

Then, relation (10.3) may be rewritten as:

$$\int \rho \delta T^{-1} \partial_t \delta e \, \mathrm{d}V = \tfrac{1}{2}\int \lambda_0 T_0^2 \delta (T_{,j}^{-1})^2 \, \mathrm{d}V + \tfrac{1}{2}\int \delta(\lambda T^2) \delta (T_{,j}^{-1})^2 \, \mathrm{d}V \tag{10.6}$$

According to equation (7.1) the l.h.s. represents the quantity $\frac{1}{2}\, \partial_t \delta^2 S$. Therefore the r.h.s. corresponds to the excess entropy production. As the stability condition is fulfilled in the present case (see 7.8 and 7.19) the sign of the excess entropy production is positive.

We also observe that in the vicinity of a steady state, the two sides of (10.1) are first-order quantities. Therefore, equation (10.6) is a relation between second-order quantities and as a result, the last term on the r.h.s. cannot be neglected with respect to the first. For the same reason, the sign of the first term alone is by no means determined by the stability condition.

Let us now investigate separately the sign of the first term. We introduce the notation

$$\Phi(T, T_0) = \int \mathscr{L}(T, T_0) \, \mathrm{d}V \tag{10.7}$$

with, as integrand, the *Lagrangian*:

$$\mathscr{L}(T, T_0) = \tfrac{1}{2}\lambda_0 T_0^2 (T_{,j}^{-1})^2 \tag{10.8}$$

The first term in (10.6) is then equal to $\delta\Phi$. The quantity Φ appears as a functional of two variables:

T_0 which is a non-varied quantity denoting the presumed solution (yet to be determined) and T, which is varied. As we shall see later, T may be considered as a *fluctuating* temperature distribution whose average is given by T_0 (cf. 10.21).

We now investigate the condition under which the integral Φ is stationary (extremum) with respect to the variations of T. This is a classic problem of variational calculus (Courant and Hilbert, 1953). The condition is given by the Euler–Lagrange equation:

$$\frac{\delta \mathscr{L}}{\delta T^{-1}} = -(\lambda_0 T_0^2 T^{-1}_{,j})_{,j} = 0 \tag{10.9}$$

However, it remains to prescribe that the solution $T^{+}(x_j)$ (say) of this equation has to coincide with the presumed value T_0. This leads *a posteriori* to the subsidiary condition:

$$T^{+} = T_0 \tag{10.10}$$

which introduced in (10.9) gives the extremal:

$$\left(\frac{\delta \mathscr{L}}{\delta T^{-1}}\right)_{T_0} = (\lambda_0 T_{0,j})_{,j} = 0 \tag{10.11}$$

that is the steady-state equation of the heat conduction problem. The integrand (10.8) of the functional (10.7) may therefore be interpreted as a Lagrangian in an extended sense. Let us now investigate the nature of this extremum. To this end, let us calculate $\Phi(T, T_0)$ around the steady state. One has:

$$\begin{aligned}\Delta\Phi &= \Phi(T, T_0) - \Phi(T_0, T_0)\\ &= \tfrac{1}{2}\int \lambda_0 T_0^2 \{[(T_0^{-1} + \theta)_{,j}]^2 - [T^{-1}_{0,j}]^2\}\, \mathrm{d}V; \quad (\theta = T^{-1} - T_0^{-1})\end{aligned} \tag{10.12}$$

Expanding the brackets in the r.h.s. and integrating by parts the linear term in θ, this term cancels due to (10.11). Hence,

$$\Delta\Phi = \tfrac{1}{2}\int \lambda_0 T_0^2 (\theta_{,j})^2\, \mathrm{d}V > 0 \tag{10.13}$$

around the steady state. Therefore, the extremum of Φ corresponds to an *absolute minimum*. Functionals which have the properties (10.11) and (10.13) will be called *local potentials* (local, with respect to the function T_0 involved). Besides (10.13) establishes the positive definite character of the first term on the r.h.s. of (10.6).

More generally, a functional such as $\Phi(y_k, y_{0k})$ of several functions y_k ($k = 1, 2, \ldots$) will be considered as a local potential each time that the following conditions are satisfied:

(i) the first-order condition for the minimum of Φ with respect to y_k

$$\delta\Phi = 0 \tag{10.14}$$

together with the subsidiary conditions:

$$y_k^{+} = y_{0k} \tag{10.15}$$

restore the conservation equations for the y_k.

(ii) the higher order condition:

$$\Delta\Phi > 0 \tag{10.16}$$

for the absolute minimum is identically verified.

Let us observe that the less restrictive second order condition (10.14)

$$\Delta\Phi = \tfrac{1}{2}\delta^2\Phi > 0 \tag{10.17}$$

is not sufficient for our purpose. Indeed, the use of the local potential as the basis of a variational technique, implies the use of test functions corresponding to deviations of arbitrary amplitude with respect to the unknown solution.

For instance, as a result of this definition, the functional:

$$F(T, T_0) = \int W_{0j} T^{-1}_{;j}\, \mathrm{d}V = \int \lambda_0 T_0^2 T^{-1}_{0;j} T^{-1}_{;j}\, \mathrm{d}V \tag{10.18}$$

which satisfies the first condition but not the second one cannot be considered as a local potential. Indeed, an integration by parts gives immediately $\Delta F = 0$ around the steady state, contrary to the minimum condition [Figure 10.1(a)].

Let us also observe that in the heat conduction problem the sign (10.16) of $\Delta\Phi$ is the same as that of (10.6). The reason is that the stability condition is here fulfilled. But this is no longer the case in problems leading to unstable solutions.† This situation is represented in Figure 10.1 where the functionals are drawn as ordinary functions.

Whatever the stability may be, the local potential has always a *minimum* value at the reference state. The variation of the local potential gives therefore in all cases a positive contribution to the excess entropy production.

For a given problem the local potential is by no means unique and several Lagrangians may be constructed in the same way, using the multipliers considered earlier (see 9.83). As an example, for the heat conduction problem, one may also introduce the following expressions in addition to the Lagrangian (10.8):

$$\tfrac{1}{2}\lambda_0 T_0[(\ln T)_{,j}]^2 \quad \text{or} \quad \tfrac{1}{2}\lambda_0 T^2_{;j} \tag{10.19}$$

We have only to multiply (10.1) by $T\delta T^{-1}$ or by $-\delta T$ instead of

† In the first derivation of the local potential (1964, 1965) our presentation was misleading on this point. One sees here that the relation $\partial_t\Phi < 0$ is only valid around stable solutions. If the system is unstable, the fluctuations will increase together with Φ[Figure 10.1(b)].

multiplying it by δT^{-1}. As a rule, the choice of the most suitable local potential for practical purposes is linked to the type of kinetic law involved. For instance, in the present case, the Lagrangian (10.8) will be more convenient if $\lambda_0 T_0^2$ is approximately constant, while the second Lagrangian (10.19) is preferable when it is λ_0 which remains practically constant.

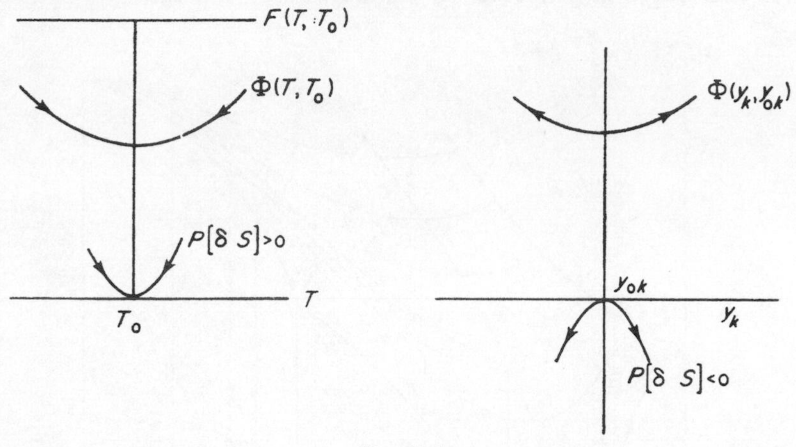

Fig. 10.1

(a) The local potential $\Phi(T, T_0)$ for a stable solution of the heat conduction problem and the functional $F(T, T_0)$.

(b) The local potential $\Phi(y_k, y_{0k})$ for an unstable solution.

Let us now add a few remarks about the meaning of the variational equations (10.9) and (10.11).

As mentioned above, the characteristic feature of the local potential $\Phi(T, T_0)$ given in (10.7), is its functional dependence on the *two* functions T and T_0. Besides, according to (7.7) and (10.8) one has:

$$2\Phi(T, T) = P(T) \tag{10.20}$$

where the r.h.s denotes the entropy production $P[S]$.† In special cases, i.e. when $\lambda \sim T^{-2}$, or when both λ and T may be treated separately as constants, as it is the case for steady states near equilibrium, Φ does not depend on T_0 but only on T and is then identical to the entropy production $P[S]$.

To visualize the properties of the functional Φ with respect to the

† This property subsists in the whole range of dissipative steady-state problems. However, it does not remain valid for time dependent problems or when convection terms appear (e.g. § 3).

functions T and T_0 we transpose on Figure 10.2 our problem to the case of a single function Φ of two variables T and T_0.

Intersection A B C of the Φ surface with the bisecting plane $T = T_0$, represents the locus of the $\Phi(T, T)$ values, that is the locus of half the entropy production, according to (10.20). When the theorem of minimum entropy production is valid, the minimum B of this curve corresponds to the stationary state of the system.

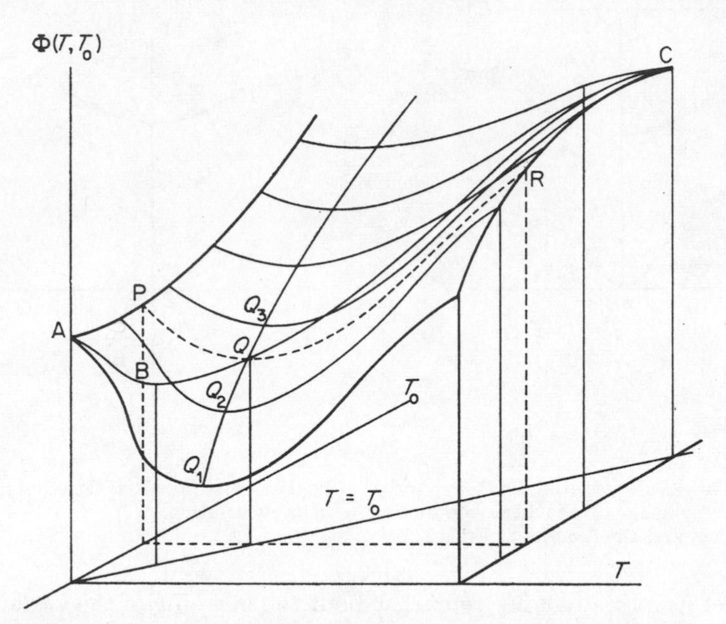

Fig. 10.2. The local potential $\Phi(T, T_0)$ as a function of both the *fluctuating* temperature T and steady state temperature T_0. Point B corresponds to the minimum of entropy production and point Q to the minimum $\Phi(T_0, T_0)$ of the local potential.

But, whatever the validity of this theorem, $\Phi(T, T_0)$ presents always a minimum as a function of T for the non varied steady state value T_0 (see curve P Q R). In other words, this minimum refers only to the class of functions which may be considered as perturbations around the given steady state. That is the reason why we now need a functional of the two temperature distributions T and T_0. Equation (10.11) cannot be interpreted as expressing a variational principle in the usual sense, because T_0 appears explicitly as a parameter in the Lagrangian. It seems more appropriate to consider equations (10.10) and (10.11) as a *variational property* which any steady state distribution $T_0(x_j)$ has to satisfy. This means that if we replace in $\Phi(T, T_0)$, T_0 by some assumed steady-state distribution, the solution of (10.9) giving the minimum of Φ, will lead in general to a value T^+ of T different from T_0. This is the case e.g. for the points Q_1, Q_2, Q_3 on Figure 10.2 which are not in the bisecting plane $T = T_0$.

On the contrary, the minimum corresponding to the point Q is in the bisecting plane and represents correctly the steady state. Of course, this point should not be confused with the minimum B of the entropy production.

As a matter of fact, T^+ is a functional of the presumed steady-state distribution and the subsidiary condition (10.10) may be rewritten more explicitly as

$$T^+(\{T_0\}) = T_0 \tag{10.21}$$

which provides a physical interpretation of the local potential method. Indeed, the function T^+ being the solution of the variational problem, corresponds to a vanishing value of the deviation δT, and therefore also of the positive definite function $-\delta^2 S$ (see 2.58). Now according to Einstein's formula discussed in Chapter VIII, there is a simple relation between the probability of a fluctuation and $\delta^2 S$. The most probable state corresponds to $\delta^2 S = 0$. The solution T^+ appears therefore as corresponding to the most probable distribution of temperature (when compared to other distributions which may be realized through small fluctuations). On the other hand, $T_0(x_j)$ is the macroscopic solution, that is the average of the temperature distribution which means that T_0 includes the effect of arbitrary fluctuations. As a result, the subsidiary condition (10.21) expresses that the *most probable temperature distribution* (with respect to small fluctuations) *has to coincide with the average temperature distribution* (with respect to all fluctuations).

3. TIME-DEPENDENT HEAT CONDUCTION PROBLEM

The local potential method may be easily extended to the time-dependent problem characterized by the temperature $T_0(x_j, t)$.

We first substract from the integrand on both sides of equation (10.3) the quantity $\rho\, \partial_t e_0 \delta T^{-1}$ ($\rho \equiv \rho_0$). We thus obtain instead of (10.6)

$$\int \rho \delta T^{-1} \partial_t \delta e \, \mathrm{d}V = \int [\tfrac{1}{2}\lambda_0 T_0^2 \delta (T_{,j}^{-1})^2 - \rho_0\, \partial_t e_0 \delta T^{-1}] \, \mathrm{d}V + \tfrac{1}{2}\int \delta(\lambda T^2) \delta (T_{,j}^{-1})^2 \, \mathrm{d}V \tag{10.22}$$

6

As the l.h.s. is the same as in equation (10.6) the two terms in the r.h.s. still corresponds to the positive excess entropy production and remain as previously, second order quantities. We construct again the local potential with the help of the first term. We have only to replace the Lagrangian (10.8) by the time dependent expression:†

$$\mathscr{L}(T, T_0) = \tfrac{1}{2}\lambda_0 T_0^2 (T_{,j}^{-1})^2 - \rho_0 T^{-1}\, \partial_t e_0 \tag{10.23}$$

and to prescribe the subsidiary condition (10.10) for all t. We thus obtain as extremal the time dependent heat equation:

$$\left(\frac{\delta \mathscr{L}}{\delta T^{-1}}\right)_{T_0} = (\lambda_0 T_{0,j})_{,j} - \rho_0 c_v^0\, \partial_t T_0 = 0 \tag{10.24}$$

On the other hand, the condition (10.13) for an absolute minimum subsists, because in $\Delta\Phi$ the new time dependent term cancels with the linear term in θ. Therefore, here also, the variation of the local potential gives a positive contribution to the excess entropy production and presents a minimum for the macroscopic evolution. Finally, all the properties established above for the steady state subsist except that (10.20) is no longer valid. There is not therefore a simple relation between local potential and entropy production.

4. RELATION WITH THE GALERKIN METHOD

Let us again consider the non-linear heat conduction problem at the stationary state. We assume that the solution $T_0(x_j)$ of the Fourier equation (10.11) may be represented everywhere in V, by a complete and linearly independent set of functions $\{\varphi_k(x_j)\}$, each of which satisfies separately the boundary conditions. We write for the sequence of order n

$$T_n^{-1} = \sum_1^n \alpha_k \varphi_k(x_j) \qquad T_{0n}^{-1} = \sum_1^n \alpha_k^0 \varphi_k(x_j) \tag{10.25}$$

where the $\{\alpha\}$ and $\{\alpha_0\}$ are two sets of parameters. We then introduce the approximations (10.25) into the local potential $\Phi(T, T_0)$. We perform the minimization as in the ordinary Rayleigh–Ritz method (e.g. Kantorovich and Krylov, 1958), with respect to each parameter α_k for constant values of the $\{\alpha_0\}$. We obtain in this way after an integration over x_j, the system of n equations:

$$f(\{\alpha\}, \{\alpha_0\}) = 0 \tag{10.26}$$

† This method was suggested to us by Prof. J. Brock (Texas University).

Using the subsidiary condition (10.10) written here in the form

$$\{\alpha\} = \{\alpha_0\} \tag{10.27}$$

the system (10.26) reduces to n equations for the parameters α_{0k}. This approach to the variational problem may be interpreted as a *self consistent method*, because it has to be understood as a consistent scheme of successive approximations.

Introducing the sequence of approximations (10.25) into the Lagrangian (10.8), we obtain after integration by parts and cancellation of the boundary terms, the n equations for the α_{0k}:

$$\int \varphi_k[\lambda_{0n} T^2_{0n} T^{-1}_{0n},_j],_j \,\mathrm{d}V = -\int \varphi_k[\lambda_{0n} T_{0n},_j],_j \,\mathrm{d}V = 0 \tag{10.28}$$

$$(k = 1,2, \ldots n)$$

One sees immediately that the n integrands appear in the form of a product of an *orthogonal* function φ_k, satisfying the boundary conditions, and the corresponding approximate value of the l.h.s. of the Fourier equation (10.11), whose solution is sought. Therefore, as far as numerical calculations go, the self-consistent method reduces to the well known Galerkin method (e.g. Kantorovich and Krylov, 1958).

Indeed, according to this method, one has to introduce the sequence of order n (10.25) into the heat equation (10.11) (dropping subscript zero). Then the n coefficients α_k are determined by the n orthogonality conditions:

$$\int (\lambda_n T_n,_j),_j \varphi_k \,\mathrm{d}V = 0$$

Clearly, the set of n conditions (10.28) is identical to these equations.

But as shown in the next section, the variational self-consistent method allows in addition to prove the convergence of the successive approximations. This proof is based on the fact that $\Phi(T, T_0)$ has a minimum for $T = T_0$ (10.16) and is therefore a direct consequence of the variational property of the local potential absent from the Galerkin method.

Moreover the local potential provides a simple physical interpretation for the Galerkin method. Indeed, as we have seen, equation

(10.28) expresses the fact that the most probable solution coincides with the average one.

5. CONVERGENCE OF THE SELF-CONSISTENT METHOD

Let us consider the value $\overline{T_n^{-1}}$ which minimizes the local potential (10.7) when the admissible functions are those of the family n (10.25) and where T_0^{-1} is replaced by its exact value. After minimization with respect to the α_k and integration by parts, we obtain the system of n equations:

$$\int \varphi_k [\lambda_0 T_0^2 \overline{T}_{n,j}^{-1}]_{,j}\, \mathrm{d}V = 0 \tag{10.29}$$

The successive minima of the local potential are certainly non increasing when n increases. Indeed, each successive family contains all the functions of the preceding one. Therefore we have the inequalities

$$\Phi(\overline{T}_1, T_0) \geqslant \Phi(\overline{T}_2, T_0) \ldots \geqslant \Phi(\overline{T}_n, T_0) \ldots \geqslant \Phi(T_0, T_0) \tag{10.30}$$

Furthermore for a complete set of functions φ_k, one may find a value of n for which $\Phi(\overline{T}_n, T_0)$ represents an arbitrary close approximation to the exact minimum $\Phi(T_0, T_0)$, namely

$$\varepsilon_n = \overline{T}_n^{-1} - T_0^{-1} \rightarrow 0 \text{ for } n \rightarrow \infty \tag{10.31}$$

Hitherto we were only dealing with the convergence properties of the classical Rayleigh–Ritz method applied to a true potential, as T_0 was treated in equation (10.29) as a simple parameter, independent of the sequence n.

Now let us go back to the solution T_{0n}^{-1} of equation (10.28) and put

$$\theta_n = T_{0n}^{-1} - \overline{T}_{0n}^{-1} \tag{10.32}$$

We want then to study the conditions under which one has for all n (Figure 10.3):

$$|\theta_n| < |\varepsilon_n| \tag{10.33}$$

or at least, the convergence in the mean:

$$\sqrt{\int \theta_n^2 \, \mathrm{d}V} < \sqrt{\int \varepsilon_n^2 \, \mathrm{d}V} \tag{10.34}$$

To introduce θ_n, into our equations we subtract (10.29) from (10.28). Then we multiply under the integral sign by $\alpha_k^0 - \bar{\alpha}_k$ and sum over k. Noting that:

$$\theta_n = \sum_{k=1}^{n} (\alpha_{0k} - \bar{\alpha}_k)\varphi_k(x_j) \tag{10.35}$$

we obtain in this way

$$\int \theta_n[\lambda_0 T_{0n}^2 T_{0n}^{-1},_j - \lambda_0 T_0^2 \bar{T}_n^{-1},_j],_j \, dV = 0 \tag{10.36}$$

After integration by parts and cancellation of the boundary term equation (10.36) becomes

$$\int \lambda_0 T_0^2 (\theta_{n},_j)^2 \, dV = -\int (\lambda_{0n} T_{0n}^2 - \lambda_0 T_0^2)\theta_{n},_j [T_0^{-1} + (\theta_n + \varepsilon_n)],_j \, dV \tag{10.37}$$

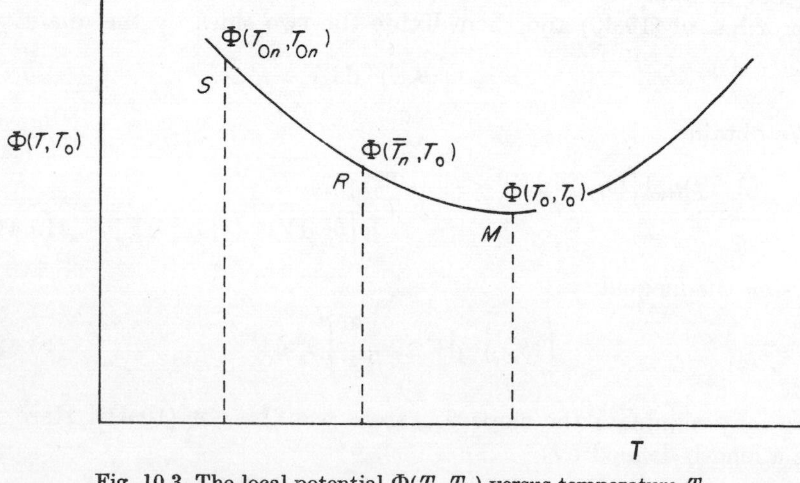

Fig. 10.3. The local potential $\Phi(T, T_0)$ versus temperature T.
M = minimum value for the exact solution.
R = minimum for the sequence n, by the Rayleigh–Ritz method.
S = Self consistent solution for the same sequence.

By manipulations similar to those used to obtain equation (10.13) one may show that the l.h.s. of (10.37) corresponds to twice the increment

$$\Phi(T_{0n}, T_0) - \Phi(\bar{T}_n, T_0)$$

For small values of θ_n and ε_n we may retain only the first order terms in the expansion:

$$\lambda_0 T_{0n}^2 = \lambda_0 T_0^2 + (\lambda_0 T_0^2)'(\theta_n + \varepsilon_n) + \ldots \tag{10.38}$$

which occur as second order terms in equation (10.37). We then obtain:

$$\int \lambda_0 T_0^2 (\theta_{n,j})^2 \, \mathrm{d}V = -\int (\theta_n + \varepsilon_n)\theta_{n,j}(\lambda_0 T_0^2)_{,j} \, \mathrm{d}V \tag{10.39}$$

This relation leads to the following inequality between moduli:

$$(\lambda_0 T_0^2)_{min} \int (\theta_{n,j})^2 \, \mathrm{d}V < |(\lambda_0 T_0^2)_{,j}|_{max} \int (|\theta_n| + |\varepsilon_n|)|\theta_{n,j}| \, \mathrm{d}V \tag{10.40}$$

The subscripts *min* and *max* denote respectively a lower and an upper limit in the given V. These limits though unknown, are independent of the sequence n.

Let us apply the Schwartz inequality† to each separate term in the r.h.s. of (10.40) and then divide the two sides by the quantity

$$\sqrt{[\int (\theta_{n,j})^2 \, \mathrm{d}V]}$$

We obtain:

$$(\lambda_0 T_0^2)_{min} [\int (\theta_{n,j})^2 \, \mathrm{d}V]^{\frac{1}{2}} < |(\lambda_0 T_0^2)_{,j}|_{max} \{[\int \theta_n^2 \, \mathrm{d}V]^{\frac{1}{2}} + [\int \varepsilon_n^2 \, \mathrm{d}V]^{\frac{1}{2}}\} \tag{10.41}$$

Using the inequality:‡

$$\int (\theta_{n,j})^2 \, \mathrm{d}V > \frac{\pi^2}{L^2} \int \theta_n^2 \, \mathrm{d}V \tag{10.42}$$

we may eliminate the slope $\theta_{n,j}$ from the l.h.s. of (10.41). Here L is a length defined by:

$$\frac{1}{L^2} = \frac{1}{a^2} + \frac{1}{b^2} + \frac{1}{c^2} \tag{10.43}$$

† $(f, g)^2 \leqslant (f, f)(g, g)$; $(f, g) = \int fg \, \mathrm{dV}$ (cf. Courant–Hilbert, 1953).

‡ Let us recall that for every piecewise smooth function $y(x)$, vanishing for $x = x_0$ and $x = x_1$, the integral

$$\int_{x_0}^{x_1} [y'^2 - \frac{\pi^2}{(x_1 - x_0)^2} y^2] \, \mathrm{d}x$$

is always positive (cf. Courant–Hilbert, 1953). The generalization to three-dimensional space is straightforward.

a, b, c, being the edges of a parallelepiped respectively along the x, y, z axis, and enclosing the volume V. Also we write

$$|(\lambda_0 T_0^2)_{,j}|_{max} = \frac{(\lambda_0 T_0^2)_{max} - (\lambda_0 T_0^2)_{min}}{L} \tag{10.44}$$

where $(\lambda_0 T_0^2)_{max}$ corresponds to an extrapolated value extended over the reference length L. In general this value will be larger than the maximum of $\lambda_0 T_0^2$ inside the volume. Finally, introducing (10.42) and (10.44) into inequality (10.41) we obtain:

$$\{(\lambda_0 T_0^2)_{min}\pi - [(\lambda_0 T_0^2)_{max} - (\lambda_0 T_0^2)_{min}]\}[\int \theta_n^2 \, \mathrm{d}V]^{\frac{1}{2}} < [(\lambda_0 T_0^2)_{max} - (\lambda_0 T_0^2)_{min}][\int \varepsilon_n^2 \, \mathrm{d}V]^{\frac{1}{2}} \tag{10.45}$$

According to (10.34) this relation establishes the convergence in the mean provided that:

$$\frac{(\lambda_0 T_0^2)_{max} - (\lambda_0 T_0^2)_{min}}{(\lambda_0 T_0^2)_{min}} < \pi \tag{10.46}$$

This inequality corresponds only to a *sufficient* condition. For instance using variable T instead of T^{-1} together with the second Lagrangian (10.19) instead of (10.8), one can derive in the same way the slightly different condition:

$$\frac{\lambda_{0max} - \lambda_{0min}}{\lambda_{0min}} < \pi \tag{10.47}$$

A more refined treatment would lead to less restrictive conditions. For suitable T dependence of λ, it may even occur that no additional condition is to be required as in the ordinary Rayleigh-Ritz method.

We see therefore that the situation is somewhat similar to that for true self-adjoint problems. The existence of a variational method permits proof of additional properties going beyond the Galerkin equations (10.28). However the local potential being not a true potential, we therefore obtain weaker conclusions than for a true self-adjoint problem. In fact the convergence of the local potential method goes much beyond the sufficient condition established above. As an example let us consider a temperature dependence of the conductivity λ of the exponential form (Thomaes, 1965):

$$\lambda(T) = \gamma \exp(-\beta T) \tag{10.48}$$

This author investigated the one-dimensional steady-state problem in an isotropic medium between walls at prescribed temperature. In this simple case, an exact solution may be calculated. It corresponds to the minimum of the potential (cf. 7.15):

$$\Phi(\beta, \gamma) = \tfrac{1}{2}\int_{x_1}^{x_2} \left(\frac{d\Theta}{dx}\right)^2 dx \tag{10.49}$$

As shown in Table 10.1, the comparison between the values of Φ calculated for the exact solution and those of the local potential

Table 10.1

$$\Phi_{(\beta)} = \tfrac{1}{2}\int_{x_1}^{x_2} \lambda^2 \left(\frac{dT}{dx}\right)^2 dx$$

$\gamma = 2;\ T_1 = 300,\ x_1 = 0;\ T_2 = 500,\ x_2 = 1$

β	0	1×10^{-4}	1×10^{-3}	5×10^{-3}
Exact solution	$8 \times 10^{+4}$	$7{\cdot}386 \times 10^4$	$3{\cdot}607 \times 10^4$	$15{\cdot}915 \times 10^2$
Local Potential	$8 \times 10^{+4}$	$7{\cdot}388 \times 10^4$	$3{\cdot}609 \times 10^4$	$15{\cdot}936 \times 10^2$

β	1×10^{-2}	6×10^{-2}	1×10^{-1}	1
Exact solution	$37{\cdot}53$	$1{\cdot}288 \times 10^{-13}$	$1{\cdot}75 \times 10^{-26}$	$5{\cdot}30 \times 10^{-261}$
Local Potential	$37{\cdot}75$	$3{\cdot}866 \times 10^{-13}$	$4{\cdot}37 \times 10^{-26}$	265×10^{-261}

method, indicates a very quick convergence of the latter in a range much larger than that corresponding to the conditions (10.46) or (10.47). The results obtained by the local potential method correspond to the first approximation, i.e. to a single term in the sequence (10.25).

In relation to the remark following (10.17), one can prove that the convergence conditions (10.46) and (10.47) remain valid without the restrictions (10.38) and (10.39) about the smallness of the deviations θ_n and ε_n (Glansdorff, 1966).

Let us finally notice that exactly as with the classical methods, the greatest difficulty here, is to prove that the set of the functions φ_k is complete.

6. THE TIME DEPENDENT PROBLEM

We shall now consider the Lagrangian (10.23) and assume that the time dependent solution $T_0(x_j, t)$ of the Fourier equation (10.24) may be represented everywhere in V and for all values of t by a complete set of functions $\{\varphi_k(x_j, t)\}$. These functions are assumed to be linearly independent. Each of them satisfies separately the boundary conditions for all values of t. In addition, the initial condition for $t = 0$ is also assumed to be satisfied by the sequences

$$T_n^{-1} = \sum_1^n \alpha_k \varphi_k(x_j, t); \quad T_{0n}^{-1} = \sum_1^n \alpha_{0k} \varphi_k(x_j, t) \tag{10.50}$$

These relations now replace (10.25).

This time dependent problem may be treated exactly as above by the variational self consistent method, using the subsidiary conditions in the form (10.27). The sufficient condition (10.47) for the convergence of the successive approximations takes now the form (Glansdorff, 1966):

$$\frac{\lambda_{0max} - \lambda_{0min}}{\lambda_{0min}} + \frac{(c_v^0)_{max} - (c_v^0)_{min}}{(c_v^0)_{max}} \frac{\pi}{2} \frac{\tau}{t} < \pi \tag{10.51}$$

with

$$\tau = \frac{L^2}{\pi^2} \frac{\rho^0 (c_v^0)_{max}}{\lambda_{0min}} \tag{10.52}$$

According to the classical linear theory of the heat conduction problem, τ appears as the maximum value of the relaxation time.

Inequality (10.51) shows that for values of t large enough, the condition of convergence becomes identical to (10.47). The proof of (10.51) is similar to that of (10.46) or (10.47) and the reader may consult the original paper for details.

7. THE ITERATION METHOD

Instead of the self-consistent method, trial functions together with iterative variational methods may also be used in conjunction with the local potential. For instance, in the stationary heat conduction problem, starting with an arbitrary function for T_0, satisfying the boundary conditions, a first approximation is calculated for T through the minimization of the local potential, exactly as in the

Rayleigh–Ritz method. Then this result for T is taken as the new T_0 and a second approximation for T is calculated in the same way, and so on. The convergence criteria (10.46), 10.47) and (10.51) obtained above for the self consistent method can also be proved in the present case whatever the choice of the first test function may be (Goche, 1971). Another slightly different criterion was obtained earlier for the special case of the one dimensional stationary heat conduction problem by Kruskal (1965).

8. GENERAL FORMULATION OF THE LOCAL POTENTIAL FOR A STEADY STATE

To obtain the general expression of the local potential, we shall start with the complete balance equations (1.28), (1.30), (1.42) for mass, momentum and energy and then proceed following the same method as in the case of the simple heat equation studied in § 2.

Let us first rewrite the balance equations in the compact form:

$$\partial_t \rho_\gamma = [\mathscr{M}_\gamma]; \quad \rho\, \partial_t \mathrm{v}_i = [\mathscr{Q}_i]; \quad \partial_t(\rho e) = [\mathscr{E}] \tag{10.53}$$

We multiply the two sides respectively by the infinitesimal increments $-\delta(\mu_\gamma T^{-1})$, $-T_0^{-1}\delta \mathrm{v}_i$, δT^{-1}, we add and integrate over the whole volume. The right hand side becomes:

$$\int \{ -[\mathscr{M}_\gamma]\delta(\mu_\gamma T^{-1}) - [\mathscr{Q}_i] T_0^{-1} \delta \mathrm{v}_i + [\mathscr{E}]\delta T^{-1}\}\, \mathrm{d}V \tag{10.54}$$

Starting from this expression, we have now to construct a local potential in the form of a functional

$$\Phi = \int \mathscr{L}[T, T_0; \mu_\gamma, \mu_{\gamma 0}; \quad \mathrm{v}_i, \mathrm{v}_{i0}]\, \mathrm{d}V \tag{10.55}$$

which generalizes (10.7) and which has to satisfy the basic conditions (10.14–10.16). Here again, T, μ_γ, v_i denote the fluctuating variables, while T_0, $\mu_{\gamma 0}$, v_{i0} represent the non varied solution, presumed to be known.

For instance, the straightforward generalization of the functional (10.18), i.e.:

$$\begin{aligned} &F[T, T_0; \quad \mu_\gamma, \mu_{\gamma 0}; \quad \mathrm{v}_i, \mathrm{v}_{i0}] \\ &\quad = \int \{ -[\mathscr{M}_\gamma]_0 \mu_\gamma T^{-1} - [\mathscr{Q}_i T^{-1}]_0 \mathrm{v}_i + [\mathscr{E}]_0 T^{-1}\}\, \mathrm{d}V \end{aligned} \tag{10.56}$$

cannot be interpreted as a local potential connected to the steady state, for the same reason as for (10.18). The extremals of (10.56) restore the conservation equations for the steady state, as

$$\frac{\delta \mathscr{L}}{\delta(\mu_\gamma T^{-1})} = [\mathscr{M}_\gamma]_0 = 0; \quad \frac{\delta \mathscr{L}}{\delta \mathrm{v}_i} = [\mathscr{Q}_i T^{-1}]_0 = 0$$

$$\frac{\delta \mathscr{L}}{\delta T^{-1}} = [\mathscr{E}]_0 = 0 \tag{10.57}$$

However, the condition (10.16) for an absolute minimum is not fulfilled, since (10.56) gives rise to

$$\Delta F = \int\{ -[\mathscr{M}_\gamma]_0 \Delta(\mu_\gamma T^{-1}) - [\mathscr{Q}_i T^{-1}]_0 \Delta \mathrm{v}_i + [\mathscr{E}]_0 \Delta T^{-1}\} \, \mathrm{d}V = 0 \tag{10.58}$$

To obtain a suitable expression of the local potential, we first integrate by parts the terms involving the flows W_j, $\rho_\gamma \Delta_{\gamma j}$, and p_{ij} in (10.54). This leads to the following explicit expression:

$$\begin{aligned}
\int\{[&-W_j \delta T^{-1} + \sum_\gamma \rho_\gamma \Delta_{\gamma j} \delta(\mu_\gamma T^{-1}) + p_{ij} T_0^{-1} \delta \mathrm{v}_i]_{,j} \\
&+ W_j \delta T^{-1}_{,j} - \sum_\gamma \rho_\gamma \Delta_{\gamma j} \delta(\mu_\gamma T^{-1}) - p_{ij} T_0^{-1} \delta \mathrm{v}_{i,j} \\
&+ \sum_\rho w_\rho \delta(A_\rho T^{-1}) - [(\rho e \mathrm{v}_j)_{,j} + P_{ij} \mathrm{v}_{i,j}] \delta T^{-1} \\
&+ \sum_\gamma (\rho_\gamma \mathrm{v}_j)_{,j} \delta(\mu_\gamma T^{-1}) + [(\rho \mathrm{v}_j \mathrm{v}_{i,j} - \rho F_i + p_{,i}) T_0^{-1} \\
&- p_{ij} T^{-1}_{0,j}] \delta \mathrm{v}_i\} \, \mathrm{d}V
\end{aligned} \tag{10.59}$$

We have assumed for simplicity that

$$F_{\gamma i} = F_i$$

The first line represents a flow term across the boundary surface Ω of the system. For fixed boundary values of T, μ_γ, v_i as well as for vanishing values of the boundary flows this term cancels. The second line of (10.59) can be transformed into an exact differential, using the usual phenomenological laws (Fourier's law, Fick's law and Newton's law) written as:

$$W_j = \lambda T^2 T^{-1}_{,j}; \quad \rho_\gamma \Delta_{\gamma j} = -D_\gamma (\mu_\gamma T^{-1})_{,j}; \quad p_{ij} = -2\eta \, \mathrm{d}_{ij}$$

$$\mathrm{d}_{ij} = \tfrac{1}{2}(\mathrm{v}_{i,j} + \mathrm{v}_{j,i}) \tag{10.60}$$

D_γ is the diffusion coefficient of component γ and η the viscosity coefficient. As in § 2, we assume that the phenomenological coefficients in (10.60) are non varied quantities, that is:

$$\lambda T^2 = \lambda_0 T_0^2; \quad D_\gamma = D_{\gamma 0}; \quad \eta = \eta_0 \tag{10.61}$$

Likewise the coefficients of the remaining terms of (10.59) will also be treated as non varied quantities. We then obtain in this way for fixed boundary conditions or vanishing boundary flows the local potential:

$$\begin{aligned}\Phi = \int \{ & \tfrac{1}{2}\lambda_0 T_0^2 (T_{,j}^{-1})^2 + \tfrac{1}{2}\sum_\gamma D_{\gamma 0}[(\mu_\gamma T^{-1})_{,j}]^2 + \eta_0 T_0^{-1} \mathrm{d}_{ij}^2 \\ & + \sum_\rho w_{\rho_0} A_\rho T^{-1} - [(\rho e \mathrm{v}_j)_{,j} + P_{ij}\mathrm{v}_{i,j}]_0 T^{-1} \\ & + \sum_\gamma [(\rho_\gamma \mathrm{v}_j)_{,j}]_0 (\mu_\gamma T^{-1}) \\ & + [(\rho \mathrm{v}_j \mathrm{v}_{i,j} - \rho F_i + p_{,i})T^{-1} - p_{ij} T_{,j}^{-1}]_0 \mathrm{v}_i \} \, \mathrm{d}V \end{aligned} \tag{10.62}$$

This expression provides the generalization of (10.7).

In order to show that functional (10.62) satisfies the conditions prescribed in § 2 (10.14 and 10.16), let us write the corresponding Euler–Lagrange equations for the extremals, taking into account the subsidiary conditions (10.15) that is

$$T^+ = T_0 \quad \mu_\gamma^+ = \mu_{\gamma 0} \quad \mathrm{v}_i^+ = \mathrm{v}_{i0} \tag{10.63}$$

We obtain successively, according to (2.18) and (10.60):

$$\begin{aligned}\left[\frac{\delta \mathscr{L}}{\delta(\mu_\gamma T^{-1})}\right]_{T_0, \mu_{\gamma 0}, \mathrm{v}_{i0}} &= [(\rho_\gamma \mathrm{v}_j)_{,j}]_0 + [(\rho_\gamma \Delta_{\gamma j})_{,j}]_0 - \sum_\rho \nu_{\gamma\rho} M_\gamma w_{\rho 0} \\ &= -[\mathscr{M}_\gamma]_0 = 0\end{aligned} \tag{10.64}$$

$$\begin{aligned}\left[\frac{\delta \mathscr{L}}{\delta \mathrm{v}_i}\right]_{T_0, \mu_{\gamma 0}, \mathrm{v}_{i0}} &= [(T^{-1} p_{ij})_{,j}]_0 + [(\rho \mathrm{v}_j \mathrm{v}_{i,j} - \rho F_i + p_{,i})T^{-1} - p_{ij} T_{,j}^{-1}]_0 \\ &= T_0^{-1}[P_{ij,j} + \rho \mathrm{v}_j \mathrm{v}_{i,j} - \rho F_i]_0 = -T_0^{-1}[\mathscr{Q}_i]_0 = 0\end{aligned} \tag{10.65}$$

$$\left[\frac{\delta \mathscr{L}}{\delta T^{-1}}\right]_{T_0, \mu_{\gamma 0}, \mathrm{v}_{i0}} = -[(\rho e \mathrm{v}_j)_{,j} + P_{ij}\mathrm{v}_{i,j}]_0 - [W_{j,j}]_0 = [\mathscr{E}]_0 = 0$$

$$(F\gamma i = Fi) \tag{10.66}$$

Therefore, as expected, we recover the steady state balance equations for mass, momentum and energy.

It remains to calculate $\Delta\Phi$, to prove that the condition (10.16) for a strict minimum is also fulfilled. We set:

$$\theta_\gamma = \mu_\gamma T^{-1} - \mu_{\gamma 0} T_0^{-1}; \quad u_i = \mathrm{v}_i - \mathrm{v}_{i0} \tag{10.67}$$
$$\theta = T^{-1} - T_0^{-1}; \quad \varepsilon_{ij} = d_{ij} - \mathrm{d}_{ij0}$$

Let us expand (10.62) in terms of these deviations. We obtain first:

$$\begin{aligned}\Phi = \Phi_0 + \int \{ & \tfrac{1}{2}\lambda_0 T_0^2(\theta_{,j})^2 + \tfrac{1}{2}\sum_\gamma D_{\gamma 0}(\theta_{\gamma,j})^2 + \eta_0 T_0^{-1}\varepsilon_{ij}^2 \\ & + \lambda_0 T_0^2 T_{0,j}\theta_{,j} + \sum_\gamma D_{\gamma 0}(\mu_\gamma T^{-1})_{0,j}\theta_{\gamma,j} + 2\eta_0 T_0^{-1}\,\mathrm{d}_{ij0}u_{i,j} \\ & - [(\rho e \mathrm{v}_j)_{,j} + P_{ij}\mathrm{v}_{i,j}]_0\theta + \sum_\gamma(\rho_\gamma \mathrm{v}_j)_{,j0}\theta_\gamma - \sum_\rho\sum_\gamma w_{\rho 0}\nu_{\gamma\rho}M_\gamma\theta_\gamma \\ & + [(\rho \mathrm{v}_j \mathrm{v}_{i,j} - \rho F_i + p_{,i})T^{-1} - p_{ij}T_{,j}^{-1}]_0 u_i\}\,\mathrm{d}V \end{aligned} \tag{10.68}$$

We then integrate by parts terms involving the gradients $\theta_{\gamma,j}$, $u_{i,j}$ and $\theta_{,j}$.

After cancellation of the boundary term, we obtain:

$$\begin{aligned}\Delta\Phi = \int \{ & \tfrac{1}{2}\lambda_0 T_0^2(\theta_{,j})^2 + \tfrac{1}{2}\sum_\gamma D_{\gamma 0}(\theta_{\gamma,j})^2 + \eta_0 T_0^{-1}\varepsilon_{ij}^2 \\ & - [\mathscr{M}_\gamma]_0\theta_\gamma - [\mathscr{Q}_i T^{-1}]_0 u_i + [\mathscr{E}]_0\theta\}\,\mathrm{d}V \end{aligned} \tag{10.69}$$

One sees that the linear part in the deviations cancels according to (10.64–10.66). The situation is thus the same as in (10.12) for the simple heat conduction problem, and the minimum condition required in (10.16) is identically satisfied since:

$$\Delta\Phi = \int \{\tfrac{1}{2}\lambda_0 T_0^2(\theta_{,j})^2 + \tfrac{1}{2}\sum_\gamma D_{\gamma 0}(\theta_{\gamma,j})^2 + \eta_0 T_0^{-1}\varepsilon_{ij}^2\}\,\mathrm{d}V > 0 \tag{10.70}$$

In conclusion, equation (10.62) provides the general expression of the local potential (10.55).

Of course, alternative formulations are also available, e.g. by using the multipliers defined in (9.84) (Glansdorff, 1966). Moreover, for the sake of simplicity, we have not performed the integration by parts of the chemical term in (10.59). This allows us to avoid supplementary manipulations postulating particular forms for the chemical kinetic laws. In any case the method remains always the same.

Let us also emphasize that, aside from the physical interpretation in terms of the most probable state (10.21), the local potential may be used as a simple variational technique of calculation, without reference to our fundamental local equilibrium assumption. In this respect, as mentioned in § 1, kinetic laws other than the linear relations (10.60) may be considered, as it is usual in the case of rheological applications (non-Newtonian fluids, . . .). A few problems of this type have been studied by Schechter (1967). Besides, interesting applications of the local potential method may be also expected in other fields than thermodynamics and hydrodynamics. An example related to the kinetic theory of gases is briefly discussed in § 11 (Prigogine, 1965; Nicolis, 1965, 1967). However, let us first consider how the local potential method may be extended to the case of time dependent processes.

9. GENERAL FORMULATION OF THE LOCAL POTENTIAL FOR THE TIME-DEPENDENT PROCESSES

We proceed exactly as in § 3.
Substitute in (10.55), the integrand

$$\mathscr{L}(T, T_0;\quad \mu_\gamma, \mu_{\gamma 0};\quad v_i, v_{i0})$$

calculated explicitly in (10.62), by the new Lagrangian:

$$\begin{aligned} &L(T, T_0;\quad \mu_\gamma, \mu_{\gamma 0};\quad v_i, v_{i0}) \\ &\quad = \mathscr{L} + \mu_\gamma T^{-1}\,\partial_t \rho_{\gamma_0} + v_i \rho_0 T^{-1}\,\partial_t v_{i0} - T^{-1}\,\partial_t(\rho e)_0 \end{aligned} \tag{10.71}$$

This time, the subscript zero refers to quantities depending both on the x coordinates and on time t.

It appears immediately that the functional

$$\Psi = \int L(T, T_0;\quad \mu_\gamma, \mu_{\gamma 0};\quad v_i, v_{i0})\,\mathrm{d}V \tag{10.72}$$

represents the general form of the local potential appropriate to time dependent processes, provided the system is subject to prescribed boundary values of T, μ_γ, v_i (time dependent or not) or to vanishing boundary flows (for all t). Indeed, using again the subsidiary conditions (10.63) the Euler–Lagrange equations for the functional Ψ, become according to (10.64–10.66):

$$\left[\frac{\delta L}{\delta(\mu_\gamma T^{-1})}\right]_{T_0, \mu_{\gamma_0}, v_{i0}} = -[\mathscr{M}_\gamma]_0 + \partial_t \rho_{\gamma_0} = 0 \tag{10.73}$$

$$\left[\frac{\delta L}{\delta v_i}\right]_{T_0, \mu_{\gamma_0}, v_{i0}} = -T_0^{-1}[\mathscr{Q}_i]_0 + \rho_0 T_0^{-1}\,\partial_t v_{i0} = 0 \tag{10.74}$$

$$\left[\frac{\delta L}{\delta T^{-1}}\right]_{T_0, \mu_{\gamma 0}, v_{i0}} = [\mathscr{E}]_0 - \partial_t(\rho e)_0 = 0 \qquad (10.75)$$

By comparison with (10.53), one sees that the extremals thus obtained are indeed the balance equations for mass, momentum and energy in the general time-dependent case.

Moreover, the condition for a true minimum

$$\Delta\Psi > 0 \qquad (10.76)$$

is also fulfilled. Indeed the additional time-dependent terms included in (10.71), cancel, as previously, the linear part in the deviations. Taking into account (10.70), we get:

$$\Delta\Psi = \Delta\Phi > 0 \qquad (10.77)$$

Clearly, the problem treated in § 3, is a particular case of this time dependent local potential. Other applications will be given in the forthcoming chapters (cf. also Schechter, 1967).

10. THE EXCESS LOCAL POTENTIAL

The problem of stability of a given steady state, based on the normal-modes analysis discussed in Chapter VI, § 8, may be solved by using the time dependent local potential method. This leads to approximate values of the frequencies ω and to approximate forms of the condition of marginal stability ($\omega_r = 0$).

To approach this problem it is often easier to proceed directly with an *excess local potential*, constructed from the excess balance equations (7.49–7.52) instead of (10.53). Around a steady state, the excess equations may be written in the compact form:

$$\partial_t \delta\rho_\gamma = [\delta\mathscr{M}_\gamma]; \quad \rho\, \partial_t \delta v_i = [\delta\mathscr{Q}_i]; \quad \partial_t \delta(\rho e) = [\delta\mathscr{E}] \qquad (10.78)$$

Likewise, the multipliers $-\delta(\mu_\gamma T^{-1})$, $-T_0^{-1}\delta v_i$, δT^{-1} have now to be replaced respectively by the increments:

$$-\delta'[\delta(\mu_\gamma T^{-1})]; \quad -T_0^{-1}\delta'[\delta v_i]; \quad \delta'[\delta T^{-1}]$$

This gives rise to the expression:

$$\int\{-[\delta\mathscr{M}_\gamma]\delta'[\delta(\mu_\gamma T^{-1})] - T_0^{-1}[\delta\mathscr{Q}_i]\delta'[\delta v_i] + [\delta\mathscr{E}]\delta'(\delta T^{-1})\}\, dV \qquad (10.79)$$

instead of (10.54). Let us observe that the time dependent solutions of the perturbation equations being denoted by δ, another symbol δ' (say) is here necessary to denote fluctuations around the most probable solution. The situation is represented schematically on Figure 10.4. We essentially consider the perturbation as a special case of macroscopic motion. It may for example correspond to the excitation of a *normal mode*. We then start from (10.79) to derive an

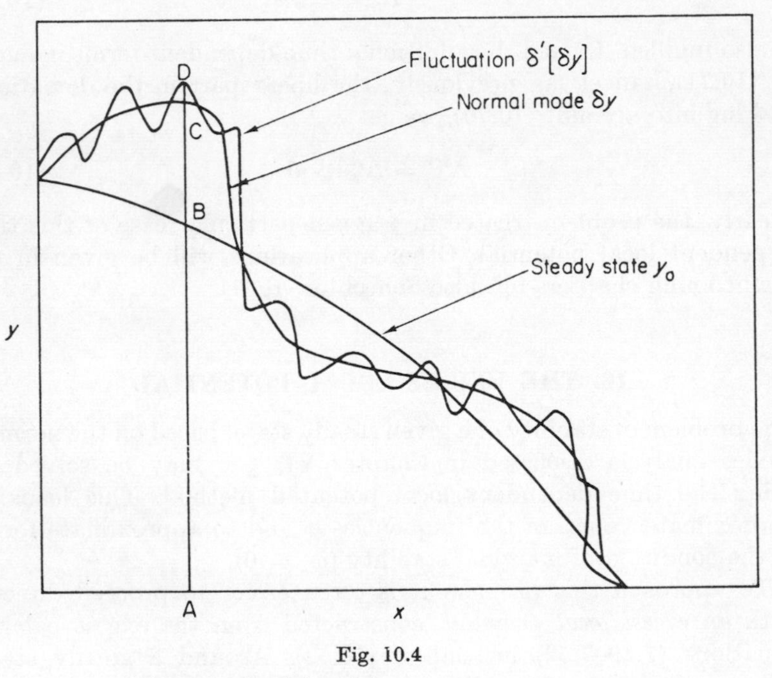

Fig. 10.4

AB = steady state $y_0(x)$.
BC = time dependent normal mode δy.
CD = fluctuation $\delta'[\delta y]$ around the normal mode.

expression for the *excess local potential* by the same manipulations as those performed previously to construct the complete time dependent local potential from (10.54). We leave this calculation as an exercise to the reader. An example is given in Chapter XII.

The main interest in using the excess local potential, is that in this case, the self-consistent method involving sequences of the type (10.25) for the perturbations δ, leads to a set of algebraic equations (10.26), which are linear and *homogeneous with respect to the* $\{\alpha_0\}$

parameters, when the subsidiary conditions (10.27) are taken into account. Indeed, these perturbation equations have to be identically satisfied for vanishing values of the perturbation δ. We may then eliminate the $\{\alpha_0\}$ through the consistency condition. Therefore, one may directly derive an approximate dispersion equation to calculate the values of ω (cf. e.g. Chap. XII).

11. LOCAL POTENTIALS IN KINETIC THEORY

The form of the distribution function, solution of the kinetic equation constitutes a central problem in statistical mechanics as well as in kinetic theory. Again, variational techniques were used quite extensively in the linear domain corresponding to small deviations from local equilibrium (Ono, 1961). It was moreover observed that when one considers non self-adjoint problems beyond the local equilibrium region (non-linear domain, system in an external field and sc on), it is no longer possible to derive the kinetic equations from a Lagrangian.

In this section we show that the concept of local potential previously introduced in macroscopic physics, may as well be extended to determine a distribution function, at least by successive approximations (Prigogine, 1965; Nicolis, 1965, 1967).

We consider a system of many particles characterized by their instantaneous positions x_i and velocities v_i. We assume that the system is dilute enough for encounters between particles to be considered as instantaneous. We also suppose that only binary collisions are involved. Such systems are completely described by the one-particle Boltzmann distribution function $f(x_i, v_i, t)$. In the presence of an external field of forces F_i (per unit mass), the distribution function f satisfies a Boltzmann type equation (Chapman and Cowling, 1939):

$$\frac{\partial f}{\partial t} + v_i \frac{\partial f}{\partial x_i} + F_i \frac{\partial f}{\partial v_i} = I\,(f, f) \tag{10.80}$$

where $I(f, f)$ is a term describing the collisional effects, on the evolution of the distribution function f. In the approximation here considered, $I(f, f)$ is independent of F_i, x_i and t. Moreover, in the particular case of a dilute one-component neutral gas of structureless molecules, $I(f, f)$ takes the well-known Boltzmann form:

$$I_B = \int \sigma(\mathbf{k}, \mathbf{k}')\, |\mathbf{g}| (f' f_1' - f f_1)\, \mathrm{d}v_1\, \mathrm{d}\mathbf{K} \tag{10.81}$$

In this expression, σ denotes the collision cross-section. The prime indicates a value after the direct collision. Moreover the integrations are performed from $-\infty$ to $+\infty$ for the velocities and over the unit sphere in respect to $\mathbf{K}$; $\mathbf{k}$ is a unit vector along the relative velocity $\mathbf{g} = \mathbf{v} - \mathbf{v}_1$.

We notice that I_B is a *non-linear* operator in f. This is an essential feature of kinetic theory, which is largely responsible for the lack of an ordinary variational principle far from equilibrium.

We now place ourselves in the framework of the fluctuation theory, extended to the x_i, v_i space. At one arbitrary point of this space, and at time t, the distribution function $f(x_i, \mathrm{v}_i, t)$ may be considered as a fluctuating function, around some reference average $\bar{f}$. Let:

$$f = \bar{f} + \delta f \tag{10.82}$$

We obtain after replacing into 10.80:

$$\frac{\partial \delta f}{\partial t} = -\frac{\partial \bar{f}}{\partial t} - \mathrm{v}_i \frac{\partial f}{\partial x_i} - F_i \frac{\partial f}{\partial \mathrm{v}_i} + I(f, f) \tag{10.83}$$

In the same way as in § 9, we now multiply both sides of (10.83) by $-f^{-1}\,\partial f$, and integrate over x_i, v_i. After similar manipulations the r.h.s. of (10.83) takes the form:

$$\begin{aligned}\mathscr{L}(f, \bar{f}) = & \int \left[\frac{\partial \bar{f}}{\partial t} + \mathrm{v}_i \frac{\partial f}{\partial x_i} + F_i \frac{\partial f}{\partial \mathrm{v}_i}\right] \delta \log f \, d\mathbf{v} \, d\mathbf{x} \\ & - \int \sigma |\mathbf{g}| \, (f' \, f_1' - f f_1) \, \delta \log f \, d\mathbf{v} \, d\mathbf{v}_1 \, d\mathbf{x} \, d\mathbf{K}\end{aligned} \tag{10.84}$$

The collision term is now transformed by symmetrization as follows:

$$\begin{aligned}\mathscr{L}_{\text{col}} &= -\frac{1}{2} \int \sigma |\mathbf{g}| (f' f_1' - f f_1) \, \delta \log f f_1 \, d\mathbf{v} \, d\mathbf{v}_1 \, d\mathbf{x} \, d\mathbf{K} \\ &= -\frac{1}{4} \int \sigma |\mathbf{g}| \, f f_1 \left(1 - \frac{f f_1'}{f \cdot f_1}\right) \delta \log \frac{f' f_1'}{f f_1} \, d\mathbf{v} \, d\mathbf{v}_1 \, d\mathbf{x} \, d\mathbf{K}\end{aligned} \tag{10.85}$$

Clearly, the quantities (10.84) and (10.85) are not exact differentials in respect to the operator δ. This means, that $\mathscr{L}$ cannot be written in the form $\delta \Psi$ of some continuous differentiable functional Ψ. We now apply the fundamental property of local potential and construct by starting from (10.84) a new functional of *two* functions:

$$\Psi(f, \bar{f}) = \int \left(\frac{\partial \bar{f}}{\partial t} + v_i \frac{\partial \bar{f}}{\partial x_i} + F_i \frac{\partial \bar{f}}{\partial v_i}\right) \log f \, d\mathbf{v} \, d\mathbf{x}$$

$$- \frac{1}{4} \int \sigma |\mathbf{g}| \bar{f} \bar{f}_1 \left(1 - \frac{f' f_1'}{f f_1} + \log \frac{f' f_1'}{f f_1}\right) d\mathbf{v} \, d\mathbf{v}_1 \, d\mathbf{x} \, d\mathbf{K} \quad (10.86)$$

One can verify by direct computations, that $\Psi(f, \bar{f})$ satisfies the basic conditions (10.14–10.16), namely:
(a) The variational equation

$$\frac{\delta \Psi(f, \bar{f})}{\delta f} = 0 \quad (10.87)$$

together with the subsidiary condition:

$$f = \bar{f} \quad (10.88)$$

restore the Boltzmann equation (10.80).

In terms of fluctuation theory this condition may again be interpreted as identifying the most probable distribution function with the average one.
(b) The functional $\Psi(f, \bar{f})$ satisfies the condition:

$$\Psi(f, \bar{f}) > \Psi(\bar{f}, \bar{f}) \quad \text{or} \quad \Delta\Psi > 0 \quad (10.89)$$

Indeed, one can easily verify that the increment $\Delta\Psi$ contains, up to second order terms, a positive definite quadratic part which depends only upon dissipation, while the linear part cancels mutually as in the foregoing examples.

The functional Ψ given in equation (10.86), is therefore a suitable local potential for equation (10.80). It can be shown, after elementary manipulations, that in the limit of small gradients and external forces, i.e. when the system is close to equilibrium, Ψ becomes functional of a single function and reduces to a Lagrangian as those used in the linear range of irreversible processes (see also § 2). Such Lagrangians are closely related to the entropy production expressed here in terms of distribution functions rather than in terms of thermodynamical averages (Ono, 1961).

However in the general case, equation (10.86) still provides us with an extended variational principle, valid for the determination of the distribution function beyond the linear domain as described by the first Chapman–Enskog approximation (see Chapman and Cowling, 1939).

The local potential method is of special interest for rarefied gases and plasmas, where the local equilibrium assumption is not satisfactory. But even in ordinary gas kinetic problems, the method could be used to calculate higher Chapman–Enskog approximations. Of course, in this case the trial functions can be chosen by starting from the local equilibrium Maxwellian distribution.

We shall not discuss here specific applications, but refer the reader to the original papers devoted to this subject (Nicolis and Sels, 1967; Babloyantz, 1968).

12. COMPARISON WITH OTHER VARIATIONAL TECHNIQUES

The local potential method enables us to solve non self-adjoint systems of differential equations using the approximation techniques of variational calculus. In the special case of self-adjoint equations, this procedure reduces simply to the classical Rayleigh–Ritz method.

Of course, other methods have already been proposed to construct functionals which take a stationary value for the solution of a given non self-adjoint system of differential equations. To this end, Lagrangians involving supplementary unknown functions not included in the original equations, have been constructed. A general survey of such methods and specially those related to the associated functions and the image systems, is given by Schechter (1967). He also discusses the difficulties which may arise.

The methods based upon associated functions should not be confused with the local potential method. As we have seen the local potential procedure is based on the distinction between fluctuating functions and their most probable values.

Other authors have already introduced variational techniques in which the unknown function was varied in some terms of the Lagrangian, and kept constant in others (Rosen, 1953, Chamber, 1956, cf. Schechter, 1967). This was however done quite *ad hoc*.

Concerning the Galerkin method, we have seen in (10.28) that as far as the Euler–Lagrangian equations of the local potential are concerned, the two methods are identical.

The main interest of the Galerkin method arises from its high degree of generality (Kantorovich and Krylov, 1958). It may be applied to non self-adjoint as well as to nonlinear systems of differential equations. Unfortunately this method being not of variational origin is deprived of any minimum property which would permit the problem of convergence of the successive approximations to be approached (§§ 5, 6 and 7). Here the local potential method adds an essential element to the Galerkin method since it introduces precisely a minimum property. Moreover in the whole range of validity of the local equilibrium assumption, this minimum property has a very appealing physical meaning. Indeed, as pointed out in Chapter VIII, this minimum corresponds to the most probable state, in agreement with the Einstein formula for fluctuations around a non-equilibrium state.

CHAPTER XI

Stability Problems in Fluids at Rest

1. INTRODUCTION

We now start the study of typical applications to illustrate the theory developed in the preceding chapters. We consider essentially situations to which we may apply the linear stability theory of Chapters VI–VII, and the evolution criterion of Chapter IX. Moreover, in Chapter XII, we apply the local potential method given in Chapter X.

As repeatedly stressed, the stability conditions we have studied, are of great generality. Investigation of various particular situations is necessary to clarify their physical content.

We shall investigate problems where *both* dissipative and convective effects determine the stability. This includes some of the usual cases of hydrodynamic stability theory, such as the onset of free convection in a layer of fluid at rest (§ 2), or the transition from laminar to turbulent flow (Chapter XII).

We also consider two important limiting situations: the case of *ideal* fluids, where dissipative processes may be neglected (§ 12, and Chapter XIII), and the case of *purely* dissipative systems, where no convection occurs (Chapters XIV–XVI).

In this chapter, we study the onset of free convection in a fluid at rest submitted to a temperature gradient. We consider mainly a thin horizontal layer of incompressible fluid heated from below (Bénard's problem). We also study briefly the stability of a vertical column of gas in § 12.

2. PERTURBATION EQUATIONS

The problem of the onset of thermal instability in horizontal layers of a fluid heated from below is well suited to illustrate many aspects of stability theory. The physical reasons for instability may be analysed in detail. Moreover, for suitable boundary conditions, the mathematical problem involved in the determination of the critical stability conditions may be solved exactly (§ 9).

A detailed description of this problem may be found in the excellent monograph by Chandrasekhar (1961) to which the reader should refer for supplementary information.

Consider a horizontal layer of fluid between two infinite parallel planes, in a constant gravitational field, and let us maintain the lower boundary at temperature T_1, and the higher boundary at T_2 with $T_1 > T_2$. Such a temperature gradient is called 'adverse' because the fluid at the bottom is then lighter than the fluid at the top, due to the thermal expansion.

For a thin layer, we may neglect the pressure dependence of the density ρ, and the equation of state may then be written in the linear form:

$$\rho = \rho^+[1 - \alpha(T - T^+)] \tag{11.1}$$

where α is the expansion coefficient (at constant pressure) and ρ^+, the density at some reference temperature T^+. For the liquids and gases we shall study, α is in the range 10^{-3}–10^{-4}. Therefore, for temperature differences of the order of 10°C, the variation in density is small, and ρ may be treated as a constant. We shall also treat all coefficients such as the specific heat c_v, thermal conductivity λ, and viscosity η, as constants.

The reference state whose stability we want to investigate is a state at rest, that is without convection:

$$\mathbf{v} = 0 \tag{11.2}$$

Therefore, the corresponding steady temperature distribution is given by the linear law:

$$T = -\beta z + T_1 \quad \text{with} \quad \beta = -\frac{\partial T}{\partial z} > 0 \tag{11.3}$$

taking the lower boundary as the x, y plane, and z in the vertical direction. We have moreover:

$$F_x = F_y = 0; \quad F_z = -g \tag{11.4}$$

We now consider small perturbations of the velocity and of the temperature distributions. We use the notations:

$$w = \delta v_z; \quad u_i = \delta v_i;$$
$$\theta = \delta T; \quad (i = 1, 2, 3) \tag{11.5}$$

These perturbations satisfy the excess balance equations (7.50–7.52) for the total mass ($\gamma = 1$) momentum and energy. In the present case we obtain in this way:

$$u_{j,j} = 0 \tag{11.6}$$

$$\rho\, \partial_t u_i = -g_i \alpha \rho \theta - (\delta P_{ij})_{,j} \quad (g_i = 0, 0, -g) \tag{11.7}$$

$$\rho c_v\, \partial_t \theta = -\delta W_{j,j} + \rho c_v \beta w \tag{11.8}$$

To derive these linearized equations we have treated the density ρ as a constant, except in the term $g\alpha\rho\theta$ of the momentum equation (11.7). This term has to be retained as it provides precisely the mechanism through which instability may arise. This is the so-called *Boussinesq approximation.*

For the contribution of $\delta\rho$ in (11.7) we used, according to (11.1) and (11.5), the equality:

$$\delta\rho = -\alpha\rho^+ \theta \simeq -\alpha\rho\theta \tag{11.9}$$

The excess momentum flow in (11.7), and the excess heat flow in (11.8), are directly related to the fluctuations θ and u_i. Indeed the usual phenomenological laws (7.11) and (7.34) give us the relations:

$$\delta W_j = -\lambda \theta_{,j}; \quad \delta p_{ij} = -\eta(u_{i,j} + u_{j,i}) \tag{11.10}$$

As the fluid is limited by the planes $z = 0$ and $z = h$, we clearly have the following boundary conditions:

$$\left.\begin{array}{l} \theta = 0 \\ w = 0 \end{array}\right\} \text{ for } \left\{\begin{array}{l} z = 0 \\ z = h \end{array}\right. \tag{11.11}$$

Moreover for a *rigid* surface on which no slip occurs, we have:

$$u_x = u_y = 0 \tag{11.12}$$

while for a *free* surface on which no tangential stresses act:

$$P_{xz} = P_{yz} = 0 \tag{11.13}$$

We shall first focus our attention on properties arising directly from the general stability criterion obtained in Chapter VII, §§ 12 and 13.

The stability conditions discussed in the following sections 3 to 5, are expressed specifically in terms of disturbances which are real quantities. The reader may remember that the corresponding expressions for complex quantities are easily obtained using (2.76). This leads to identical conclusions. In § 6 to 8, we shall use explicitly the complex form of the stability conditions.

3. STABILITY CONDITIONS FOR A FLUID LAYER

We first apply separately the thermodynamic and the hydrodynamic stability conditions (7.98) and (7.102) to the Bénard problem. As a consequence of the boundary conditions (11.11–11.13), the surface terms (7.99) and (7.103) vanish. Therefore, we are only concerned with stability conditions involving volume integrals, namely:

$$P[\varepsilon^2 \delta S] > 0 \qquad P[\tau^2 \delta E_{kin.}] < 0 \tag{11.14}$$

The explicit expressions of the corresponding source terms are given in the excess balance equations (7.96), (7.97) and (7.101). They may be considerably simplified using condition (11.2) and the Boussinesq approximation adopted in § 2. Introducing also the phenomenological laws (11.10) and taking for convenience $\tau^2 = 1$, we obtain the following equalities:

$$\delta p_{ij} \delta \mathrm{v}_{i,j} = \delta p_{ij} u_{i,j} = -\eta(u_{i,j} + u_{j,i}) u_{i,j} = -2\eta\, \mathrm{d}_{ij}^2 \tag{11.15}$$

$$F_i \delta\rho \delta \mathrm{v}_i = g\alpha\rho\theta w \tag{11.16}$$

$$\delta W_j \delta(\varepsilon^2 T^{-1})_{,j} = \lambda \varepsilon^2 T^{-2} (\theta_{,j})^2 + \tfrac{1}{2}\lambda(\theta^2)_{,j} (T^{-2}\varepsilon^2)_{,j} \tag{11.17}$$

$$-\delta(p_{ij} T^{-1}) \delta(\varepsilon^2 \mathrm{v}_{i,j}) + \varepsilon^2 T^{-1} \delta p_{ij} \delta \mathrm{v}_{i,j} = 0 \tag{11.18}$$

$$-\varepsilon^2 \delta \mathrm{v}_j [T^{-1}_{,j} \delta(\rho e) - (\mu T^{-1})_{,j} \delta\rho] = -\rho c_v \beta w \theta \varepsilon^2 T^{-2} \tag{11.19}$$

We adopt in (11.15) the definition:

$$\mathrm{d}_{ij} = \tfrac{1}{2}(u_{i,j} + u_{j,i}) \tag{11.20}$$

Also it is convenient to take

$$\varepsilon = T_0 \tag{11.21}$$

where the subscript zero denotes the state of reference. Therefore, in equation (11.17–11.19), we have: $\delta\varepsilon^2 = 0$. As a result of (11.21), the second term on the r.h.s. of (11.17) vanishes and this equality reduces to:

$$-\delta W_j \delta T_{,j} = \lambda(\theta_{,j})^2 \tag{11.22}$$

We have here a typical example of the algebraic simplifications which may be expected from a suitable choice of weighting functions. Let us now introduce these equalities into the source terms of the balance equations (7.96), (7.101). We obtain in this way the explicit form of the stability conditions for the Bénard problem.

Hydrodynamic stability condition.

The second inequality (11.14) becomes:

$$\int [2\nu\, \mathrm{d}_{ij}^2 - g\alpha\theta w - v u_{j,j}\delta p]\, \mathrm{d}V > 0 \tag{11.23}$$

where ν denotes the *kinematic viscosity*:

$$\nu = \frac{\eta}{\rho} \tag{11.24}$$

As the layer is unbounded along the horizontal plane, the disturbances θ, u_i, δp are periodic functions of x and y. Therefore an integration over x and y amounts to averaging over the horizontal plane. Denoting such mean values by acute brackets $\langle\ \rangle$, inequality (11.23) may be rewritten as:

$$\int_0^h [2\nu\langle \mathrm{d}_{ij}^2\rangle - g\alpha\langle\theta w\rangle - v\langle u_{j,j}\delta p\rangle]\, \mathrm{d}z > 0 \tag{11.25}$$

Moreover, using the incompressibility condition (11.6), the last term vanishes.

Also, by subtracting the zero value $\nu(u_{j,j})^2$ under the sign of (11.25) and integrating by parts the first term, we obtain after reduction, the equivalent expression:

$$\int_0^h [\nu\langle(u_{i,j})^2\rangle - g\alpha\langle\theta w\rangle]\, \mathrm{d}z > 0 \tag{11.26}$$

Thermodynamic stability condition.

Likewise, the first inequality (11.14) becomes successively

$$\int \left[\kappa(\theta_{,j})^2 - \beta\theta w + \frac{T}{c_v}\, v u_{j,j}\delta p\right] \mathrm{d}V > 0 \tag{11.27}$$

and

$$\int_0^h [\kappa\langle(\theta_{,j})^2\rangle - \beta\langle\theta w\rangle]\, \mathrm{d}z > 0 \tag{11.28}$$

Here, κ denotes the *thermal diffusivity*:

$$\kappa = \frac{\lambda}{\rho c_v} \tag{11.29}$$

The inequality (11.26) is well known. Its content has been interpreted by Chandrasekhar in the following terms:†

'*Instability occurs at the minimum temperature gradient at which a balance can be steadily maintained between the kinetic energy dissipated by viscosity and the internal energy released by the buoyancy force*'.

Therefore, inequality (11.26) appears as the expression of a competition between two mechanisms generating internal or kinetic energy.

The interpretation of the inequality (11.28) is slightly different. The first term corresponds to the generation of entropy, due to temperature fluctuations, while the second term represents the flow of the entropy fluctuations carried away by the fluctuations of the vertical velocity. We may therefore express the second inequality in the following terms:

Instability occurs at the minimum temperature gradient at which a balance can be steadily maintained between the entropy generated through heat conduction by the temperature fluctuations and the corresponding entropy flow carried away by the velocity fluctuations.

If this entropy flow overcomes the entropy produced by heat conduction, the fluctuation will penetrate far into the layer and the state at rest will become unstable.

It is worth while to stress that inequalities (11.26) and (11.28), both refer to properties of fluctuations. In (11.26) also, the dissipation of kinetic energy results from the velocity fluctuations. However, as we study the stability of a system *at rest*, this dissipation is equal to the total dissipation of the kinetic energy in the system. On the contrary, in (11.28) the entropy production arising from the heat conduction, due to the temperature fluctuations, should not be confused with the entropy production due to the adverse temperature

† See especially, Chandrasekhar 1961 (Chapter II, § 19 and Bibliographical Notes). On page 73 of this monograph, references are given to earlier works on the thermal instability by Jeffreys, 1956 and Malkus, 1954.

It may be mentioned that it is the beautiful original paper by Chandrasekhar (1958), which suggested to the authors the possibility of a generalization of non-equilibrium thermodynamics to include hydrodynamic processes.

gradient (11.3). We shall come back to this problem in § 11, where we consider briefly the Bénard problem for binary mixtures. As we shall see, instability may then occur, even when the lighter fluid is at the top!

4. BÉNARD INSTABILITY AND ENTROPY PRODUCTION

In § 3, the stability of the layer was discussed starting from the two separate hydrodynamic and thermodynamic conditions. Let us now investigate the complete thermo-hydrodynamic conditions (6.39). More specifically, we shall use the corresponding expressions (6.40), (6.41) which allows us to select suitable weighting functions. We adopt here the values:

$$\varepsilon^2 = T^{+^{-1}} T^2; \quad \tau^2 = T \quad (\zeta = z') \tag{11.30}$$

where T^+ denotes a positive constant having the dimension of a temperature which will be fixed later (11.32).

As a result, the corresponding stability condition will be slightly different from (7.94), which is associated to the values $\varepsilon^2 = \tau^2 = 1$.

Taking into account the Boussinesq approximation, as well as the incompressibility of the fluid, we obtain the inequality (compare with (11.26), (11.28); see also (7.57)):

$$P[\delta Z'] = \int [\lambda T^{+^{-1}} (\theta_{,j})^2 + \eta (u_{i,j})^2 - (\beta \rho c_v T^{+^{-1}} + \rho g \alpha) \theta w] \, \mathrm{d}V > 0 \tag{11.31}$$

We recognize in the r.h.s. the dissipative terms associated with the thermodynamic flows and forces, and the two terms arising from the velocity fluctuations already involved in (11.26) and (11.28).

Let us adopt the value:

$$T^+ = \frac{c_v \beta}{\alpha g} \tag{11.32}$$

In this way the last two terms in (11.31) combine to give a single term. Moreover let us introduce the usual dimensionless *Rayleigh number*, which characterizes the slope β:

$$\mathscr{R}a = \frac{g \alpha h^4}{\kappa \nu} \beta > 0 \tag{11.33}$$

The complete stability condition (11.31) then becomes:

$$P[\delta Z'] = \int \rho[(\mathscr{R}a)^{-1} \frac{(g\alpha)^2 h^4}{\nu} (\theta_{,j})^2 + \nu(u_{i,j})^2 - 2g\alpha\theta w] \, \mathrm{d}V > 0 \quad (11.34)$$

The r.h.s. may be interpreted as the increment of the function:

$$\mathscr{F} = \int \rho \left[(\mathscr{R}a)^{-1} \frac{(g\alpha)^2 h^4}{\nu} (T_{,j})^2 + \nu(\mathrm{v}_{i,j})^2 - 2g\alpha T \mathrm{v}_z \right] \mathrm{d}V \quad (11.35)$$

This increment is taken between the perturbed state and the state at rest. To prove it we expand (11.35) in terms of:

$$T = T_0 + \theta \quad \mathrm{v}_i = 0 + u_i \quad (11.36)$$

where the subscript zero refers to the state at rest, and remember that the mean values $\langle\theta\rangle$ and $\langle u_i\rangle$ vanish identically. After averaging over the x, y plane we obtain in this way the equality:

$$\langle\mathscr{F}\rangle = \langle\mathscr{F}_0\rangle + \langle P[\delta Z']\rangle \quad (11.37)$$

Besides, applying (11.35) to the state at rest one finds:

$$\langle\mathscr{F}_0\rangle = \frac{\rho\nu\kappa^2}{h^3} \mathscr{R}a \quad (11.38)$$

As a result, the stability condition (11.31) for the state at rest may be interpreted as a minimum condition for the function $\langle\mathscr{F}\rangle$. We shall come back to this property in § 6. One also observes that near thermodynamic equilibrium, that is when $\mathscr{R}a \to 0$, one has $\langle\mathscr{F}_0\rangle \to 0$ whereas $\langle\mathscr{F}\rangle \to \infty$ and $\langle P[\delta Z']\rangle \to \infty$ for all perturbed states. This situation is represented schematically on Figure 11.1.

For normal modes such as curve (4), the increment $\langle\mathscr{F}_4\rangle - \langle\mathscr{F}_0\rangle$ i.e. $\langle P_4[\delta Z']\rangle$ is always positive. Therefore, the stability condition (11.34) is always fulfilled. On the contrary, for cases such as (2), (3), instability occurs respectively beyond $(\mathscr{R}a)_2$, $(\mathscr{R}a)_3$. The Bénard point corresponds to the lowest limit, called the critical Rayleigh number $(\mathscr{R}a)_c = (\mathscr{R}a)_1$.

Instability occurs when $\langle P[\delta Z']\rangle$ vanishes. The function $\langle\mathscr{F}\rangle$ takes then the same value both for the state at rest and the excited normal mode (see 11.37). Therefore, the instability corresponds to the *degeneracy* of $\langle\mathscr{F}\rangle$. The analogy with phase transition is striking and will be further developed in § 5.

It is also important to observe that for all fluctuations leading to an instability, the function $\langle \mathscr{F} \rangle$ derived from (11.35), appears as the difference between two *positive* quantities, namely:

$$\langle \mathscr{F} \rangle = (\text{dissipative effects}) - (\text{convective effects}) \quad (11.39)$$

In other words, the stability occurs as the result of a competition

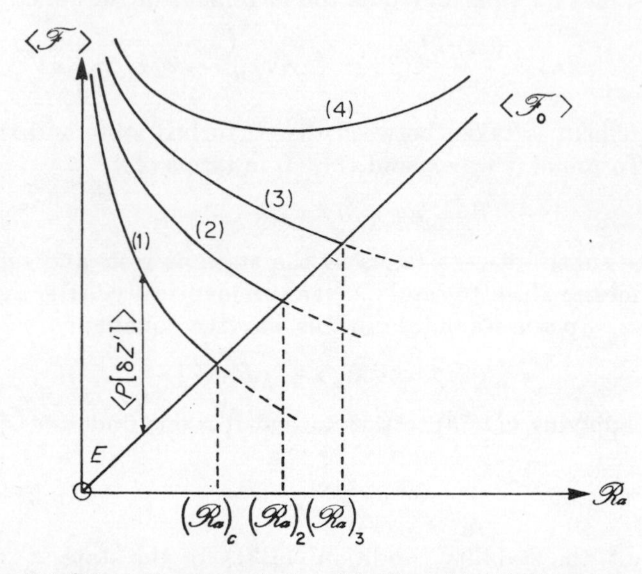

Fig. 11.1. Schematic representation of the Bénard instability in terms of (11.37) at $t = t_0$. The straight line $\langle \mathscr{F}_0 \rangle$ corresponds to the state at rest. Curves (1) to (4) are related to normal modes in the x, y plane; the dashed part represents the range of instability, namely

$$\langle P[\delta Z'] \rangle < 0 \quad \text{or} \quad \langle P_m[\delta Z'] \rangle < 0$$

for complex modes, see end of § 2. The onset of free convection corresponds to the critical Rayleigh number $(\mathscr{R}_a)_c = (\mathscr{R}_a)_1$.

between two opposite tendencies, that is, the stabilizing dissipative effects and the destabilizing convective effects.

Such a competitive aspect is quite general in problems involving the possibility of an instability. This is in agreement with our discussion in Chapter VII, § 6.

We also observe on (11.34) that in the range of the small Rayleigh numbers, dissipative effects generated by the temperature fluctuations are dominant, while the importance of the velocity fluctuations appears only for higher values of the Rayleigh number.

5. THERMODYNAMIC INTERPRETATION AND DISSIPATIVE STRUCTURE

In § 2, we have assumed constant values of the phenomenological coefficients such as heat conductivity and viscosity.

Therefore, the state at rest below the critical Rayleigh number (Figure 11.1) belongs to the region of the strictly linear non-equilibrium thermodynamics. This implies the validity of the theorem of minimum entropy production (Chapter III, § 4; Chapter VII, § 9).

When we reach the marginal state, the entropy production varies suddenly with the appearance of the first unstable normal mode

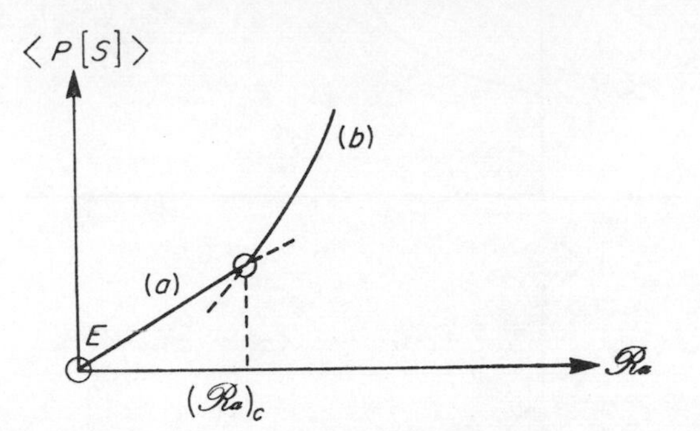

Fig. 11.2. (*a*) Thermodynamic branch including the equilibrium state *E*. (*b*) Branch generated by the first critical normal mode. Schematic representation of the entropy production.

(§ 10). The new solution generated by this normal mode leads to a *discontinuity in the slope* of the entropy production $\langle P[S]\rangle$ as schematically represented on Figure 11.2. This is not surprising since at the critical Rayleigh point, a new mechanism occurs due to viscous dissipation generated by convection.

There is no discontinuity for $\langle P[S]\rangle$ itself, as the amplitude of the critical normal mode remains infinitesimal at the marginal state. To obtain a finite amplitude we have to consider $\mathscr{R}a$ values slightly higher than $(\mathscr{R}a)_c$. For values of $\mathscr{R}a$ beyond $(\mathscr{R}a)_c$, the description of the system in terms of linear thermodynamics of irreversible processes breaks down. New coupling effects are introduced since a temperature gradient gives rise to a convective current. This coupling, not included in the phenomenological laws, arises from the stationary perturbation equations (Chapter III, § 3).

We have already mentioned the analogy between the Bénard

problem and a phase transition. Let us investigate this question more thoroughly. Below the critical Rayleigh number, the perturbation equations (11.6–11.8), have only the simple trivial zero value steady state solution, corresponding to the system at rest (also § 10). All the normal modes are damped: $\omega_r < 0$ (and all the ω_i vanish as pointed out in § 7). At $\mathcal{R}a = (\mathcal{R}a)_c$, we obtain *in addition* a new non-trivial solution. As shown in § 4, the function $\langle \mathcal{F} \rangle$ introduced in (11.35), takes then the same value for these two solutions. This

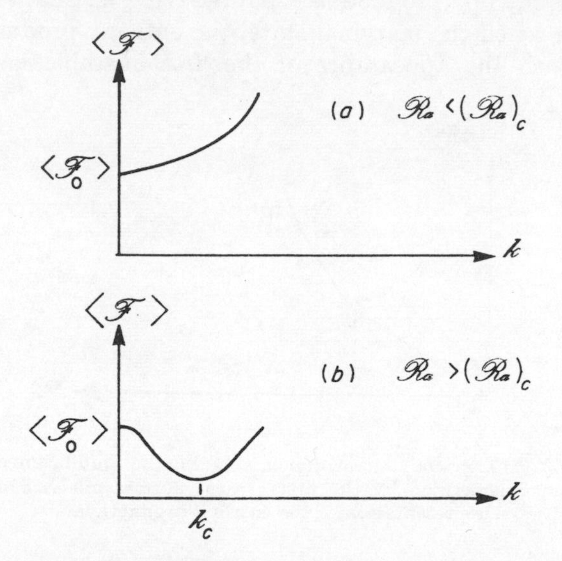

Fig. 11.3. (*a*) The state at rest is stable; $\langle \mathcal{F}_0 \rangle$ is a minimum. (*b*) The state at rest is unstable just after the transition point. The minimum of $\langle \mathcal{F} \rangle$ corresponds to the new solution assumed to be stable.

function depends on the amplitude of the perturbations, and on parameters such as the wavelength or the wave vector of the normal mode considered (also, § 10).

Let us plot $\langle \mathcal{F} \rangle$ as a function of these parameters. To simplify the visualization we use a single parameter, such as the wave number k.

The Figure 11.3(a) corresponds to the case of the system at rest $(\mathcal{R}a < (\mathcal{R}a)_c)$. The corresponding value $\langle \mathcal{F}_0 \rangle$ given by (11.38) is smaller than $\langle \mathcal{F} \rangle$ for all perturbed situations.

On the contrary, for $\mathcal{R}a$ slightly larger than $(\mathcal{R}a)_c$, the values of $\langle \mathcal{F} \rangle$ for the critical normal mode will be slightly lower than $\langle \mathcal{F}_0 \rangle$ and correspond to a minimum if the new solution is stable as shown on Figure 11.3(b).

The manifold of solutions corresponding to the system at rest will be called the *thermodynamic branch*. At the Bénard point, the thermodynamic branch becomes unstable. We have then a transition to a new 'branch' (Figure 11.2).

This transition involves the appearance of a '*dissipative structure*'. Indeed, at the critical point, the system uses part of its thermal energy to build up the kinetic energy necessary to maintain the macroscopic stationary cellular motion which occurs at the onset of free convection. Then the layer appears as formed by juxtaposed cells aligned to form a regular hexagonal pattern in the horizontal plane (Chandrasekhar, 1961, Chapter II). The stationary behaviour of this cellular motion is studied in § 7.

The function $\langle \mathscr{F} \rangle$ plays here a role similar to that of the Helmholtz free energy, in ordinary phase transitions. Parameters such as the wave number of the disturbances (included in k in Figure 11.3), correspond to the intensive variables.

The two time independent states ($k = 0$ and $k = k_c$) considered in Figure 11.3(b) are separated by a line of time dependent processes. However, an important difference with the *van der Waals type of phase transition* is apparent on Figure 11.3. We do not have *two* stable or metastable equilibrium states separated by *one* unstable equilibrium state. Here, till the Bénard point, only one steady state exists, and just above the Bénard point, we observe one stable and one unstable steady state. If we increase the Rayleigh number beyond $(\mathscr{R}a)_c$, then the stable state will correspond to a superposition of an increasing number of normal modes.

In Chapter XVI, we shall meet some examples closer to the *van der Waals type*, as they will involve two stable steady states separated by an unstable one.

6. NEUTRAL STABILITY CONDITION

We now consider the layer in the marginal state of *neutral stability*, that is at the border between the stable and the unstable states. In the kinetic stability theory based on the normal mode analysis, the marginal state is reached as soon as the real part of a normal mode frequency vanishes.

According to (6.34), (6.39), the thermo-hydrodynamic condition of neutral stability may be written as:

$$\partial_t \delta_m^2 Z' = 2\omega_r \delta_m^2 Z' = 0 \tag{11.40}$$

The prime corresponds again to the weighting functions (11.30).

As shown in Chapter VII (cf. 7.94, 7.95) one has for non-varied boundary conditions:

$$\tfrac{1}{2}\,\partial_t \delta_m^2 Z' = P_m[\delta Z'] \tag{11.41}$$

Therefore, the condition (11.40) becomes:

$$P_m[\delta Z'] = 0 \tag{11.42}$$

Applying the rule (2.76) for complex quantities, we derive from (11.34) the following explicit neutral stability condition:

$$\int \rho \left[(\mathscr{R}a)^{-1} \frac{(g\alpha)^2 h^4}{\nu}\ \theta_{,j}\theta^*_{,j} + \nu u_{i,j}u^*_{i,j} - g\alpha(\theta w^* + \theta^* w) \right] \mathrm{d}V = 0 \tag{11.43}$$

As a result, the corresponding Rayleigh number is given, after averaging over the x, y-plane, by the ratio:

$$\mathscr{R}a = \frac{(g\alpha)^2 h^4 \displaystyle\int_0^h \langle \theta_{,j}\theta^*_{,j}\rangle\, \mathrm{d}z}{g\alpha\nu \displaystyle\int_0^h (\langle \theta w^*\rangle + \langle \theta^* w\rangle)\, \mathrm{d}z - \nu^2 \displaystyle\int_0^h \langle u_{i,j}u^*_{i\,j}\rangle\, \mathrm{d}z} \tag{11.44}$$

The lowest value of this ratio for non-trivial zero solutions, of the perturbation equations (11.6–11.8), represents the critical Rayleigh number, which characterizes the onset of free convection.

This property opens the way to a variational approach for its determination, as shown in §§ 8, 9. But let us first establish the value of ω_i (the imaginary part of ω) at the marginal state.

7. THE PRINCIPLE OF EXCHANGE OF STABILITIES AND THE EVOLUTION CRITERION

We now show that in the Bénard problem, convection sets in as a stationary motion, or in other words, that the so-called *principle of the exchange of stabilities* is satisfied (Chandrasekhar, 1961, Chapter I, II). This means that in the marginal state ($\omega_r = 0$), ω is real ($\omega_i = 0$). In fact, we shall prove more, i.e. $\omega_i = 0$, even for non vanishing values of ω_r.

We start from our general evolution criterion and we first extend the properties derived in Chapter IX, § 6, to situations involving convection. Following the same method, we multiply the two sides of the excess momentum equation (11.7) by $-\partial_t u_i^*$ and those of the excess energy equation (11.8), by

$$-\frac{\alpha g}{c_v \beta}\,\partial_t \theta^*$$

according to the weighting functions selected in (11.30), (11.32). We then proceed in the same way for the complex conjugate mode and we add. Performing in the r.h.s. the integration by parts described e.g. in § 2 of Chapter IX, and using (11.6), we obtain, after cancellation of the surface term, the evolution criterion in the form:

$$\begin{aligned}
-2(\omega_r^2 + \omega_i^2) \int \rho \left(\frac{\alpha g}{\beta}\,\theta\theta^* + u_i u_i^*\right) \mathrm{d}V \\
= (\omega_r - i\omega_i) \int \rho \left[(\mathscr{R}a)^{-1} \frac{(g\alpha)^2 h^4}{\nu}\,\theta_{,j}\theta^*_{,j} + \nu u_{i,j} u^*_{i,j} \right. \\
\left. - g\alpha(w\theta^* + w^*\theta) \right] \mathrm{d}V \\
+ (\omega_r + i\omega_i) \int \rho \left[(\mathscr{R}a)^{-1} \frac{(g\alpha)^2 h^4}{\nu}\,\theta^*_{,j}\theta_{,j} + \nu u^*_{i,j} u_{i,j} \right. \\
\left. - g\alpha(w^*\theta + w\theta^*) \right] \mathrm{d}V \leqslant 0
\end{aligned} \tag{11.45}$$

This is the straightforward generalization of (9.52) to the present problem involving convective effects. Indeed (11.45) may be written in the compact form:

$$(\omega_r^2 + \omega_i^2)\delta_m^2 Z' = \omega_r P_m[\delta Z'] + \omega_i \Pi_m[\delta Z'] \leqslant 0 \tag{11.46}$$

As in the proof of (9.57), the equality (cf. 11.40, 11.41):

$$\omega_r \delta_m^2 Z' = P_m[\delta Z'] \tag{11.47}$$

allows us to write *separately*:

$$\omega_i^2 \delta_m^2 Z' = \omega_i \Pi_m[\delta Z'] \leqslant 0 \tag{11.48}$$

But at the same time, one observes directly using (11.45), that in this problem $\Pi_m[\delta Z']$ vanishes identically. Therefore, one has:

$$\omega_i = 0 \tag{11.49}$$

since $\delta_m^2 Z'$ is a negative quadratic form (cf. 6.39). We may therefore conclude that oscillating disturbances are excluded around the steady state of the layer at rest, up to and including the marginal state.

We recover in this way through our evolution criterion, a property previously derived by Pellew and Southwell (1940) from a detailed analysis of the perturbation equations at the marginal state (Chandrasekhar, 1961).

The interest of the present method lies again in its generality. Every time the antisymmetric contribution Π in the evolution criterion may be eliminated by a suitable choice of the weighting functions, the principle of the exchange of stabilities is valid.

8. A VARIATIONAL FREE MINIMUM PRINCIPLE FOR THE CRITICAL RAYLEIGH NUMBER

We now consider the value of the Rayleigh number in the marginal state as given by equation (11.44) in the form of a ratio of two integrals:

$$\mathscr{R}a = \frac{I_1}{I_2} \tag{11.50}$$

The critical value characterizing the onset of instability, corresponds to the lowest value of this ratio, when the functions involved in the r.h.s. are evaluated in terms of the steady state solutions of the perturbation equations.

Let us investigate if the critical value of the Rayleigh number corresponds to a *free minimum* of the r.h.s. of (11.50). As well known, free minimum means that all the functions involved in the r.h.s. of (11.50) may be varied *independently* and *arbitrarily*. They only have to satisfy the boundary conditions.

In this respect we would like to underline the distinction between the present formulation and the variational principles previously established by Pellew and Southwell (1940) and by Chandrasekhar (1954) (Chandrasekhar, 1961). Indeed these principles imply a preliminary elimination of all the unknown functions in terms of one of them, with the help of the perturbation equations themselves (Chandrasekhar, 1961, Chapter II). On the contrary we vary independently all the functions involved.

The ratio (11.50) was originally derived from these perturbation equations, starting from (11.6–11.8). We now use it as the basis of a variational principle to recover the marginal perturbation equations in the form of Euler–Lagrange equations (steady state of § 7). Such a reciprocity may only be expected if no additional condition were to be introduced to derive equation (11.43). Otherwise we would have to reintroduce such additional conditions through appropriate Lagrange multipliers.

This is however what we have done using the incompressibility condition $u_{j,j} = 0$ (e.g. 11.25–11.26). As we want now to recover this condition among the Euler–Lagrange equations, without the use of Lagrange multipliers, we first rewrite equation (11.43) in its complete form as:

$$\int \rho \left[(\mathscr{R}a)^{-1} \frac{(g\alpha)^2 h^4}{\nu} \theta_{,j}\theta^*_{,j} + 2\nu \, \mathrm{d}_{ij} \, \mathrm{d}^*_{ij} - g\alpha(\theta w^* + \theta^* w) - v(\varpi u^*_{j,j} + \varpi^* u_{j,j}) \right] \mathrm{d}V = 0$$

$$(\varpi \equiv \delta p) \tag{11.51}$$

In this way the incompressibility condition is only implied in the basic perturbation equations (11.6) and (11.8), but not in (11.7).† The integrals in (11.50) now become explicitly:

$$I_1 = (g\alpha)^2 h^4 \int \theta_{,j}\theta^*_{,j} \, \mathrm{d}V \tag{11.52}$$

and

$$I_2 = \int [g\alpha\nu(\theta w^* + \theta^* w) - 2\nu^2 \, \mathrm{d}_{ij} \, \mathrm{d}^*_{ij} + \nu v(\varpi u^*_{j,j} + \varpi^* u_{j,j})] \, \mathrm{d}V \tag{11.53}$$

Let us now minimize the ratio (11.50).

We deduce from (11.50):

$$\delta\mathscr{R}a = \frac{1}{I_2}\left(\delta I_1 - \frac{I_1}{I_2}\delta I_2\right) = \frac{1}{I_2}(\delta I_1 - \mathscr{R}a\delta I_2)$$

This provides the condition for the minimum:

$$\delta I_1 - \mathscr{R}a\delta I_2 = 0 \tag{11.54}$$

The increments δI_1 and δI_2 may be derived from (11.52), (11.53) as:

$$\delta I_1 = -(g\alpha)^2 h^4 \int [(\theta_{,j})_{,j}\delta\theta^* + (\theta^*_{,j})_{,j}\delta\theta] \, \mathrm{d}V \tag{11.55}$$

† To avoid this distinction, it is easier to derive directly (11.51) starting from equations (11.6–11.8). We have then to follow the general method of Chapter VII, § 5, instead of introducing specific special restrictions (such as the incompressibility condition, the Boussinesq approximation, . . .) as we did to derive (11.31) from the general expression of the stability condition.

and

$$\delta I_2 = \int \{g\alpha\nu[\theta\delta w^* + w^*\delta\theta + \theta^*\delta w + w\delta\theta^*]$$
$$+ 2\nu^2[(\mathrm{d}_{ij}),_j\delta u_i^* + (\mathrm{d}_{ij})^*,_j\delta u_i]$$
$$- \nu v[\varpi,_j\delta u_j{}^* + \varpi^*,_j\delta u_j]$$
$$- \nu v[u_j,_j\delta\varpi^* + u_j^*,_j\delta\varpi]\} \,\mathrm{d}V \tag{11.56}$$

We introduce these two expressions into (11.54). The Euler–Lagrange equations corresponding to this variational problem are obtained by cancellation of the coefficients of the arbitrary increments:

$$\delta\theta^*,\ \delta\theta,\ \delta u_i^*,\ \delta u_i,\ \delta\varpi^*,\ \delta\varpi.$$

Grouping the terms in (11.54) and using the definition (11.33) of the Rayleigh number, we get successively:

1°/For the coefficient of $\delta\theta^*$:

$$\lambda(\theta,_j),_j + \rho c_v \beta w = 0 \tag{11.57}$$

The conjugate expression for $\delta\theta$ is derived in a similar way.

As expected we recover here the steady state perturbation equation for energy, according to the first equality (11.10) (see 11.8).

2°/For the coefficient of δu_i^*:

$$2\eta(\mathrm{d}_{ij}),_j - \varpi,_j\delta_{ij} - g_i\alpha\rho\theta = 0$$
$$(g_i = 0, 0, -g); \quad \delta_{ij} = \begin{cases} 0 & i \neq j \\ 1 & i = j \end{cases} \tag{11.58}$$

This is indeed the steady state perturbation equation for momentum, according to the second equality (11.10) used together with (11.20) (see 11.7).

3°/For the coefficient of $\delta\varpi^*$:

$$u_j,_j = 0 \tag{11.59}$$

One recognizes the condition of incompressibility as expressed by the perturbation equation for mass (11.6).

It remains to establish that the lowest value of $\mathcal{R}a$, i.e. the critical Rayleigh number, corresponds indeed to a true minimum of the ratio (11.50).

As already observed when we derived (11.43) from (11.34), one has:

$$\rho^{-1}\mathcal{R}a\nu P_m[\delta Z'] = I_1 - \mathcal{R}aI_2 \tag{11.60}$$

Therefore, according to (11.50), the difference for a given marginal Rayleigh number, between the generalized excess entropy production due to an arbitrary mode and that of the marginal mode, is given by:

$$\rho^{-1}\mathscr{R}a\nu P_m[\delta Z'] = \rho^{-1}\mathscr{R}a\nu\Delta P_m[\delta Z'] = \Delta I_1 - \mathscr{R}a\Delta I_2 \qquad (11.61)$$

We have now to remember that for the lowest value $(\mathscr{R}a)_c$ all disturbances except the critical mode satisfy the stability condition (11.34). We get therefore:

$$\rho^{-1}\mathscr{R}a\nu P_m[\delta Z'] = \Delta I_1 - (\mathscr{R}a)_c\Delta I_2 > 0 \qquad (11.62)$$

Taking into account this stability condition, the corresponding increment of the ratio I_1/I_2 gives us:

$$\Delta\left(\frac{I_1}{I_2}\right) = \frac{I_1 + \Delta I_1}{I_2 + \Delta I_2} - \frac{I_1}{I_2} = \frac{1}{I_2 + \Delta I_2}\left(\Delta I_1 - \frac{I_1}{I_2}\Delta I_2\right)$$

$$\simeq \frac{(\mathscr{R}a)_c}{I_1}[\Delta I_1 - (\mathscr{R}a)_c\Delta I_2] > 0 \qquad (11.63)$$

In the last term we have assumed small values of the increment ΔI_2 namely,

$$\Delta I_2 < I_2 \qquad (11.64)$$

We also notice that $(\mathscr{R}a)_c$ and I_1 are positive quantities (11.52). Therefore as expected, the critical Rayleigh number corresponds to a true minimum. Let us make a few supplementary remarks.

The reader will observe on (11.63) that the variational principle based on the minimization of the Rayleigh ratio I_1/I_2, is quite similar to those which could be based on the minimization of either the generalized excess entropy production $\langle P_m[\delta Z']\rangle$ (11.61), or of the function $\langle \mathscr{F} \rangle$ (11.37), after averaging over the x, y plane.

With this formulation, we no longer have a ratio as in (11.50) and the restriction (11.64) is not required. We then obtain an *absolute* minimum. This additional property may be useful for numerical computation, since the amplitude of the error introduced by test functions is a priori unknown (see Chapter X).

Moreover, as the time dependence exp (ωt), disappears from the ratio I_1/I_2, the corresponding variational principles may be based as well on $\langle P_m[\delta Z']\rangle_{t=0}$ or on $\langle \mathscr{F} \rangle_{t=0}$ which depend on the value of the disturbances at $t = 0$. To illustrate our variational principle we shall calculate in Section 10 the approximate value of $(\mathscr{R}a)_c$ in a simple case.

9. NORMAL MODE APPROACH TO THE BÉNARD PROBLEM

Let us now study the Bénard problem, starting with the kinetic theory of stability based on the normal mode analysis. This approach of the problem is carefully developed by Chandrasekhar (1961, Chapter II). We therefore only summarize it briefly here.

We want to show how the marginal state may be derived as the solution of an eigenvalue problem.

We first eliminate the perturbation ϖ of the hydrostatic pressure from the excess momentum equation (11.7), by taking the curl of the two sides (curl grad $\equiv 0$). Taking the curl once more one obtains after simple manipulations:

$$\partial_t \nabla^2 w = g\alpha \left(\frac{\partial^2 \theta}{\partial x^2} + \frac{\partial^2 \theta}{\partial y^2} \right) + \nu \nabla^4 w \tag{11.65}$$

together with:

$$\partial_t \zeta = \nu \nabla^2 \zeta \quad \left(\zeta = \frac{\partial u_y}{\partial x} - \frac{\partial u_x}{\partial y} \right) \tag{11.66}$$

The quantity ζ denotes the z-component of the vorticity.

One has also the equation (cf. 11.8, 11.10, 11.27):

$$\partial_t \, \theta = \beta w + \kappa \nabla^2 \theta \tag{11.67}$$

We now analyse an arbitrary perturbation in terms of a complete set of normal modes. As we suppose the layer contained between two infinite horizontal planes, we may develop an arbitrary disturbance in terms of two-dimensional periodic waves. This gives:

$$w = W(z) \exp\left[i(k_x x + k_y y) + \omega t\right] \tag{11.68}$$

$$\theta = \Theta(z) \exp\left[i(k_x x + k_y y) + \omega t\right] \tag{11.69}$$

where

$$k = (k_x^2 + k_y^2)^{\frac{1}{2}} \tag{11.70}$$

is the wave number of the disturbance and ω the frequency which in principle, may be a complex quantity. The vorticity ζ plays no role and will no longer be considered (Chandrasekhar, 1961, p. 32).

For functions, such as (11.68–11.69), one has:

$$\partial_t = \omega; \quad \frac{\partial^2}{\partial x^2} + \frac{\partial^2}{\partial y^2} = -k^2; \quad \nabla^2 = \frac{\mathrm{d}^2}{\mathrm{d}z^2} - k^2 \tag{11.71}$$

Therefore, equations (11.65) and (11.66) now become:

$$\omega\left(\frac{\mathrm{d}^2}{\mathrm{d}z^2}-k^2\right)W=-g\alpha k^2\Theta+\nu\left(\frac{\mathrm{d}^2}{\mathrm{d}z^2}-k^2\right)^2 W \tag{11.72}$$

$$\omega\Theta=\beta W+\kappa\left(\frac{\mathrm{d}^2}{\mathrm{d}z^2}-k^2\right)\Theta \tag{11.73}$$

The corresponding boundary conditions are (cf. 11.11–11.13):

$$\Theta=W=0 \quad \text{for} \quad z=0 \quad \text{and} \quad z=h \tag{11.74}$$

and

$$\left.\begin{aligned}&\frac{\mathrm{d}W}{\mathrm{d}z}=0, \text{ on a rigid surface}\\ &\text{or}\\ &\frac{\mathrm{d}^2W}{\mathrm{d}z^2}=0, \text{ on a free surface}\end{aligned}\right\} \tag{11.75}$$

It is useful to introduce non-dimensional variables such as

$$Z=\frac{z}{h} \quad \text{and} \quad \tau=\frac{k^2}{\nu}t \tag{11.76}$$

Let

$$a=kh \quad \text{and} \quad \sigma=\frac{\omega h^2}{\nu} \tag{11.77}$$

be the wave number and the frequency in these units. We also introduce the symbol

$$\mathrm{D}=\frac{\mathrm{d}}{\mathrm{d}Z}$$

Equations (11.72), (11.73), now take the form:

$$(\mathrm{D}^2-a^2)(\mathrm{D}^2-a^2-\sigma)W=\left(\frac{g\alpha}{\nu}h^2\right)a^2\Theta \tag{11.78}$$

$$\left(\mathrm{D}^2-a^2-\frac{\nu}{\kappa}\sigma\right)\Theta=-\left(\frac{\beta}{\kappa}h^2\right)W \tag{11.79}$$

The ratio $\frac{\nu}{\kappa}$ is the *Prandtl number.*

A marginal state separating stable from unstable states is characterized by the condition $\sigma_r=0$. We have already proved, using our evolution criterion that σ is a real quantity (cf. 11.49). This theorem can also be derived directly from the perturbation equations (Chandrasekhar, 1961, p. 24).

As a result, the equations governing the marginal state are obtained simply by setting $\sigma = 0$ in equations (11.78–11.79), namely:

$$(\mathrm{D}^2 - a^2)^2 W = \left(\frac{g\alpha}{\nu} h^2\right) a^2 \Theta \tag{11.80}$$

$$(\mathrm{D}^2 - a^2)\Theta = -\left(\frac{\beta}{\kappa} h^2\right) W \tag{11.81}$$

Eliminating Θ between these equations, we get:

$$(\mathrm{D}^2 - a^2)^3 W = -(\mathscr{R}a) a^2 W \tag{11.82}$$

where $\mathscr{R}a$ is the Rayleigh number defined in (11.33).

The boundary conditions (11.74), (11.75), take here the form:

$$W = 0; \quad (\mathrm{D}^2 - a^2)^2 W = 0 \quad \text{at} \quad Z = 0 \quad \text{and} \quad Z = 1 \tag{11.83}$$

and

$$\mathrm{D}W = 0 \text{ (rigid surface)}$$

or

$$\mathrm{D}^2 W = 0 \text{ (free surface)} \tag{11.84}$$

An identical equation can be derived for Θ.

We are facing here an *eigenvalue problem* for $\mathscr{R}a$ at the marginal state. In other words, only for particular values of $\mathscr{R}a$ shall we find a non-vanishing solution W, satisfying the boundary conditions (11.83) and (11.84) for a given a^2.

As an example, let us consider the solution for two *free boundaries*. The functions

$$W = A \sin n\pi Z$$

where A is a constant and n an integer, satisfy equation (11.82) and the corresponding boundary conditions, (11.83), (11.84). Substitution in (11.82) leads to:

$$\mathscr{R}a = \frac{(n^2\pi^2 + a^2)^3}{a^2} \tag{11.85}$$

For a given a^2, the lowest value of $\mathscr{R}a$ occurs for $n = 1$. Therefore:

$$\mathscr{R}a = \frac{(\pi^2 + a^2)^3}{a^2} \tag{11.86}$$

which is still a function of a^2. The critical Rayleigh number corresponds to the *minimum* of $\mathscr{R}a$ as a function of a^2 and satisfies therefore the condition:

$$\frac{\partial \mathscr{R}a}{\partial a^2} = \frac{1}{a^4}(\pi^2 + a^2)^2(2a^2 - \pi^2) = 0 \tag{11.87}$$

that is:

$$a^2 = \frac{\pi^2}{2} \tag{11.88}$$

The critical value of $\mathscr{R}a$ beyond which instability will set in, is given by:

$$(\mathscr{R}a)_c = \frac{27}{4}\pi^4 = 657{\cdot}5 \tag{11.89}$$

It corresponds to the critical wavelength (cf. 11.77)

$$\lambda_c = \frac{2\pi}{a} h = 2^{3/2} h \tag{11.90}$$

The simplicity of the stability theory based on normal mode analysis should be emphasized. However, this simplicity is due in part to the possibility of eliminating all the variables except one. For instance in equation (11.82), only W remains, Θ has been eliminated. In problems where such an elimination is no longer possible, a variational method such as the one described in § 8, and based on the existence of a free minimum may be useful.

10. APPROXIMATE DETERMINATION OF THE CRITICAL RAYLEIGH NUMBER BY THE FREE MINIMUM METHOD

Let us illustrate briefly by a simple example, the variational principle derived in § 8 and compare it with the exact result derived in section 9.

After integration over the x, y plane, we write (11.51) in the form:

$$\mathscr{R}a \int_0^1 [g\alpha\nu\langle \theta w^* + \theta^* w\rangle - 2\nu^2\langle \mathrm{d}_{ij}\,\mathrm{d}_{ij}^*\rangle + v\nu\langle \varpi^* u_{j,j} + \varpi u_{j,j}{}^*\rangle]\,\mathrm{d}Z - (g\alpha)^2 h^4 \int_0^1 \langle \theta_{,j}\theta_{,j}^*\rangle\,\mathrm{d}Z = 0 \tag{11.91}$$

We use the dimensionless quantities

$$Z = \frac{z}{h}; \quad Y = \frac{y}{h}; \quad X = \frac{x}{h}; \quad a = kh$$

Let us introduce into (11.91) the following ten trial functions:

$$\begin{aligned}
\theta &= \alpha_\theta [\exp i(Xa_X + Ya_Y)]\phi(Z) \\
u_1 &= \alpha_1 [\exp i(Xa_X + Ya_Y)] \frac{d\phi}{dZ} \\
u_2 &= \alpha_2 [\exp i(Xa_X + Ya_Y)] \frac{d\phi}{dZ} \\
u_3 = w &= \alpha_3 [\exp i(Xa_X + Ya_Y)]\phi(Z) \\
\varpi &= \alpha_\varpi [\exp i(Xa_X + Ya_Y)]F(Z)
\end{aligned} \tag{11.92}$$

together with the five complex conjugates θ^*, u_1^*, . . . We also adopt the same Z-dependence $\phi(Z)$ for θ and w, and $d\phi/dZ$ for u_1 and u_2, in order to satisfy both the boundary conditions as well as simple properties derived from the perturbation equations (e.g. cancellation of the vorticity). Accordingly, $\phi(Z)$ is taken to be the odd function:

$$\phi(Z) = A_1 Z + A_2 Z^3 + A_3 Z^5 + \cdots$$

For the sake of simplicity we limit ourselves in (11.92), to test functions containing only a single arbitrary independent parameter, i.e. α_θ, α_θ^* . . . The other parameters are determined by the boundary conditions. This limits $\phi(Z)$ to a polynomial of degree five.

Indeed let us consider two free limiting surfaces. The boundary conditions (11.11–11.13) then yield:

$$\phi(0) = \phi(1) = 0; \quad \phi''(0) = \phi''(1) = 0$$

Therefore apart from a multiplicative factor absorbed in the α's we get

$$\phi(Z) \sim 7Z - 10Z^3 + 3Z^5$$

On the other hand $F(Z)$ may be chosen arbitrarily, but satisfying the boundary conditions. Indeed, $F(z)$ is finally eliminated from the result given in (11.94).

We can now perform the integration over Z in (11.91). This leads to a relation between the ten unknown complex parameters α_1, α_1^*, . . .

Minimizing with respect to each parameter α_1^*, α_2^* . . . we get the following relations:†

$$\frac{\nu}{h}[A\alpha_1 a^2 + A\alpha_1 a_x^2 + Aa_x a_y \alpha_2 + B\alpha_1] + i\nu\, a_x\, \alpha_{\varpi} K = 0$$

$$\frac{\nu}{h}[A\alpha_2 a^2 + A\alpha_2 a_y^2 + Aa_x a_y \alpha_1 + B\alpha_2] + i\nu\alpha_{\varpi}\, a_y K = 0$$

$$(g\alpha)\alpha_\theta C - \frac{\nu\alpha_3}{h^2}(2A + a^2C + 2D) + \frac{\nu K\alpha_{\varpi}}{h} = 0$$

$$R\nu\alpha_3 C - (g\alpha)h^2\alpha_\theta(A + a^2C) = 0$$

$$\alpha_3 = -i(\alpha_1 a_x + \alpha_2 a_y) \tag{11.93}$$

and similarly for α_1, α_2 . . .

In (11.93) we have set:

$$A = \int_0^1 \left(\frac{d\phi}{dZ}\right)^2 dZ; \qquad B = \int_0^1 \left(\frac{d^2\phi}{dZ^2}\right) dZ$$

$$C = \int_0^1 \phi^2(Z)\, dZ; \qquad D = \int_0^1 \phi(Z)\frac{d^2\phi}{dZ^2}\, dZ$$

$$K = \int_0^1 \frac{d\phi}{dZ} F(Z)\, dZ$$

Using the above value of $\phi(Z)$, we get,

$$A = 27{\cdot}43, \quad B = 274{\cdot}3, \quad C = 2{\cdot}752, \quad D = 27{\cdot}4$$

Let us now observe that (11.93) occurs as a set of homogeneous equations with respect to the parameters α. The consistency condition for the existence of a non-trivial solution (det $= 0$) gives rise to:

$$\mathcal{R}a = \frac{(A + a^2C)\left(4A + \dfrac{B}{a^2} + a^2C + 2D\right)}{C^2} \tag{11.94}$$

One sees that K is eliminated and then also $F(Z)$. For the special choice $F(Z) = \phi(Z)$ one finds $K = 0$.

† We have integrated over Z before minimizing. One might equivalently invert these two operations as it is usual e.g. in the Rayleigh–Ritz method. But it must be kept in mind that our purpose here is slightly different, as we do not need the separate value of each parameter. This means that the Z–dependence of the pressure perturbation does not influence the critical value of the Rayleigh number.

As in (11.87), we now calculate the critical Rayleigh number by minimization of $\mathscr{R}a$:

$$a^6 + a^4 \left(\frac{5A + 2D}{2C}\right) - \frac{AB}{C^2} = 0 \qquad (11.95)$$

Solving for a^2, (11.95) gives us successively:

$$a_c^2 = 4{\cdot}99 \quad (\mathscr{R}a)_c = 672{\cdot}15$$

instead of the exact values (11.87), (11.88):

$$a_c^2 = 4{\cdot}935 \quad (\mathscr{R}a)_c = 657{\cdot}5$$

Moreover as the equation (11.94) depends on the boundary conditions only through the numerical values of, $A \ldots D$, the present method may be immediately applied to the case of two rigid surfaces. We thus obtain for the same choice of test functions:

$$a_c^2 = 10{\cdot}415; \quad (\mathscr{R}a)_c = 1821{\cdot}8$$

instead of the exact value (Chandrasekhar, Chapter II):

$$a_c^2 = 9{\cdot}716; \quad (\mathscr{R}a)_c = 1707{\cdot}76$$

Remembering that these results have been obtained with the help of only one arbitrary parameter for the functions u_1, u_2, u_3 and θ, we may consider the agreement as satisfactory.

Henceforth we have at our disposal various techniques, since we can construct a Lagrangian involving either all the unknown functions as in the free minimum method, or only some of them, as done by Chandrasekhar. In this last method the steady state perturbation equations are used for this pre-elimination.

Besides, for a given sequence of test functions all these methods lead practically to the same result.

Of course, the free minimum method implies more lengthy calculations, but in turn remains valid when the eliminations become cumbersome.

11. ONSET OF INSTABILITY IN THE TWO-COMPONENT BÉNARD PROBLEM

As already pointed out in §§ 4–6, the onset of an instability is directly connected with the vanishing of the generalized excess entropy production $P[\delta Z]$ (or of $P_m[\delta Z']$ used in 11.42).

In the case of the Bénard problem for a single fluid, the instability has a simple mechanical interpretation as given in § 3. However, such an interpretation is no longer possible for the two-component Bénard problem which includes the additional effect of thermodiffusion.

Let us indeed consider a binary mixture submitted to temperature gradient. We shall enter into a minimum of details; supplementary informations are given in the original paper by Schechter, Hamm and Prigogine (1970).

Let us first recall that, according to the definition (1.20) of the barycentric velocity, the momentum balance equation (1.30), as well as the excess momentum balance equation (7.51) or (11.7), remain valid for a multicomponent system. Therefore, keeping also the viscosity coefficient constant, we have only to replace the equation of state (11.1) by:

$$\rho = \rho^+[1 - \alpha(T - T^+) + \gamma(N_1 - N_1^+)] \qquad (11.96)$$

Here N_1 denotes the mass fraction of the denser component 1 (say). We shall therefore assume that γ is a positive constant. The cross stands as in (11.1), for an arbitrary reference state. Usually, equation (11.96), does not introduce more than 1 per cent of error.

We now obtain for the density perturbation:

$$\delta\rho = \rho^+(-\alpha\theta + \gamma\Gamma) \quad (\Gamma \equiv \delta N_1) \qquad (11.97)$$

This equation replaces (11.9) Then, introducing the Boussinesq approximation and the incompressibility condition as in § 3, we derive the hydrodynamic stability condition for a binary mixture as:

$$\int_0^h [\nu\langle(u_{i,j})^2\rangle - g\alpha\langle\theta w\rangle + g\gamma\langle\Gamma w\rangle]\, dz \geqslant 0 \qquad (11.98)$$

This relation generalizes (11.26).

The main feature is that the onset of instability for which (11.98) vanishes, depends now on the competition between three different effects, instead of two as it was the case in (11.26). These three effects are respectively: the dissipation of kinetic energy, the energy released by the buoyancy forces due both to the temperature gradient and to the concentration gradient. The two last effects may be separately either stabilizing or destabilizing.

In a binary mixture the diffusion flow is given by (e.g. de Groot and Mazur, 1961; Prigogine, 1967):

$$J_{dif1} = \rho_1\Delta_1 = -\rho D\nabla N_1 - \rho D' N_1 N_2 \nabla T \qquad (11.99)$$

Here D is the isothermal diffusion coefficient, and D', the thermal diffusion coefficient.

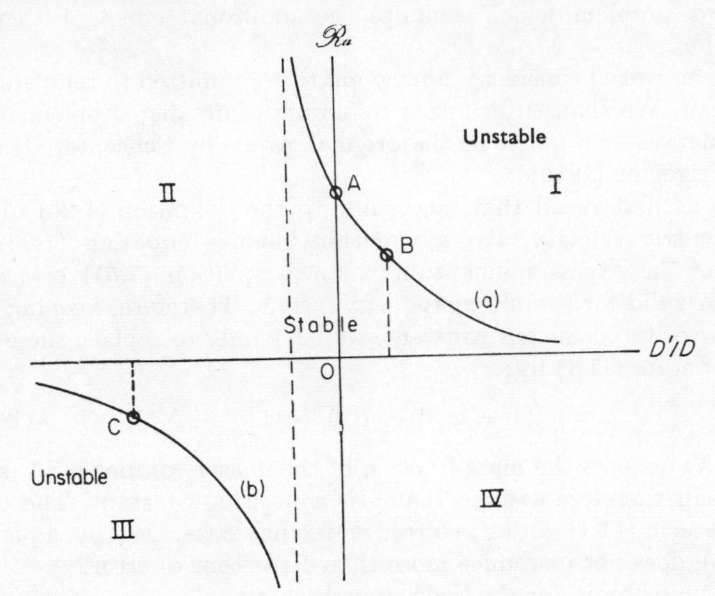

Fig. 11.4. Schematic representation of the stability regions for the two component Bénard problem.

(a), (b): curves of critical stability.

A = critical Rayleigh number for a single fluid ($D'/D = 0$).

B = layer heated from below. Thermodiffusion reinforces the adverse temperature gradient

$$(\mathscr{R}_a)_B < (\mathscr{R}_a)_A.$$

C = layer heated from the top. The denser component migrates to the top.

Relation (11.99) is an example of the phenomenological linear laws we studied in Chapter III.

At the marginal state, we may have both diffusion according to (11.99) and hydrodynamic convection.

The situation is summarized schematically in Figure 11.4, where the Rayleigh number is plotted versus the ratio D'/D. The curves (a) and (b) represent the locus of the critical states and the region of stability is located between these curves (Schechter, Hamm and Prigogine, 1970).

Moreover one has:

$$\mathscr{R}a > 0 \quad (<0)$$

for a fluid heated from the bottom (the top). Also:

$$D'/D > 0 \quad (<0)$$

for a migration of the more dense component towards the cold (the hot) boundary.

Therefore, according to (11.96), for $\mathcal{R}a > 0$ and $D'/D > 0$, the density of the fluid increases from the bottom to the top ($\rho_{,z} > 0$).

On the contrary, for $\mathcal{R}a < 0$ and $D'/D < 0$, the density decreases from the bottom to the top ($\rho_{,z} < 0$) for sufficiently small values of (D'/D) not represented on Figure 11.4.

Instability phenomena may occur in regions I, III but not in II and IV. This is quite natural. For instance in II, where $\mathcal{R}a > 0$, $D'/D < 0$ and sufficiently large in modulus, thermodiffusion overcompensates the thermal dilatation, and the layer will remain stable.

In sector I (positive values of D'/D) thermal diffusion destabilizes the system. Therefore $(\mathcal{R}a)_B < (\mathcal{R}a)_A$. The rate of change is often important.

An unexpected feature appears at point C in sector III: the density is decreasing from the bottom to the top of the layer. Still the system is unstable. As we have already mentioned this situation shows clearly that a simple mechanical interpretation of the instability based on the state of the system is impossible in the present case (cf. § 3). The third term in (11.98) is now destabilizing (therefore negative) whereas both the first and the second terms are stabilizing (therefore positive). Instability occurs when the energy released by the buoyancy forces due to the concentration gradient overcomes the two first effects. Free convection then appears. However the hydrodynamic patterns of motion are quite different from the one component Bénard problem (for more details, see Schechter and Hamm, 1970).

The calculations of Schechter, Hamm and Prigogine (1970), on which Figure 11.4 is based, have been performed on the assumption that the principle of exchange of stability is still valid here (§ 7).

In a very recent paper Legros and Platen (1970) have taken into account complex frequencies ($\omega_i \pm 0$) using the local potential technique (cf. Chapter XII). The qualitative conclusions agree with those of Schechter, Hamm and Prigogine.

The problem of thermodiffusion in binary mixtures, and the study of its stability has recently attracted more attention, partly for its interest in oceanographic problems (Turner, Veronis, 1967–1970; Hurle and Jakeman, 1970).

12. STABILITY OF A VERTICAL COLUMN OF FLUID

In the Bénard problem, dissipative processes play an essential role. It is also interesting to consider the opposite case of a perfect fluid, where by definition all dissipative processes may be neglected. Therefore the entropy production vanishes identically according to (2.21). Nevertheless, the stability problem retains its full significance, because the basic local equilibrium assumption remains valid in the case of a perfect fluid, as pointed out in Chapter II, § 2.

As a simple classical example, we shall consider the stability of a vertical column of fluid. This example may be also considered as an introduction to the more elaborated problem concerning the stability of wave propagation in a perfect fluid as studied in Chapter XIII.

The equation of motion (1.29) takes the simple form, known as the *Euler equation*:

$$\rho(\partial_t \mathrm{v}_i + \mathrm{v}_j \mathrm{v}_{i,j}) = \rho F_i - p_{,i} \qquad (11.100)$$

On the other hand, the entropy balance equation (2.19) reduces to:

$$\partial_t(\rho s) + [\rho s \mathrm{v}_j]_{,j} = 0 \qquad (11.101)$$

or alternatively, using (1.17):

$$\mathrm{d}s = 0 \qquad (11.102)$$

Let us now consider the case of a fluid at rest in a uniform gravitational field. Then, the Euler equation (11.100), written for the vertical direction z, becomes

$$\rho = -\frac{1}{g} p_{,z} \qquad (11.103)$$

We suppose a given steady distribution of temperature $T(z)$. Therefore, using in addition the equation of state, we have:

$$p = p(z); \quad \rho = \rho(z); \quad T = T(z) \qquad (11.104)$$

where $p(z)$ is a decreasing function according to (11.103). We want to investigate the stability of this steady state.

Suppose a small vertical fluctuation w of the velocity at height z. A fluid element will then undergo a displacment δz. As the motion is adiabatic, this fluid element carries along its own entropy. On the other hand, it adjusts its pressure to the hydrostatic pressure of the new level $z + \delta z$ reached. Therefore:†

$$\delta p = 0 \tag{11.105}$$

and

$$\delta \rho = -\frac{1}{v^2}\,\delta v = -\frac{1}{v^2}\left(\frac{\partial v}{\partial s}\right)_p \delta s \tag{11.106}$$

The relation between $\delta \rho$ and the velocity fluctuation w is given by the Euler equation (11.100).

Hence:

$$\partial_t w = -gv\delta\rho = \rho g\left(\frac{\partial v}{\partial s}\right)_p \delta s \tag{11.107}$$

As the entropy remains constant along the motion (11.102) we have:

$$\delta s = -s_{,z}\delta z \tag{11.108}$$

whatever the sign of δz.

Using (2.33), (2.37) as well as (11.103), we get:

$$\delta s = -\left(\frac{c_p}{T}\,T_{,z} + \alpha g\right)\delta z \tag{11.109}$$

where α denotes as in (11.1), the expansion coefficient at constant pressure.

We also observe that

$$\rho g\left(\frac{\partial v}{\partial s}\right)_p = \frac{gT\alpha}{c_p} \tag{11.110}$$

and, for slow motion:

$$\partial_t w = \mathrm{d}_t w = \mathrm{d}_t^2(\delta z) \tag{11.111}$$

Let us introduce these three relations into (11.107). We then obtain the explicit form:

$$\mathrm{d}_t^2(\delta z) = -\alpha g\left(T_{,z} + \frac{\alpha g}{c_p}\,T\right)\delta z \tag{11.112}$$

This equation describes a harmonic oscillation around the local

† This means that the wave propagation generated by the perturbation in the fluid at rest, with the sound velocity $c^2 = (\partial p/\partial \rho)_s$ is here neglected. Wave propagation is studied separately in Chapter XIII.

equilibrium value ($\delta z = 0$), when the following conditions are satisfied:

$$\alpha > 0 \quad \text{and} \quad T_{,z} > -\frac{\alpha g T}{c_p} \tag{11.113}$$

or

$$\alpha < 0 \quad \text{and} \quad T_{,z} < -\frac{\alpha g T}{c_p} \tag{11.114}$$

Otherwise δz will grow in time and the state at rest becomes *unstable*.

These stability conditions have a very simple mechanical meaning. Indeed, using (11.107), (11.109) and (11.110), one sees that they may be written alternatively:

$$\delta\rho\delta z > 0 \tag{11.115}$$

Therefore, if the fluid element is displaced upwards ($\delta z > 0$), the density increases and gravitation will tend to restore the initial situation (and vice versa for $\delta z < 0$). If this condition is not satisfied we expect the appearance of convection.

The existence of critical temperature gradients is of course well known (e.g. Landau and Lifshitz, 1959, § 4). Dissipative processes play no role in the stability conditions (11.113), (11.114). We are still in the stable region in spite of undamped oscillations. The real part ω_r of the frequency ω vanishes here, in the absence of dissipative processes (cf. 6.35).

Let us consider the balance equations for this simple situation.

First, the balance equation for the excess kinetic energy is readily obtained by multiplying the two sides of (11.112) by ρw. Using also (11.111) we thus obtain:

$$\tfrac{1}{2}\rho\, \partial_t w^2 = -\alpha g \rho \left(T_{,z} + \frac{\alpha g}{c_p} T\right) w\delta z \tag{11.116}$$

If (11.113) is satisfied, the kinetic energy associated to the perturbation oscillates in time.

Similarly, taking into account (11.105) the excess entropy balance equation (7.96) becomes for perfect fluid:

$$\tfrac{1}{2}\, \partial_t \delta^2(\rho s) = -w[T^{-1}_{,z}\delta(\rho e) - (\mu T^{-1})_{,z}\delta\rho] \tag{11.117}$$

Using (2.47) we get:

$$(\mu T^{-1})_{,z} = e T^{-1}_{,z} + v(pT^{-1})_{,z} \tag{11.118}$$

Therefore (11.117) takes the form:

$$\tfrac{1}{2}\,\partial_t\delta^2(\rho s) = \rho T^{-1}w[T_{,z}\delta s - p_{,z}\delta v] \qquad (11.119)$$

Let us expand δv in terms of the variables s and p using (2.33), (2.37) as well as (11.105) and (11.110). We obtain in this way, taking into account (11.109):

$$\tfrac{1}{2}\,\partial_t\delta^2(\rho s) = \frac{\rho}{c_p}\,s_{,z}w\delta s = -\frac{\rho c_p}{T^2}\left(T_{,z} + \frac{\alpha g}{c_p}\,T\right)^2 w\delta z \qquad (11.120)$$

Therefore we have always:

$$\tfrac{1}{2}\,\partial_t\delta^2(\rho s) < 0 \quad \text{for} \quad t = 0 \qquad (11.121)$$

as $w\delta z$ is a positive quantity for $t = 0$.

However, if the stability condition (11.113) is fulfilled $w\delta z$ will subsequently change sign. Exactly as w^2, $\delta^2(\rho s)$ oscillates in time.

In the Liapounoff stability theory (Chapter VI, § 2), *asymptotic stability* corresponds to situations where the perturbation vanishes for long times, while *stability* implies only the weaker condition that the perturbed system remains in some neighbourhood of the reference state. This more general type of stability is also often referred to as the Poisson–Poincaré stability.

We may thus conclude that for dissipative systems our stability theory corresponds to asymptotic stability, whereas in the case of a perfect fluid, we only deal with non asymptotic stability in this extended sense.

It is worth while to point out the difference between the balance equation for entropy as expressed here by (11.101), and the balance equation for the excess entropy (11.120): while the source term is absent in (11.101), according to the definition of an ideal fluid, there remains a source term in (11.120). This can be easily understood. As a result of the fluctuations we have a periodic exchange between kinetic and internal energy. The thermodynamic state changes in time at a given point. The excess entropy source terms contain *both* reversible and dissipative contributions. Therefore it will not vanish even for perfect fluids.

CHAPTER XII

Application of the Local Potential to Stability Problems of Laminar Flow

1. INTRODUCTION

In this chapter, we investigate the thermal and mechanical stability of a steady laminar flow with respect to small disturbances. More specifically, we consider the plane Poiseuille flow of an incompressible fluid, between two infinite horizontal plates each maintained at a prescribed constant temperature.

The stability region of the Poiseuille flow is derived from the excess local potential in the neighbourhood of the basic flow, as described in Chapter X, § 10.

The two main dimensionless parameters characterizing this problem are respectively:

The *Rayleigh number* (11.33):

$$\mathcal{R}a = \frac{g\alpha \Delta T h^3}{\kappa \nu} \tag{12.1}$$

where ΔT denotes the temperature difference between the lower and the upper boundary. In the present problem, $\mathcal{R}a$ will be either a positive or a negative quantity, according to the sign of ΔT (Figure 12.1). As in Chapter XI, h denotes the height of the layer.

The *Reynolds number* is defined as:

$$\mathcal{R}e = \frac{U^+ h}{\nu} \tag{12.2}$$

where U^+ is some reference velocity, usually chosen as half the maximum velocity for the Poiseuille flow (Figure 12.1a).

The general expression of the excess local potential we shall derive in § 3 includes two limiting situations. First of all, for $\mathscr{R}e = 0$ it corresponds to the Bénard problem as discussed in Chapter XI. On the other hand, for $\mathscr{R}a = 0$ it corresponds to the usual transition from laminar to turbulent flow, in a fluid whose temperature is assumed to be uniform. As already observed in Chapter VII, § 3 in connection with the Helmholtz's theorem, this is a consistent assumption for sufficiently slow motion. Then dissipation may be neglected as a second order quantity in the energy balance equation (1.42). We shall adopt here this assumption in the whole range of the laminar flow, up to the onset of turbulence. This also implies that in problems where $\mathscr{R}a \neq 0$, the transverse temperature gradient still

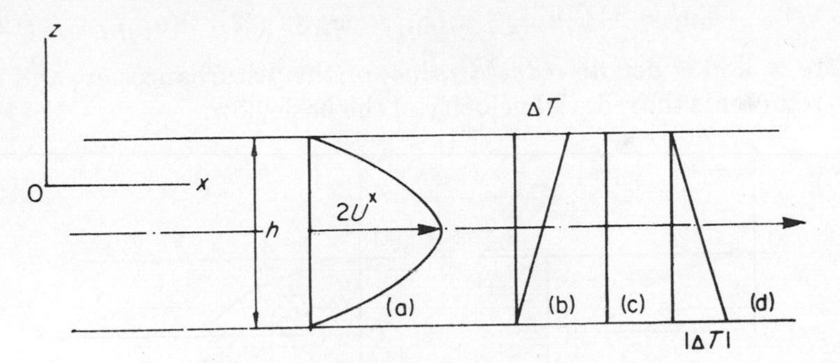

Fig. 12.1. The plane Poiseuille flow, with a transverse temperature gradient.

(a) Parabolic velocity distribution.
(b) Fluid heated from above ($\Delta T > 0$).
(c) Uniform temperature flow.
(d) Fluid heated from below ($\Delta T < 0$).

remains constant as for a fluid at rest (considering the viscosity ν and thermal conductivity λ as constants). The velocity and temperature distributions of the basic flow may then be represented as in Figure 12.1.

In the next section we first study the case of a flow at uniform temperature. More details may be found in the excellent monograph by C. C. Lin (1955). In § 4, we study the influence of a transverse temperature gradient. Numerical results are presented and discussed in §§ 5 to 8.

For the sake of simplicity we limit ourselves to a two dimensional perturbation problem, i.e. along the axis x and z (Figure 12.1). Indeed, as proved by Squire (1933, e.g. Lin, 1955), the link with the more realistic three-dimensional perturbation problem, is straightforward, at least for flow at uniform temperature.

2. THE EIGENVALUE PROBLEM FOR HYDRODYNAMIC STABILITY

Let us start with the linearized excess balance equations for mass and momentum as derived from equations (7,50), (7.51) in the absence of external forces and for an incompressible fluid. We obtain the following equations written in a dimensionless form:

$$0 = u_{i,i} \tag{12.3}$$

$$\partial_t u_i = -\bar{U}_j u_{i,j} - u_j \bar{U}_{i,j} - \varpi_{,i} + (\mathscr{R}e)^{-1}(u_{i,j})_{,j} \tag{12.4}$$

Here u_i and ϖ denote *reduced values* of the disturbances δv_i and δp. $\bar{U}_j$ represents the reduced velocity of the basic flow.

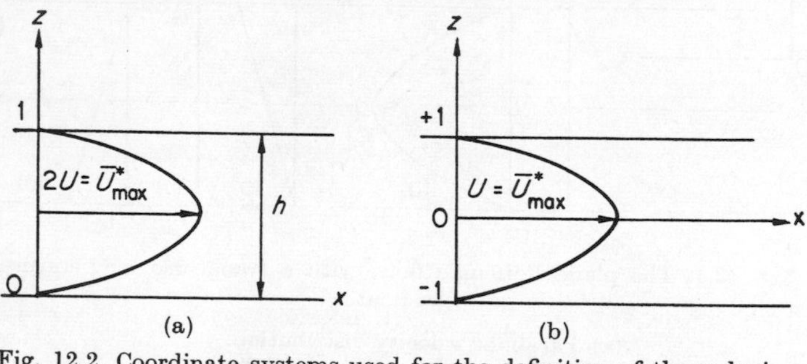

Fig. 12.2. Coordinate systems used for the definition of the reducing factors.

Let us observe that different sets of reducing factors may be used to obtain (12.4). We shall mainly use the two following sets:

$$U^+ = \bar{U}_{max}/2,\ h,\ h(U^+)^{-1},\ \rho(U^+)^2,$$

[Figure 12.2(a)] or

$$U^+ = \bar{U}_{max},\ h/2,\ \tfrac{1}{2}h(U^+)^{-1},\ \rho(U^+)^2,$$

[Figure 12.2(b)], for velocity, length, time and pressure respectively. The first choice corresponds to the coordinates used in Chapter XI for the Bénard problem. The second corresponds to the coordinates used by Lin (1955).

Other sets of reducing factors will be also introduced below (e.g. (12.17), with $\theta = 0$).

Let u and w be respectively the x and z component of an arbitrary disturbance u_i†. As such a disturbance is necessarily periodic in the indefinite x direction, we may write for one single Fourier's component:

$$\begin{aligned} u &= U(z)\, e^{i\alpha(x-ct)} \\ w &= W(z)\, e^{i\alpha(x-ct)} \\ \varpi &= \Pi(z)\, e^{i\alpha(x-ct)} \end{aligned} \tag{12.5}$$

Here α is a real quantity denoting the wave number; $U(z)$, $W(z)$ and $\Pi(z)$ are dimensionless amplitude functions, generally complex.

In the case of the plane Poiseuille flow, one has:

$$\bar{U}_j = \bar{U}_x = \bar{U}(z); \quad \bar{U}_y = \bar{U}_z = 0 \tag{12.6}$$

Then, the balance equations (12.3) and (12.4) take the simpler form:

$$i\alpha U + DW = 0 \tag{12.7}$$

$$[D^2 - \alpha^2 - i\alpha\mathscr{R}e(\bar{U} - c)]U = \mathscr{R}e(D\bar{U})W + i\alpha\mathscr{R}e\Pi \tag{12.8}$$

$$[D^2 - \alpha^2 - i\alpha\mathscr{R}e(\bar{U} - c)]W = \mathscr{R}eD\Pi; \ (D \equiv \mathrm{d}/\mathrm{d}z) \tag{12.9}$$

Also, c is an eigenvalue generally complex. Its imaginary part, when negative, describes the damping of the fluctuation. Between c and the reduced frequency σ defined by (11.77) one has the equality:

$$\sigma = -i\alpha c \tag{12.10}$$

Eliminating U and Π, among (12.7), (12.8) and (12.9), we finally obtain a single equation for W:

$$(D^2 - \alpha^2)^2 W = i\alpha\mathscr{R}e[(\bar{U} - c)(D^2 - \alpha^2)W - (D^2\bar{U})W] \tag{12.11}$$

together with the boundary conditions:

$$W = DW = 0 \quad (\text{for } z = z_1,\ z = z_2) \tag{12.12}$$

Equation (12.11) is the well known *Orr–Sommerfeld equation* (Lin, 1955). Together with (12.12), this equation introduces an eigen-value problem similar to that arising from equations (11.82), (11.83) for the Bénard problem. Let us consider the marginal state for which the imaginary part c_i of c vanishes. Then for a given wave number α, only particular values of $\mathscr{R}e$ give rise to non-vanishing solutions $W(z)$ of (12.11), satisfying (12.12).

In fact, it is more convenient here to approach this problem in a

† Note that the vertical component z is a reduced variable.

slightly different way. For a given real α and a given real $\mathscr{R}e$, we determine the eigenvalue c. This is the method generally adopted (Lin, 1955, Platten, 1970). If c_i is a positive quantity, the flow is unstable in the frame of the linearized theory. If c_i is a negative quantity, the perturbation is damped and the flow is stable.

The relation between α and $\mathscr{R}e$ for $c_i = 0$, in the $\alpha - \mathscr{R}e$ plane, is usually called the curve of marginal or of neutral stability (cf. Figure 12.3). The smallest Reynolds number in respect to all possible values of the wave number α, for which c_i changes sign, is called the *critical Reynolds number*.

3. THE EXCESS LOCAL POTENTIAL FOR HYDRODYNAMIC STABILITY

In Chapter X, § 10 we have used the operator δ to denote a time dependent solution of the perturbation equations, as e.g. $\delta \mathrm{v}_i$, δp. Also we have used the operator δ' to represent increments of these quantities (e.g. Figure 10.4). The distinction having been made once and for all, we shall introduce simpler symbols and write

$$u_i = \delta \mathrm{v}_i; \quad \varpi = \delta p$$

for the reduced perturbations, and

$$\delta u_i = \delta'(\delta \mathrm{v}_i); \quad \delta\varpi = \delta'(\delta p)$$

for their increments.

Accordingly, we now multiply the two sides of equation (12.3) by $-\delta\varpi$, those of equation (12.4) by $-\delta u_i$ and we add. We also take into account the equality:

$$(\partial_t u_i)(\delta u_i) = \tfrac{1}{2}\, \partial_t(\delta u_i)^2 + (\partial_t u_i^0)(\delta u_i) \tag{12.13}$$

The index zero used here refers to a non-varied quantity, i.e. a quantity denoting a solution of the perturbation equations, as pointed out in Chapter X. We then obtain:

$$\begin{aligned} -\tfrac{1}{2}\, \partial_t(\delta u_i)^2 = {} & [\bar{U}_j u_i \delta u_i + u_j \bar{U}_i \delta u_i - (\mathscr{R}e)^{-1} u_{i,j} \delta u_i \\ & + \varpi \delta u_j + u_j \delta\varpi]_{,j} - \bar{U}_j u_i \delta u_{i,j} - \bar{U}_{j,j} u_i \delta u_i \\ & - \bar{U}_i u_j \delta u_{i,j} - \bar{U}_i u_{j,j} \delta u_i - \varpi \delta u_{i,i} \\ & + (2\mathscr{R}e)^{-1} \delta(u_{i,j})^2 - u_i \delta\varpi_{,i} + \partial_t u_i^0 \delta u_i \end{aligned} \tag{12.14}$$

Let us integrate over the volume. The divergence vanishes due to the fixed boundary conditions. Using the method developed in Chapter X, we obtain the excess local potential:

$$\Phi = \int [-\bar{U}_j^0 u_i^0 u_{i,j} - \bar{U}_{j,j}^0 u_i^0 u_i - \bar{U}_i^0 u_j^0 u_{i,j} - \bar{U}_i^0 u_{j,j}^0 u_i - u_i^0 \varpi_{,i} - \varpi^0 u_{i,i} + (2\mathscr{R}e)^{-1}(u_{i,j})^2 + u_i\, \partial_t u_i^0]\, dV \quad (12.15)$$

Indeed, it is easy to verify that the Euler–Lagrange equations of the functional (12.15) restore the excess balance equations (12.3), (12.4), when used together with the time-dependent subsidiary conditions (Chapter X, § 9):

$$u_i^0 = u_i^+; \quad \varpi^0 = \varpi^+ \quad (12.16)$$

Let us also emphasize that in this free minimum variational method, all the functions u_i and ϖ are varied independently. We may also eliminate some of the unknown functions between the balance equations prior to the construction of the local potential. In fact, an excess local potential involving only one single unknown function is often the most appropriate starting point for calculations (e.g. the end of § 10, Chapter XI). We shall give an example in the next section (12.34).

4. THE EXCESS LOCAL POTENTIAL FOR STABILITY OF FLOW WITH A TRANSVERSE TEMPERATURE GRADIENT

We now consider the complete set of the excess balance equations (7.50), (7.51) and (7.52). Again we use dimensionless quantities but other reducing factors than those of Section 2. We write this time for the excess momentum and energy equations:

$$\partial_t u_i = -\mathscr{R}e \bar{U}_j u_{i,j} - \mathscr{R}e u_j \bar{U}_{i,j} - \varpi_{,i} + (u_{i,j})_{,j} + \mathscr{R}a \theta \alpha_i; \; (\alpha_i = 0, 0, 1) \quad (12.17)$$

$$\mathscr{P}r \partial_t \theta = -\mathscr{P}\acute{e} \bar{U}_j \theta_{,j} + w + (\theta_{,j})_{,j} \quad (12.18)$$

The space coordinates are reduced by $d = h/2$, the time by d^2/ν, the disturbances of velocity, pressure and temperature (θ), respectively by κ/d, $d^2/\rho\kappa\nu$ and ΔT (12.1). Also the basic flow is reduced by the characteristic velocity U^+ (Figure 12.2(a)).

Four dimensionless quantities appear in the excess balance equations (12.17–12.18): the Rayleigh number (12.1), the Reynolds number (12.2), the Prandtl number, and the Péclet number, namely:

$$\mathscr{P}r = \frac{\nu}{\kappa}; \quad \mathscr{P}é = \frac{U^{+}h}{\kappa} \tag{12.19}$$

However these four quantities are not independent as:

$$\mathscr{P}é = \mathscr{R}e \times \mathscr{P}r \tag{12.20}$$

For the two dimensional plane Poiseuille flow, the perturbation equations (12.17), (12.18) become:

$$\partial_t w = -\mathscr{R}e\bar{U}w,_x - \varpi,_z + \mathscr{R}a\theta + \nabla^2 w \tag{12.21}$$

$$\partial_t u = -\mathscr{R}e\bar{U}u,_x - \varpi,_x - \mathscr{R}ew\bar{U},_z + \nabla^2 u \tag{12.22}$$

$$\mathscr{P}r\,\partial_t\theta = -\mathscr{P}é\bar{U}\theta,_x + w + \nabla^2\theta \tag{12.23}$$

Again, to establish the corresponding expression of the excess local potential, we multiply the two sides of (12.21), (12.22), (12.23), respectively by infinitesimal increments $-\delta w$, $-\delta u$, $-\delta\theta$ and we add. After an integration by parts over z, we integrate over the x, z plane. The boundary terms vanish on the upper and the lower limits. According to the method followed in Chapter X, §§ 8, 9, we derive in this way a time dependent excess local potential in the form: $\Psi(u, u^0, w, w^0, \theta, \theta^0; \varpi^0, \mathscr{R}e, \mathscr{R}a, \mathscr{P}r)$, together with the basic minimum property (10.77).

The subsidiary conditions are here:

$$u^{+} = u^0; \quad w^{+} = w^0; \quad \theta^{+} = \theta^0 \tag{12.24}$$

We now adopt as in (12.5) for the disturbances the following expressions:

$$\begin{aligned} w &= W(z)\,\mathrm{e}^{i\alpha x}\,\mathrm{e}^{\sigma t} \\ u &= \frac{i}{\alpha}\,DW(z)\,\mathrm{e}^{i\alpha x}\,\mathrm{e}^{\sigma t} \\ \theta &= \Theta(z)\,\mathrm{e}^{i\alpha x}\,\mathrm{e}^{\sigma t} \\ \varpi &= \Pi(z)\,\mathrm{e}^{i\alpha x}\,\mathrm{e}^{\sigma t} \end{aligned} \tag{12.25}$$

σ denotes a reduced eigenvalue.

With the set (12.25), the continuity equation (12.3) is identically fulfilled and we can now write an excess local potential in the form:

$$\Psi(W, W^0, \Theta, \Theta^0, \Pi^0; \mathscr{R}e, \mathscr{R}a, \mathscr{P}r, \alpha, \sigma),$$

containing only two unknown functions: W, Θ, instead of three as above.

Moreover, we easily eliminate Π^0 using equations (12.22) and (12.25). The excess local potential finally becomes:

$$\Psi = \int_{z_1}^{z_2} \Big\{ \Big[\alpha^2 + \sigma + i\alpha\mathscr{R}e\bar{U}\Big]\Big[W^0 W + \frac{1}{\alpha^2}(DW^0)(DW)\Big] + [\alpha^2 + \mathscr{P}r(\sigma + i\alpha\mathscr{R}e\bar{U})]\,\Theta^0\Theta + \tfrac{1}{2}[(DW)^2 + (D\Theta)^2] - \mathscr{R}a\Theta^0 W - W^0\Theta - \frac{1}{2\alpha^2}(D^2W)^2 + \frac{2}{\alpha^2}(D^2W^0)(D^2W) - \frac{i\mathscr{R}e}{\alpha}(D\bar{U})(DW)W^0 \Big\}\,\mathrm{d}z \quad (12.26)$$

Let us also recall that the dimensionless quantities involved in (12.26), as well as the limits z_1 and z_2, depend on the choice of co-ordinates (Figure 12.2).

The time dependent excess local potential (12.26) has two well defined limits:

(i) In the absence of flow ($\mathscr{R}e = 0$), the extremals of (12.26) together with the subsidiary conditions $W^0 = W^+$; $\Theta^0 = \Theta^+$, give us the Euler–Lagrange equation:

$$\left(\frac{\delta\Psi}{\delta\Theta}\right)_{W^0,\Theta^0;\,\mathscr{R}e=0} = 0 \quad (12.27)$$

or

$$(D^2 - \alpha^2 - \sigma\mathscr{P}r)\Theta = -W \quad (12.28)$$

and on the other hand:

$$\left(\frac{\delta\Psi}{\delta W}\right)_{W^0,\Theta^0;\,\mathscr{R}e=0} = 0 \quad (12.29)$$

or

$$(D^2 - \alpha^2)(D^2 - \alpha^2 - \sigma)W = \mathscr{R}a\alpha^2\Theta \quad (12.30)$$

Equations (12.28) and (12.30) restore the energy and momentum equations (11.79, 11.78) for the Bénard problem, written in reduced variables used here.

(ii) In the absence of temperature gradient ($\mathscr{R}a = 0$) and disregarding temperature perturbations ($\Theta = 0$), one obtains, with the subsidiary condition $W^0 = W^+$, the Euler–Lagrange equation:

$$\left(\frac{\delta \Psi}{\delta W}\right)_{W^0, \mathscr{R}a = 0} = 0 \tag{12.31}$$

or

$$(D^2 - \alpha^2)^2 W = i\alpha\mathscr{R}e\left[\left(\bar{U} + \frac{\sigma}{i\alpha\mathscr{R}e}\right)(D^2 - \alpha^2)W - (D^2\bar{U})W\right] \tag{12.32}$$

To recover the notations used in Sections 2 and 3, we have to write equation (12.10) as:

$$\sigma = -i\alpha\mathscr{R}ec \tag{12.33}$$

again in the reduced variables adopted in the present section.

We observe immediately that equation (12.32) is indeed the Orr–Sommerfeld equation (12.11). Likewise, for $\Theta = 0$, and using once more (12.33), the excess local potential Ψ obtained in (12.26), takes after multiplication by α^2, the reduced form:

$$\begin{aligned}\Phi = \int_{-1}^{+1} \Bigg\{ & \left[\alpha^4 + i\alpha^3\mathscr{R}e(\bar{U} - c)\right]\left[W^0 W + \frac{1}{\alpha^2}(DW^0)(DW)\right] \\ & + \frac{\alpha^2}{2}(DW)^2 - \tfrac{1}{2}(D^2W)^2 + 2(D^2W^0)(D^2W) \\ & - i\alpha\mathscr{R}e(D\bar{U})(DW)W^0\Bigg\} d\xi\end{aligned} \tag{12.34}$$

This is a suitable local potential for the study of hydrodynamic stability considered in §§ 2 and 3.

To avoid confusion, we use in (12.34) the symbol ξ to denote the z coordinate in the coordinate system chosen by Lin (1955) (Figure 12.2(b)).

It must be kept in mind, that the general expression (12.26) of the excess local potential, which enables us to investigate the stability problem of flows involving transverse temperature gradient, depends of only two independent unknown functions W and Θ.

In the subsequent sections, we shall consider three examples. The first corresponds to $\mathscr{R}a = 0$ (critical Reynolds number). The second to $\mathscr{R}e = 0$ (critical Rayleigh number) and the third to the mixed problem.

5. THE CRITICAL REYNOLDS NUMBER FOR THE PLANE POISEUILLE FLOW

Following the procedure developed in Chapter X, we use the variational technique of successive approximations. First of all we introduce into (12.34), the sequence of order n of the trial functions, that is:

$$W_n = a_1\varphi_1 + \ldots + a_n\varphi_n; \quad W_n^0 = a_1^0\varphi_1 + \ldots + a_n^0\varphi_n \qquad (12.35)$$

each φ_k satisfying the boundary conditions. We then minimize in respect to a_k, and we introduce in the result the subsidiary conditions $a_k = a_k^0$ (compare e.g. with equations 10.25–10.27). The n linear homogeneous equations in a_k obtained in this way have to admit a non trivial solution for the a_k, when the system reaches a state of neutral stability. This implies that the determinant of the coefficients vanishes (Lee and Reynolds, 1964, 1967):

$$\text{Det}\,|A_{kl} - cB_{kl}| = 0 \quad (k, l = 1, 2, \ldots n) \qquad (12.36)$$

One has:

$$\begin{aligned} A_{kl} &= I_{kl}^{(2)} + 2\alpha^2 I_{kl}^{(1)} + \alpha^4 I_{kl}^{(0)} \\ &\quad + i\alpha\mathscr{Re}(J_{kl}^{(2)} + J_{kl}^{(1)} + \alpha^2 J_{kl}^{(0)}) \\ B_{kl} &= i\alpha\mathscr{Re}(I_{kl}^{(1)} + \alpha^2 I_{kl}^{(0)}) \end{aligned} \qquad (12.37)$$

with

$$\begin{aligned} I_{kl}^{(0)} &= \int_{-1}^{+1} \varphi_k\varphi_l\,\mathrm{d}\xi & J_{kl}^{(0)} &= \int_{-1}^{+1} \bar{U}\varphi_k\varphi_l\,\mathrm{d}\xi \\ I_{kl}^{(1)} &= \int_{-1}^{+1} (D\varphi_k)(D\varphi_l)\,\mathrm{d}\xi & J_{kl}^{(1)} &= \int_{-1}^{+1} (D^2\bar{U})\varphi_k\varphi_l\,\mathrm{d}\xi \\ I_{kl}^{(2)} &= \int_{-1}^{+1} (D^2\varphi_k)(D^2\varphi_l)\,\mathrm{d}\xi & J_{kl}^{(2)} &= -\int_{-1}^{+1} \bar{U}\varphi_k(D^2\varphi_l)\,\mathrm{d}\xi \end{aligned}$$

$$D^n = \mathrm{d}^n/\mathrm{d}\xi^n \qquad (12.38)$$

In the system of reference used by Lin (Figure 12.2(b)), the basic flow is

$$\bar{U} = 1 - \xi^2 \qquad (12.39)$$

We choose as trial functions the set introduced by Lee and Reynolds (1964, 1967):

$$\varphi_k = (1 - \xi^2)^2\xi^{2(k-1)} \qquad (12.40)$$

They satisfy identically the boundary conditions. Moreover the integrals (12.38) are easy to evaluate. The eigenvalue problem for c, corresponding to equations (12.36), (12.37), has been solved by Platten (1970) on a computer, for 450 values of α and $\mathscr{R}e$. This gives enough information to determine with a fair accuracy, the stable ($c_i < 0$) and the unstable region ($c_i > 0$) in the α, $\mathscr{R}e$ plane.

For each point α, $\mathscr{R}e$ the convergence of the value obtained for the fundamental mode was investigated successively up to a 20×20 matrix. For small values of the product $\alpha\mathscr{R}e$ a suitable convergence is obtained, but for increasing values of $\alpha\mathscr{R}e$ the convergence becomes less and less satisfactory.

Fortunately, near the critical point, that is for small c_i (Figure 12.3), the numerical approximation on c_i becomes of the same order

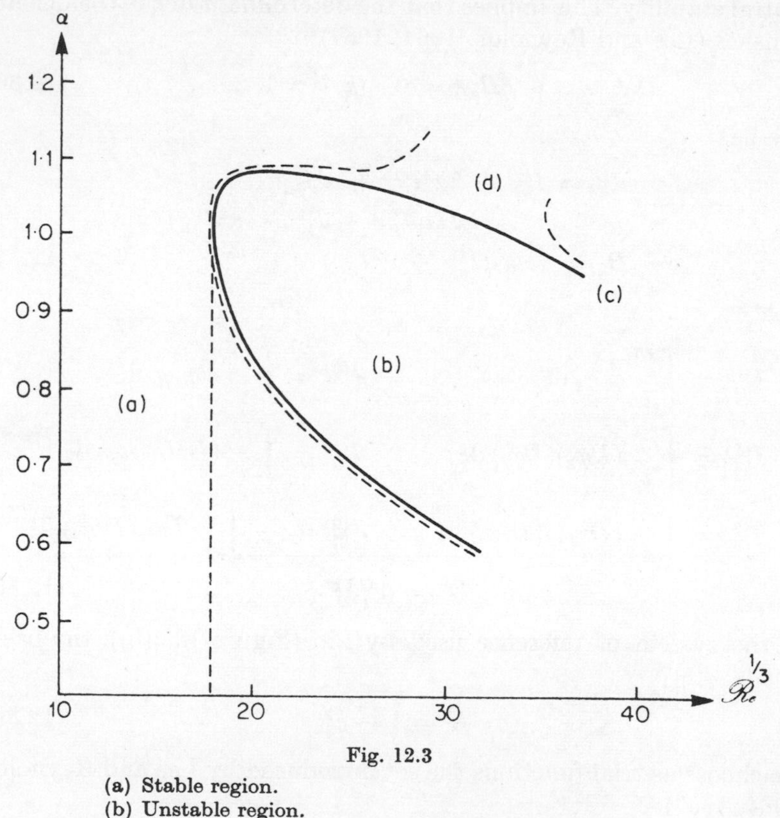

Fig. 12.3

(a) Stable region.
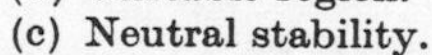
(b) Unstable region.
(c) Neutral stability.
(d) Indetermination due to the numerical approximation.

of magnitude as c_i itself. However for very large values of $\alpha\mathscr{R}e$ (e.g. $\alpha\mathscr{R}e > 50{,}000$), the method fails due to a lack of accuracy in this region.

As represented on Figure 12.3, the curve of neutral stability calculated in this way, is surrounded by a region corresponding to the limit of accuracy obtained by the local potential technique.

According to Platten (1970) the lack of accuracy observed for large values of the product $\alpha\mathscr{R}e$, could be due to the *non self-adjoint* character of the Orr-Sommerfeld equation (12.10) (and not to the numerical method). In fact, Platten observes that the *non self-adjoint* contribution of the Orr-Sommerfeld equation, is represented by the quantity $i\alpha\mathscr{R}e\bar{U}D^2W$, whose importance increases with $\alpha\mathscr{R}e$. In this respect, it should be emphasized that the convergence studied in Chapter X, § 5 for a non self-adjoint problem, was only solved for a situation not too far from an equilibrium state.

Besides, the same type of computations applied to particular cases for which the Orr–Sommerfeld equation reduces to a self-adjoint equation (e.g. unidimensional flow, $\bar{U}$ = constant) leads again to *uniform* convergence. Indeed, for $\bar{U}$ = constant, the eigenvalue spectrum may be calculated exactly and then compared with the results arising from our approximate method. In each case, Platten observes an excellent agreement.

To illustrate the method, we give below some typical results corresponding to $\alpha = 1$.

For $\mathscr{R}e = 100$ (Table 12.1), there is an excellent convergence to the final value $c_i = -0{\cdot}1629$. As c_i is found to be negative, the flow is stable for this wave number. The discrepancies of the last digits are due to the round-of errors in the computer calculations.

For $\mathscr{R}e = 2500$ (Table 12.2), the flow remains stable for $\alpha = 1$ ($c_i < 0$), but the convergence is already weaker than in the foregoing case.

For $\mathscr{R}e = 5900$ (Table 12.3), c_i changes sign continuously. We are near the critical value. Indeed, for higher values of $\mathscr{R}e$, c_i becomes positive and instability occurs (Table 12.4).

With the 450 points calculated in the α, $\mathscr{R}e$ plane, Platten derives the critical value of the Reynolds number with an accuracy of ± 150, and obtains finally:

$$5600 < (\mathscr{R}e)_c < 5900 \tag{12.41}$$

This result is in good agreement with the value $(\mathscr{R}e)_c = 5780$, given by Thomas and Lin (1952, 1953).

Table 12.1

$\alpha = 1, \mathscr{Re} = 100$			
n (sequence)	c_i	n (sequence)	c_i
2	−0·172161	12	−0·162942
3	−0·164498	13	−0·162915
4	−0·162195	14	−0·162957
5	−0·162967	15	−0·162944
6	−0·162946	16	−0·162950
7	−0·162944	17	−0·162962
8	−0·162944	18	−0·162947
9	−0·162944	19	−0·162944
10	−0·162944	20	−0·162942
11	−0·162945		

Table 12.2

$\alpha = 1, \mathscr{Re} = 2500$			
n (sequence)	c_i	n (sequence)	c_i
2	−0·011960	12	−0·012464
3	−0·040551	13	−0·017830
4	+0·008418	14	−0·014178
5	−0·000085	15	−0·020270
6	−0·017376	16	−0·005494
7	−0·011408	17	−0·015388
8	−0·015762	18	−0·013269
9	−0·013779	19	−0·014132
10	−0·014613	20	−0·014132
11	−0·012342		

Table 12.3

$\alpha = 1, \mathscr{Re} = 5900$			
n (sequence)	c_i	n (sequence)	c_i
2	−0·005068	12	+0·000475
3	−0·017068	13	+0·000073
4	+0·020529	14	−0·001761
5	+0·024528	15	−0·002231
6	+0·001250	16	+0·001458
7	−0·000029	17	+0·004265
8	+0·002102	18	+0·000719
9	−0·001550	19	−0·000175
10	+0·000774	20	+0·000229
11	+0·004750		

Table 12.4

$\alpha = 1, \mathscr{R}e = 8000$			
n (sequence)	c_i	n (sequence)	c_i
2	−0·003737	12	−0·001712
3	−0·012580	13	−0·002749
4	+0·022941	14	+0·004114
5	+0·029698	15	+0·001853
6	+0·020298	16	+0·002457
7	+0·006242	17	+0·003975
8	+0·004185	18	+0·002177
9	+0·002306	19	+0·003267
10	+0·002263	20	+0·002704
11	+0·002065		

6. THE CRITICAL RAYLEIGH NUMBER FOR THE BÉNARD PROBLEM

Now we consider the situation where $\mathscr{R}e$ and $\bar{U}$ vanish in (12.26) and $\mathscr{R}a > 0$.

As in Chapter XI, we use here the reference system adopted on Figure 12.2(a). We choose also as boundary conditions, those corresponding to rigid surfaces (cf. (11.11), (11.12)):

$$\left.\begin{matrix} DW = W = 0 \\ \Theta = 0 \end{matrix}\right\} \text{ for } \begin{cases} z = 0 \\ z = 1 \end{cases} \tag{12.42}$$

We introduce the simple trial functions for W and Θ respectively:

$$\varphi_k = (1 - z)^2 z^2 (2z - 1)^{2(k-1)} \tag{12.43}$$

$$\psi_k = (1 - z) z (2z - 1)^{(k-1)} \tag{12.44}$$

They satisfy identically (12.42). As in (12.35) the sequence of order n is

$$W_n = a_1\varphi_1 + \ldots + a_n\varphi_n \tag{12.45}$$

$$\Theta_n = b_1\psi_1 + \ldots + b_n\psi_n \tag{12.46}$$

and likewise for W_n^0 and Θ_n^0.

We proceed then as in Section 5. One obtains this time a satisfactory uniform convergence for the eigenvalues σ. An example is given in Table 12.5. This result is not surprising since the equations (12.28), (12.30) of the Bénard problem are self-adjoint. For the same reason, the local potential used here reduces to a true potential as already observed in Chapter XI, § 10, and the variational technique becomes identical to the well-known Rayleigh–Ritz method.

With a 'one-term' approximation, Schechter and Himmelblau (1965) found:

$$(\mathcal{R}a)_c = 1750$$

while the free minimum method used in Chapter XI yields for the same problem and for the same approximation:

$$(\mathcal{R}a)_c = 1822$$

With a 'three-term' approximation, we obtain from Table 12.5, the value:

$$(\mathcal{R}a)_c = 1710$$

which is very near to the exact value 1708, calculated by Chandrasekhar (Chapter XI, § 10).

Table 12.5

	σ minimum for $\alpha = 3{\cdot}117$		
	$\sigma < 0$ stable		$\sigma > 0$ unstable
n (sequence)	$\mathcal{R}a = 1000$	$\mathcal{R}a = 1700$	$\mathcal{R}a = 1710$
1	−1·15692088	−0·075660939	−0·060019552
2	−1·15605572	−0·073235449	−0·058178305
3	−1·11377581	−0·013082648	+0·002232301
4	−1·11377581	−0·013082643	+0·002232306
5	−1·11377517	−0·011895156	+0·003428226
6	−1·11377517	−0·011895156	+0·003428226
7	−1·11377515	−0·011894804	+0·003428587
8	−1·11377515	−0·011894804	+0·003428587
9	−1·11377515	−0·011894804	+0·003428587
10	−1·11377514	−0·011894792	+0·003428599

7. THE BENARD PROBLEM FOR LAMINAR FLOW

To approach this problem, we need the complete form (12.26) of the excess local potential, with a positive Rayleigh number (heating from below).

We want to investigate the influence of a slow Poiseuille flow on the formation of the Bénard cells. We may then expect that a 'one-term' approximation will suffice to derive the influence of the Reynolds number on the critical value $(\mathcal{R}a)_c$.

However, it must be kept in mind that the local potential (12.26) has been written only for two-dimensional disturbances, and that the link with the three dimensional problem is no longer given by Squire's theorem quoted in § 1, when thermal gradients occur.

Using the system of reference $0 \leqslant z \leqslant 1$ (Figure 12.2(a)), the basic flow $\bar{U}$ in the x–direction may be written as:

$$\bar{U} = U^{++}(z - z^2) \tag{12.47}$$

Here, U^{++} corresponds to four times the maximum velocity at $z = 1/2$ (that is 8 times the reference velocity U^+ defined in (12.2)). Therefore, the Reynolds number occurring in equation (12.26) is equal to 8 times the usual Reynolds number considered in § 5. To avoid confusion we introduce the symbol:

$$(\mathscr{Re})' = 8\mathscr{Re} \tag{12.48}$$

together with (cf. 12.20):

$$(\mathscr{P}\acute{e})' = 8\mathscr{P}\acute{e} \tag{12.49}$$

Applying now the local potential method to this problem, one obtains easily the dispersion equation (for details see Platten, 1970):

$$\left(\frac{\sigma}{630} + \frac{2\sigma}{105\alpha^2} + \frac{\alpha^2}{630} + \frac{4}{105} + \frac{4}{5\alpha^2} + \frac{i\alpha(\mathscr{Re})'}{2772} + \frac{i(\mathscr{Re})'}{630\alpha}\right) \times \left(\frac{\sigma\mathscr{Pr}}{30} + \frac{1}{3} + \frac{\alpha^2}{30} + \frac{i\alpha\mathscr{Pr}(\mathscr{Re})'}{140}\right) = \mathscr{Ra}\,\frac{1}{(140)^2} \tag{12.50}$$

As the principle of exchange of stabilities was only established for the simple Bénard problem (Chapter XI, § 7), we can no longer consider here σ as a real quantity. We then separate (12.50) into a real and an imaginary part. Moreover, at the marginal state of neutral stability, σ_r vanishes in both relations. Therefore, these relations permit the elimination of σ_i and give rise to the equation of the neutral curve. This yields (Platten, 1970):

$$\left(\frac{\alpha^2}{630} + \frac{4}{105} + \frac{4}{5\alpha^2}\right)\left(\frac{1}{3} + \frac{\alpha^2}{30}\right) - \left[\frac{\alpha(\mathscr{Re})'}{2772} + \frac{(\mathscr{Re})'}{630\alpha} - \left(\frac{1}{630} + \frac{2}{105\alpha^2}\right)C\right] \times \left[\frac{\alpha(\mathscr{Re})'}{140} - \frac{1}{30}C\right]\mathscr{Pr} = \mathscr{Ra}\,\frac{1}{(140)^2} \tag{12.51}$$

where

$$C = \frac{\left(\frac{\alpha^2}{630} + \frac{4}{105} + \frac{4}{5\alpha^2}\right)\frac{\alpha \mathscr{P}r}{140} + \left(\frac{\alpha}{2772} + \frac{1}{630\alpha}\right)\left(\frac{1}{3} + \frac{\alpha^2}{30}\right)}{\left(\frac{\alpha^2}{630} + \frac{4}{105} + \frac{4}{5\alpha^2}\right)\frac{\mathscr{P}r}{30} + \left(\frac{1}{630} + \frac{2}{105\alpha^2}\right)\left(\frac{1}{3} + \frac{\alpha^2}{30}\right)} (\mathscr{R}e)' \tag{12.52}$$

The minimum $(\mathscr{R}a)_c$ of this neutral stability curve can be computed for various values of the Reynolds and Prandtl numbers. One obtains then $(\mathscr{R}a)_c$, as well as the associate critical values of the wave number α, versus $\mathscr{R}e$ and $\mathscr{P}r$. This dependence is given on Tables 6 and 7, as well as on Figure 12.4.

One may conclude that the critical Rayleigh number increases for increasing values of the Reynolds number. Therefore, the region of stability of the solution corresponding to the state at rest is extended when a slow horizontal motion is prescribed to the fluid. In other words, the Poiseuille flow tends to destroy the Bénard cells and to stabilize the temperature distribution which prevails at rest. Also the critical wavelength and therefore the size of the Bénard cells decrease.

As shown on Figure 12.4 this effect is important because it appears in the range of realistic values of the Prandtl number (6·75 for liquid water). A preliminary series of experiments performed by Legros, Platten and others (1970), show an order of magnitude consistent with the theoretical results.

Table 12.6

	$\mathscr{P}r = 7$	
$(\mathscr{R}e)' = 8\mathscr{R}e$	α_c	$(\mathscr{R}a)_c$
0	3·12	1750
100	3·27	2190
200	4·04	3401
300	5·20	4895
400	6·05	6320
500	6·68	7620
600	7·19	8792
700	7·60	9843
800	7·95	10784
900	8·25	11626
1000	8·51	12378
2000	9·90	16725
3000	10·41	18368
4000	10·64	19109
5000	10·76	19494

Table 12.7

$\mathscr{P}r = 0{\cdot}7$		
$(\mathscr{R}e)' = 8\mathscr{R}e$	α_c	$(\mathscr{R}a)_c$
0	3·12	1750
100	3·24	1917
200	3·61	2374
300	4·15	3004
400	4·69	3698

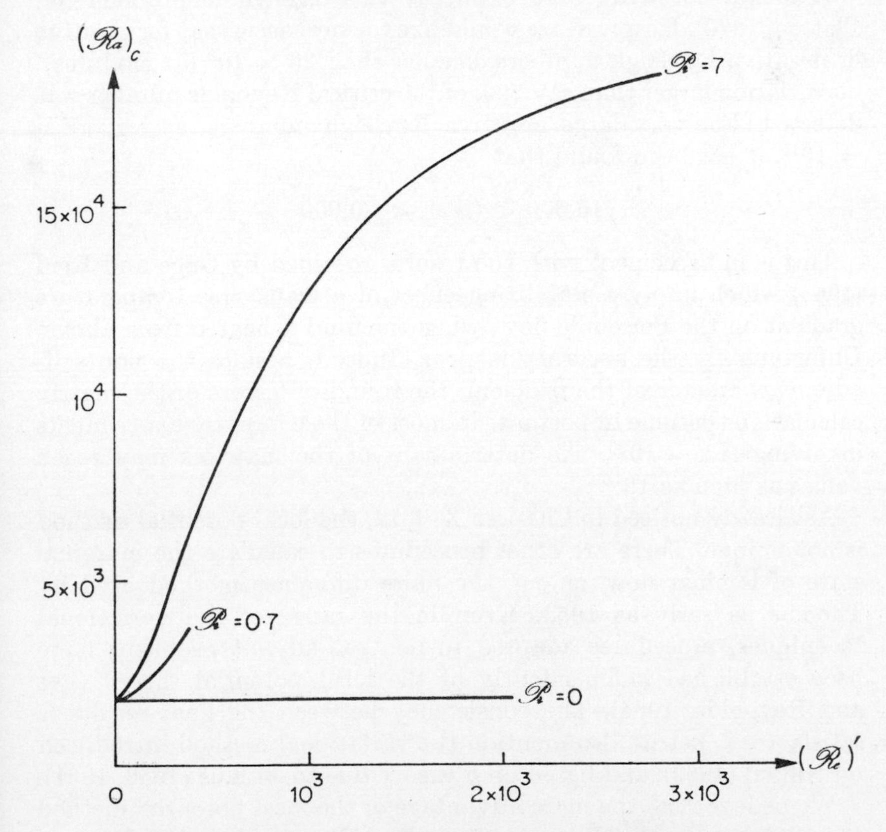

Fig. 12.4. The critical Rayleigh number versus the Reynolds number for various values of the Prandtl number.

8. INFLUENCE OF A TRANSVERSE TEMPERATURE GRADIENT ON TURBULENCE

This problem can be investigated by the same method and with the help of the same complete form (12.26) of the excess local potential. The main difference is that we now consider a negative value of the Rayleigh number (heating from the top).

In contrast with the case studied in § 7, we have now to retain many terms in the sequences (12.45), (12.46).

The successive steps are now quite obvious and lead to a $2n \times 2n$ eigenvalue problem. As a particular case, in absence of temperature disturbances, it reduces to the system (12.36–12.38).

We shall not write here explicitly this eigenvalue problem (cf. Platten, 1970). However, we would like to emphasize that for positive or negative Rayleigh numbers, smaller than 20×10^3 (in modulus), no variation larger than say 200, of the critical Reynolds number was detected. For very large negative Rayleigh numbers, as e.g. $|\mathscr{R}a| = 10^{11}$, it has been found that

$$16{,}000 < (\mathscr{R}e)_c < 20{,}000$$

This is in agreement with the results obtained by Gage and Reid (1968) which imply a stabilizing effect of a transverse temperature gradient on the Poiseuille flow, when the fluid is heated from above. Unfortunately the accuracy is poor. Indeed, besides the non self-adjoint character of the problem, the round-off errors of the matrix calculations become important. In most of the computer experiments involving $\mathscr{R}a \sim 10^{10}$, the determinant of the matrices may reach values as high as 10^{100}.

As already noticed in Chapter X, § 12, the local potential method is not unique. There are other procedures to calculate the marginal state of laminar flow, as e.g. the finite difference method used by Thomas as early as 1952. Even in the range of the variational techniques, procedures adapted to non self-adjoint problems have been established independently of the local potential theory (Lee and Reynolds, 1964). The consistency between the final results is satisfactory. Let us also mention the variational method introduced by Nihoul (1969), and based on a theory due to Malkus (1956, 1961).

We believe that the main advantage of the local potential method consists in its commodity and generality (Chapter X, § 12). No supplementary *ad hoc* adjoint problem has to be introduced.

Let us also emphasize the importance of proper test functions φ_k, as introduced into (12.35). Too restrictive test functions have to be rejected. As an example, the set

$$(1 - y)y^{(k-1)} \quad \text{vanishing for} \quad 0 \text{ or } 1$$

has a *vanishing* first derivative at $y = 0$, and therefore is too restrictive in a situation where such additional condition is not required by the boundary conditions.

On the contrary the set

$$(1 - y)y^{(2k-1)} \quad \text{vanishing for} \quad 0 \text{ or } 1$$

avoids this restriction (see Lee and Reynolds, 1964). A too restrictive set of trial functions will generally lead to bad convergence. Let us finally observe that numerous applications of the local potential method to different stability problems in the field of hydrodynamics and magneto hydrodynamics may be found in the literature. Let us specially quote the interesting contributions due to Butler and Hayes 1967–1970, Nigam 1967, Himmelblau (1965), Roberts (1965), Schechter (1965, 1967) and others.

It may finally be useful to recall here that different analytical expressions of the local potential are generally available with the help of multiplicators, among which one has to choose (Chapter X, § 8).

CHAPTER XIII

Stability of Finite Amplitude Waves

1. INTRODUCTION

In the last example of Chapter XI, we have considered the stability problem of a column of ideal fluid, with respect to perturbations carried along with the matter ($\delta s \neq 0$; $\delta p = 0$; no travelling perturbations).

We now consider the opposite situation corresponding to an isentropic flow of an ideal fluid, subject to small isentropic perturbations ($\delta s = 0$; $\delta p \neq 0$). We then have only travelling perturbations such as e.g. sound waves (cf. also Chapter XI, § 12).

As a typical example, we investigate the stability of a one dimensional isentropic 'simple wave' both for compression and rarefaction (Glansdorff and Banaï, 1970). The main interest of this problem is that we are now concerned with the stability problem of a time dependent process in a compressible fluid, while the examples treated till now were all devoted to steady state stability problems.

Therefore, the kinetic stability method based on the normal mode analysis, is no longer valid, while our stability criterion remains available.

Before approaching this subject it may be useful to recall briefly some basic properties of travelling waves. This includes sound waves corresponding to infinitesimal perturbations and finite amplitude waves. More detailed information is available in the excellent monographs by Landau and Lifshitz (1959), or by Zeldovich and Raizer (1966).

2. SOUND WAVES

For an ideal fluid, in the absence of external forces and density gradient, the mass and momentum equations for small perturbations become (7.50, 7.51):

$$\partial_t \delta\rho + \rho u_{i,i} = 0 \tag{13.1}$$

$$\rho\, \partial_t u_i + \varpi_{,i} = 0 \quad (i = 1, 2, 3) \tag{13.2}$$

For isentropic perturbations, we have:

$$\varpi = \left(\frac{\partial p}{\partial \rho}\right)_s \delta\rho \tag{13.3}$$

We obtain therefore five equations relating the five functions $\delta\rho$, u_i, ϖ. It is convenient to introduce a velocity potential Φ defined by (Landau and Lifshitz, 1959, § 63):

$$u_i = \Phi_{,i} \tag{13.4}$$

Equation (13.2) then leads to:

$$\varpi = -\rho\, \partial_t \Phi \tag{13.5}$$

and equation (13.1) shows that the velocity potential satisfies the wave equation:

$$\partial_t^2 \Phi - c^2(\Phi_{,i})_{,i} = 0; \quad c^2 = \left(\frac{\partial p}{\partial \rho}\right)_s \tag{13.6}$$

The solution of this equation corresponding to plane waves is:

$$\Phi = f_1(x + ct) + f_2(x - ct) \tag{13.7}$$

where c denotes the velocity of the wave. One obtains the same wave equation for the disturbances $\delta\rho$, u_i and ϖ respectively, by taking the derivative of equation (13.6), with respect to either x_i or t.

The unperturbed reference state is stable in respect to sound waves, in the extended sense introduced in Chapter XI, § 12, for an ideal fluid. If in addition, dissipative phenomena are taken into account we have *asymptotic* stability.

However the situation changes completely when the basic reference state is no longer homogeneous. The equations become then non-linear. Even when the form of the wave equations (13.6) remains valid, the velocity c changes from point to point.

The investigation of stability under such conditions is considerably more difficult. For this reason it appears of particular interest to show that our stability criterion enables us to solve this problem at least in the one dimensional case.

3. COMPRESSION AND RAREFACTION WAVES RIEMANN INVARIANTS

We consider a one-dimensional gas flow. Initially the gas is at rest in a pipe. At its left the pipe is provided with a mobile piston; to the right it extends over a sufficiently large distance to be considered as infinite (Figure 13.1(1)).

We may now move the piston either to the left (Figure 13.1(2)), or to the right (Figure 13.1(3)). In the first case the velocity of the piston

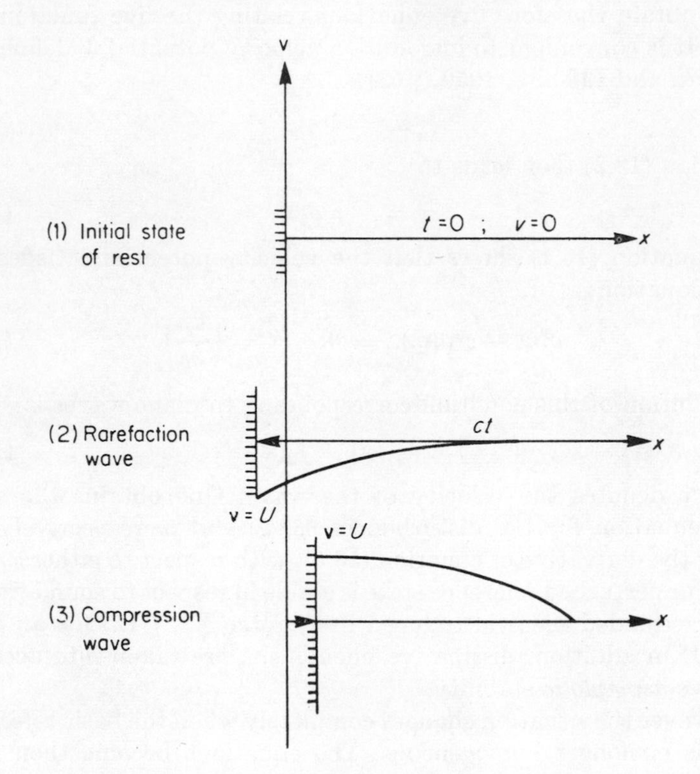

Fig. 13.1. Production of rarefaction and compression waves.

$U(x, t)$ is negative and the gas expands (the gas velocity v is also negative). The disturbance produced by the motion of the piston is propagated to the right with the velocity of sound c. We then have a *rarefaction wave.* If the motion of the piston is directed to the right we obtain a *compression wave.*

For completeness let us present a brief summary of the hydrodynamic theory (for more detail see e.g. Landau and Lifshitz, 1959, § 92–97; Zeldovich and Raizer, in Hayes and Probstein, 1966).

The continuity equation (1.12) gives us for isentropic one dimensional motion, the equality:

$$\frac{1}{\rho c}\frac{\partial p}{\partial t} + \frac{\mathrm{v}}{\rho c}\frac{\partial p}{\partial x} + c\frac{\partial \mathrm{v}}{\partial x} = 0 \quad (\mathrm{v} = \mathrm{v}_x) \tag{13.8}$$

On the other hand, Euler's equation (11.100) becomes here:

$$\frac{\partial \mathrm{v}}{\partial t} + \mathrm{v}\frac{\partial \mathrm{v}}{\partial x} + \frac{1}{\rho}\frac{\partial p}{\partial x} = 0 \tag{13.9}$$

Combining these two equations one gets immediately:

$$\left[\frac{\partial \mathrm{v}}{\partial t} \pm \frac{1}{\rho c}\frac{\partial p}{\partial t}\right] + (\mathrm{v} \pm c)\left[\frac{\partial \mathrm{v}}{\partial x} \pm \frac{1}{\rho c}\frac{\partial p}{\partial x}\right] = 0 \tag{13.10}$$

By definition, *characteristics* are lines in the x, t plane, whose slopes represent the velocity of propagation of disturbances, relative to a fixed system of coordinates.

As the gas is moving with velocity v, and disturbances are propagated with the speed $\pm c$ in respect to the gas, we obtain two families of characteristics C_+ and C_-, defined by the equations:

$$\frac{\mathrm{d}x}{\mathrm{d}t} = \mathrm{v} + c \quad (C_+); \quad \frac{\mathrm{d}x}{\mathrm{d}t} = \mathrm{v} - c \quad (C_-) \tag{13.11}$$

By comparison with (13.10) we see that:

$$\mathrm{dv} + \frac{1}{\rho c}\,\mathrm{d}p = 0 \quad \text{on} \quad C_+ \tag{13.12}$$

$$\mathrm{dv} - \frac{1}{\rho c}\,\mathrm{d}p = 0 \quad \text{on} \quad C_- \tag{13.13}$$

where the operator d denotes derivation along the corresponding characteristic.

Moreover, as the flow is isentropic (§ 1), each of the variables ρ, c may be considered as a function of the single variable p. This permits us to interpret (13.12) and (13.13) as exact differentials. The functions:

$$J_+ = \mathrm{v} + \int \frac{\mathrm{d}p}{\rho c}; \quad J_- = \mathrm{v} - \int \frac{\mathrm{d}p}{\rho c} \tag{13.14}$$

are called the Riemann invariants (e.g. Landau and Lifshitz, 1959). Indeed one has:

$$\left.\begin{aligned} \mathrm{d}J_+ = 0 \quad \text{i.e.} \quad J_+ = \text{constant on } C_+ \\ \mathrm{d}J_- = 0 \quad \text{i.e.} \quad J_- = \text{constant on } C_- \end{aligned}\right\} \tag{13.15}$$

The existence of these invariants is essential to the theory.

In case of a perfect gas for which $\gamma = c_p/c_v$ is considered as a constant, one has the adiabatic equation:

$$p \sim \rho^{\gamma} \tag{13.16}$$

and according to (13.6):

$$c^2 \sim \gamma\rho^{\gamma-1} \tag{13.17}$$

Therefore:

$$J_{\pm} = \mathrm{v} \pm \frac{2}{\gamma - 1} c \tag{13.18}$$

We may now use the quantities J_+ and J_-, to eliminate two variables such as for example v and c. Therefore the equations of the characteristics (13.11) now become

$$\begin{aligned} \frac{\mathrm{d}x}{\mathrm{d}t} &= F_+(J_+, J_-) \text{ on } (C_+) \\ \frac{\mathrm{d}x}{\mathrm{d}t} &= F_-(J_+, J_-) \text{ on } (C_-) \end{aligned} \tag{13.19}$$

The functions F_+, F_-, depend on the boundary as well as the initial conditions. As an example we may consider the rarefaction wave of Figure 13.1(2). We suppose that the piston is accelerated from the state at rest at $t = 0$ till $t = t_A$. For $t > t_A$ the velocity is then maintained constant.

The distribution of characteristics is represented on Figure 13.2. At the initial time, the characteristics C_- originate in the region where the gas is still in equilibrium (at the right of point 0). As $J_-(x, 0)$, has the same value for all $x > 0$ and moreover, as this value is then conserved along the characteristics C_-, we conclude that J_- reduces to a constant at all points of the x, t diagram. But then *both* invariants along C_+ are constant and (13.19) gives us for the characteristics C_+ the simple equation:

$$x = F_+(J_+, J_-)t + \varphi(J_+) \text{ on } (C_+) \tag{13.20}$$

The characteristics C_+ are therefore straight lines.

Let us discuss some consequences of this important result. Comparing with (13.11) we see that:

$$x = [\mathrm{v} + c(\mathrm{v})]t + \varphi(\mathrm{v}) \quad (C_+) \tag{13.21}$$

the dependence $c(\mathrm{v})$ arising from the invariance of J_+ along C_+ (see 13.18).

Alternatively

$$\mathrm{v} = f\{x - [\mathrm{v} + c(\mathrm{v})]t\} \tag{13.22}$$

or

$$c = g\{x - [\mathrm{v} + c(\mathrm{v})]t\} \tag{13.23}$$

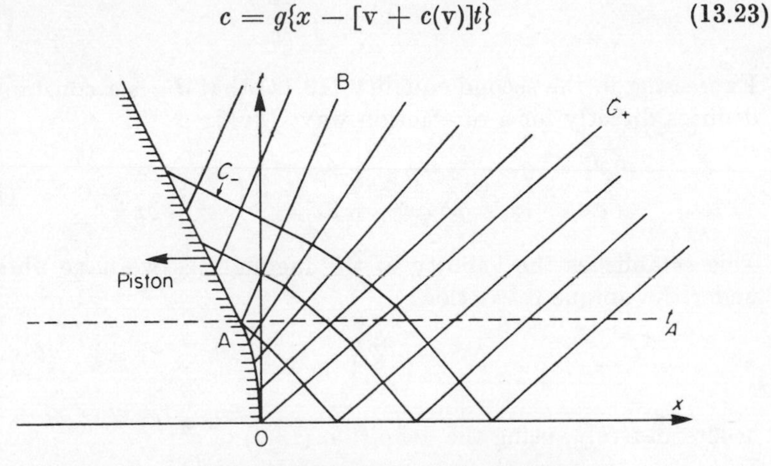

Fig. 13.2. Characteristics of a rarefaction wave.

The explicit form of functions f and g, depends again on the initial and boundary conditions. We observe the analogy with the plane wave $f_1(x - ct)$ in (13.7). However in the present case, points corresponding to different values of the velocity v, travel with different speeds. Therefore the profile of the wave changes in time.

Waves of finite amplitude defined by equations (13.21–13.23), are called *simple waves*, (e.g. Landau and Lifshitz, 1959, or Zeldovich and Raizer, 1966) i.e. waves consistent with the state of homogeneous equilibrium.

As the invariant J_- is constant for all values of x and t, we deduce for a perfect gas, from (13.18), the relation:

$$J_- = \mathrm{v} - \frac{2}{\gamma - 1}\, c = -\frac{2}{\gamma - 1}\, c_e \tag{13.24}$$

where c_e denotes the speed of sound in the equilibrium region, i.e. for the fluid at rest. This yields:

$$\mathrm{v} = -\frac{2}{\gamma - 1}(c_e - c); \quad c = c_e + \frac{\gamma - 1}{2}\mathrm{v} \tag{13.25}$$

At the piston, we have $\mathrm{v} = U < 0$; hence, according to (13.16) and (13.25): $c < c_e, \rho < \rho_e, p < p_e$.

The validity of these inequalities is by no means restricted to the case of perfect gases. Indeed, consider the function:

$$\Psi(p) = \int \frac{\mathrm{d}p}{\rho c} \tag{13.26}$$

Expressing in the second equality (13.14) that J_- is a constant, one deduces directly for a rarefaction wave ($\partial \mathrm{v}/\partial x > 0$):

$$\frac{\partial \Psi}{\partial x} = \frac{\partial \mathrm{v}}{\partial x} = \frac{1}{\rho c}\frac{\partial p}{\partial x} = \frac{c}{\rho}\frac{\partial \rho}{\partial x} = \frac{2}{\rho}\left(\frac{\partial p}{\partial c^2}\right)_s \frac{\partial c}{\partial x} > 0 \tag{13.27}$$

This establishes the validity of the inequalities we have obtained, under the unique restriction:

$$\left(\frac{\partial p}{\partial c^2}\right)_s > 0 \tag{13.28}$$

or alternatively, using the definition (13.6) of c^2:

$$\left(\frac{\partial^2 \rho}{\partial p^2}\right)_s < 0 \tag{13.29}$$

We shall adopt this condition which is identically satisfied for a perfect gas. It is similar but not identical to the inequality

$$(\partial^2 v/\partial p^2)_s > 0,$$

considered as usually satisfied by Landau and Lifshitz in their theory of weak shock waves (1959, § 83).

Of course, for a compression wave ($\partial \mathrm{v}/\partial x < 0$), the same conclusions may be drawn but with the opposite sign in (13.27), namely:

$$\frac{\partial \mathrm{v}}{\partial x} < 0, \quad \frac{\partial p}{\partial x} < 0, \quad \frac{\partial \rho}{\partial x} < 0, \quad \frac{\partial c}{\partial x} < 0 \tag{13.30}$$

Coming back now to perfect gases and using (13.11) and (13.25), we see that the C_+ characteristics which originate from the piston are given by:

$$\frac{dx}{dt} = c + v = c_e + \frac{\gamma + 1}{2} v = c_e - \frac{\gamma + 1}{2} |U|; (C_+) \quad (13.31)$$

Therefore, during the time interval $0 < t < t_A$ (see Figure 13.2), the characteristics C_+ form a family of *divergent* straight lines. On the contrary, for a compression wave, we find a family of *convergent* straight lines as represented on Figure 13.3.

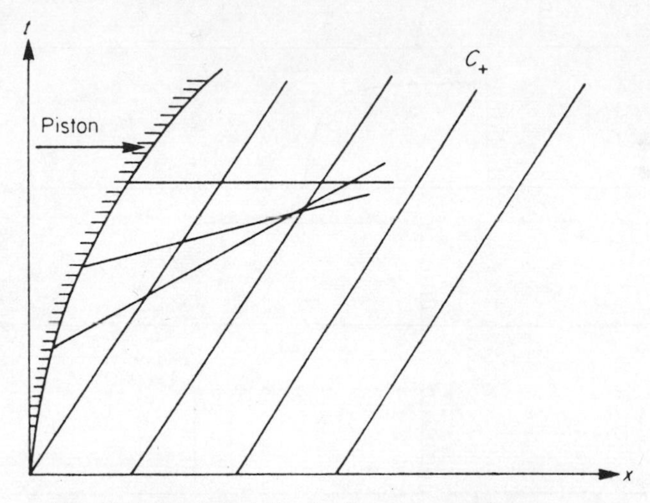

Fig. 13.3. Characteristics C_+ for a compression wave.

It is important to notice that for simple compression waves, characteristics intersect. As each characteristic corresponds to a well defined velocity v, this means that the hydrodynamic velocity is no longer a uniform function. We then have formation of a shock wave. This is illustrated on Figure 13.4. The time scale for the formation of such a shock wave is of the order of

$$\tau = \frac{c}{a} \quad (13.32)$$

where the quantity a denotes the acceleration of the piston (e.g. Landau and Lifshitz, 1959, § 94). For small accelerations this time may be arbitrarily large.

Again, all these results are not limited to perfect gases but we shall not go here into more details.

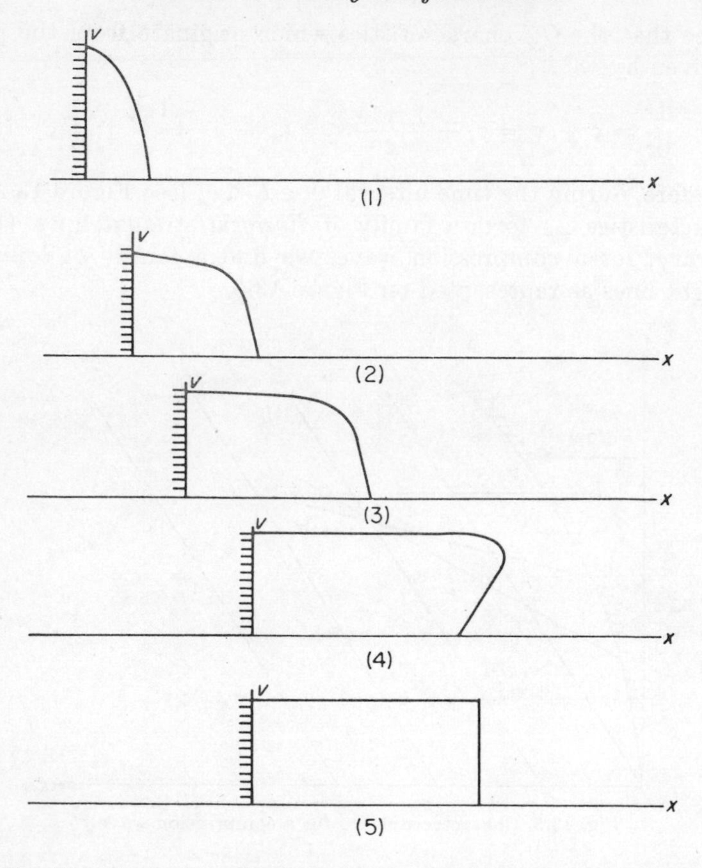

Fig. 13.4. Successive velocity profiles in a simple compression wave. Formation of a shock wave.

4. SMALL DISTURBANCES OF TRAVELLING WAVES

To fluctuations ϖ, of the pressure, and u, of the velocity, correspond, according to (13.14), fluctuations of the Riemann invariants:

$$\delta J_{\pm} = u \pm \frac{1}{\rho c}\varpi \tag{13.33}$$

In this relation, we have neglected higher order terms in the fluctuations. The fluctuations δJ_{+} are therefore propagated along the characteristics C_{+} and δJ_{-} along the characteristics C_{-}. Indeed, we may still adopt as the total derivative for the transport of $\delta J_{\pm}$, the operators used in (13.12), (13.13):

$$\mathrm{d} = \partial_t + (\mathrm{v} \pm c) \frac{\partial}{\partial x} \tag{13.34}$$

A perturbation of either v or c in (13.34), would correspond to second order effects.

It is convenient to study separately the effect of the perturbation of each of the invariants. We first suppose:

$$\delta J_{-} = 0 \tag{13.35}$$

This may occur when there are no disturbances generated in the region of homogeneous equilibrium. Indeed, (13.35) is then valid at time $t = 0$, and the unperturbed value $J_{-}(x, t)$, is propagated along the characteristics C_{-}, which leads directly to (13.35) in the whole region x, t we have to consider. Therefore, according to (13.33), (13.35), we have everywhere for the perturbations δJ_{+}:

$$\varpi = \rho c u \tag{13.36}$$

Similarly, using the definition (13.6) of c, one has:

$$\delta \rho = \frac{\rho}{c} u \tag{13.37}$$

and for the corresponding (i.e. isentropic) disturbance θ of temperature†:

$$\theta = \left(\frac{\partial T}{\partial p}\right)_s \varpi = \frac{T}{c_p}\left(\frac{\partial v}{\partial T}\right)_p \varpi = \frac{\alpha T c}{c_p} u \tag{13.38}$$

We see that one is finally left with a single independent fluctuating variable, which may be chosen to be u.

Let us now suppose the opposite case:

$$\delta J_{+} = 0 \tag{13.39}$$

We obtain by the same line of reasoning, the corresponding equalities:

$$\varpi = -\rho c u; \quad \delta \rho = -\frac{\rho}{c} u; \quad \theta = -\frac{\alpha T c}{c_p} u \tag{13.40}$$

We are now ready to undertake the study of the stability problem.

† One deduces from 2.31 that:

$$\left(\frac{\partial T}{\partial p}\right)_s = \left(\frac{\partial v}{\partial s}\right)_p = \frac{T}{c_p}\left(\frac{\partial v}{\partial T}\right)_p = \frac{vT\alpha}{c_p}$$

5. INSTABILITY OF THE SIMPLE COMPRESSION WAVE

As a rule, instability is easier to prove than stability. Indeed, the existence of one single unstable disturbance suffices to establish instability, while investigation of all possible disturbances is necessary to prove stability. In this respect, the difference between the instability problem of a compression wave, considered in this section, and the stability problem of the rarefaction wave treated in § 6, is striking.

Let us apply our hydrodynamic stability criterion (7.102) to the compression wave considered on Figure 13.1(3). For unperturbed boundary conditions (the motion of the piston is assumed to be given), the boundary term (7.103) vanishes. Moreover the dissipative effects are absent in this one-dimensional isentropic motion ($p_{ij} = 0$). Taking $\tau^2 = 1$ in (7.101), and using (13.36), (13.37) along the characteristics C_+, we obtain for the hydrodynamic stability criterion:

$$P[\delta E_{kin}] = - \int_{x_P}^{\infty} u^2 \left[\tfrac{1}{2} \frac{\partial(\rho \mathrm{v} + \rho c)}{\partial x} + \frac{\rho}{c} (\mathrm{v} + c) \frac{\partial \mathrm{v}}{\partial x} \right] \mathrm{d}x < 0 \quad (13.41)$$

According to inequalities (13.30), it appears clearly that the condition (13.41) is violated, as x_P, v and c, are positive quantities. Therefore, *the simple compression wave is unstable.*

What is the physical meaning of such an instability? As mentioned in (13.32), the unperturbed motion has already a life time limited to the formation of a shock wave. We now observe that this life time should be even smaller due to the instability of the basic flow. Such an instability may appear everywhere in the wave and at any moment, as the result of a fluctuation. Then the simple wave is replaced by another secondary flow pattern.

Unfortunately we lack experimental evidence in this field. Is this reduction of the life time realistic, or is it a simple consequence of the too simplified isentropic one-dimensional pattern of flow adopted? This question, as well as the role of dissipation (thermal and viscous) in this stability problem, remains to be elucidated.

Anyway, we are here concerned with an interesting example of a *macroscopic time dependent evolution*, which is unstable in respect to small fluctuations. Such a situation, does not occur in thermal problems as described in Chapter VII, § 2. Indeed, as we have seen, the stability conditions are then satisfied, and the macroscopic evolution described by the time-dependent Fourier equation is stable.

We may stress also that the instability of the compression waves, only occurs for finite amplitude. Again, we need a finite distance from equilibrium to make the thermodynamic branch unstable (see Chapter XI, § 5).

6. STABILITY OF THE SIMPLE RAREFACTION WAVE

We now investigate the stability problem of the rarefaction wave considered in Figure 13.1(2). As in the foregoing example, we assume unperturbed boundary conditions. On the other hand, for isentropic one dimensional flows, one has ($\mu = h - Ts$):

$$s_{,j} = 0 \quad \text{i.e.} \quad (\mu T^{-1})_{,j} = (hT^{-1})_{,j} \tag{13.42}$$

Also

$$\delta s = 0 \quad \text{i.e.} \quad \delta(\mu T^{-1}) = \delta(hT^{-1}) \tag{13.43}$$

for the small isentropic disturbances we consider. Accordingly, all dissipative effects vanish, and therefore the excess entropy production $\sum_\alpha \delta J_\alpha \delta X_\alpha$ disappears from the balance equations (7.56), (7.57).

We then obtain explicitly for the stability criterion (6.38) applied to the present problem, the inequality:

$$\begin{aligned} P[\delta Z] = \int_{x_P}^{+\infty} \Bigg\{ &-[u\delta(\rho e) + u\varpi + \tfrac{1}{2}\rho \mathrm{v} u^2] \frac{\partial T^{-1}}{\partial x} + u\delta\rho \frac{\partial}{\partial x}(hT^{-1}) \\ &- [\varpi\delta T^{-1} - \rho T^{-1}u^2 - \mathrm{v}T^{-1}u\delta\rho] \frac{\partial \mathrm{v}}{\partial x} + \tfrac{1}{2}T^{-1}u^2 \frac{\partial}{\partial x}(\rho \mathrm{v}) \\ &+ [\delta(\rho e) \frac{\partial}{\partial x} \delta T^{-1} - \delta\rho \frac{\partial}{\partial x} \delta(hT^{-1})]\, \mathrm{v} \Bigg\} \, dx \geqslant 0 \end{aligned} \tag{13.44}$$

Conditions (13.42), (13.43) enable us to express the changes of p, ρ, T, e, h, (gradients or disturbances) in terms of one of them, say the pressure p. The reduction which follows contains a lot of elementary manipulations, which are given separately in the next section to simplify the representation. The integrand of (13.44) becomes in this way

$$\begin{aligned} \sigma = &\frac{\mathrm{v}}{2\rho^2 c^2 T}\left(\frac{3\alpha}{c_p} + \frac{1}{c^2}\right)\frac{\partial p}{\partial x}\varpi^2 + \frac{1}{2\rho c^2 T}\left(\frac{\partial \mathrm{v}}{\partial x} - 2\frac{\mathrm{v}}{c}\frac{\partial c}{\partial x}\right)\varpi^2 \\ &+ \frac{1}{\rho T}\left(\frac{\alpha}{c_p} + \frac{1}{c^2}\right)\frac{\partial p}{\partial x}u\varpi + \frac{1}{\rho T}\left(\frac{\alpha}{c_p}\varpi^2 + \frac{3}{2}\rho^2 u^2 + \frac{\rho \mathrm{v}}{c^2}u\varpi\right)\frac{\partial \mathrm{v}}{\partial x} \\ &+ \frac{\mathrm{v}}{2T}\left(\frac{\alpha}{c_p} + \frac{1}{c^2}\right)\frac{\partial p}{\partial x}u^2 \end{aligned} \tag{13.45}$$

Let us now introduce the relations (13.36) and (13.40) between the disturbances u and ϖ. Eliminating ϖ, equation (13.45) takes the form:

$$\sigma = \frac{u^2}{T}\left\{\left[\frac{\alpha}{c_p}(2\mathrm{v} \pm c) + \frac{1}{c^2}(\mathrm{v} \pm c)\right]\frac{\partial p}{\partial x} + \left(2 + \frac{\alpha c^2}{c_p} \pm \frac{\mathrm{v}}{c}\right)\rho\frac{\partial \mathrm{v}}{\partial x} - \frac{\rho \mathrm{v}}{c}\frac{\partial c}{\partial x}\right\}; (C_\pm) \qquad (13.46)$$

The link between the three remaining gradients implies the choice of an equation of state. For a perfect gas, at constant γ one derives from equation (13.24):

$$\frac{\partial \mathrm{v}}{\partial x} = \frac{2}{\gamma - 1}\frac{\partial c}{\partial x} \qquad (13.47)$$

and (13.46) becomes ($\alpha = T^{-1}$):

$$\sigma = \frac{cu^2}{c_p T^2}\left\{\frac{2\gamma - 1}{\gamma - 1}\cdot\frac{\mathrm{v}}{c} \pm \frac{\gamma}{\gamma - 1}\right\}\frac{\partial p}{\partial x} + \frac{\rho u^2}{T}\left\{2\frac{\gamma + 1}{\gamma - 1} + \frac{\mathrm{v}}{c}\left(\pm\frac{2}{\gamma - 1} - 1\right)\right\}\frac{\partial c}{\partial x}; (C_\pm) \qquad (13.48)$$

Due to the relation

$$p \sim c^{2\gamma/(\gamma-1)} \qquad (13.49)$$

which is a consequence of equations (13.16), (13.17), one has also:

$$\frac{1}{p}\frac{\partial p}{\partial x} = \frac{2\gamma}{\gamma - 1}\frac{1}{c}\frac{\partial c}{\partial x} \qquad (13.50)$$

Therefore elimination of the pressure gradient from the integrand (13.48), permits us to write the stability condition (13.44) in the form of the two inequalities:

$$P_+[\delta Z] = \int_{x_P}^{+\infty} \frac{\rho u^2}{(\gamma - 1)T}\left[(3\gamma + 1)\frac{\mathrm{v}}{c} + 4\gamma + 2\right]\frac{\partial c}{\partial x}\,\mathrm{d}x \geqslant 0 \qquad (13.51)$$

$(t > t_0)$ on C_+

$$P_-[\delta Z] = \int_{x_P}^{+\infty} \frac{\rho u^2}{(\gamma - 1)T}\left[3(\gamma - 1)\frac{\mathrm{v}}{c} + 2\right]\frac{\partial c}{\partial x}\,\mathrm{d}x \geqslant 0 \qquad (13.52)$$

$(t > t_0)$ on C_-

Let us recall that in case of a rarefaction wave one has (see Figure 13.1(2) and equation (13.27)):

$$x_P < 0 \quad v < 0 \quad \frac{\partial c}{\partial x} > 0 \tag{13.53}$$

Moreover the sound velocity has to remain positive at the piston. Therefore, according to the second relation (13.25) we have the inequality

$$(c)_{x_P} = c_e - \frac{\gamma - 1}{2}|U| > 0 \quad \text{i.e.} \quad |U| < \frac{2}{\gamma - 1}c_e = U_{max} \tag{13.54}$$

For larger velocities imposed to the piston, the fluid could not follow and the validity of continuous hydrodynamics becomes questionable.

To be sure that we remain in the region of validity of our basic local equilibrium assumption, we shall suppose that the velocity of the piston is lower than the local sound velocity. Consequently, we have at every point:

$$|v| < c \tag{13.55}$$

The flow is then everywhere subsonic.

We also assume

$$\gamma \leqslant \frac{5}{3} \tag{13.56}$$

The upper limit corresponds to a monoatomic gas. Using (13.53), (13.55), (13.56) we easily check that both the conditions (13.51), (13.52) are satisfied.

Therefore *the rarefaction simple wave is a stable time-dependent process in the subsonic region*. However, this conclusion does not exclude the possibility of some instabilities arising from anomalous thermodynamic properties leading to a change of sign, either in (13.28), (13.29), (Zeldovich and Raizer, 1966; Kahl and Mylin, 1969, on the possiblity of rarefaction shock), or in (13.56) (perfect gases with, e.g. $\gamma = 3$, as considered by Landau, 1965). Also instabilities are possible in the supersonic region, but we shall not go into further considerations here.

7. REDUCTION OF $P[\delta Z]$

We establish here briefly the main equalities we used, to reduce the integrand of (13.44) to the simpler form (13.45). Again we suppose unperturbed boundary conditions.

First of all we need, besides equations (13.3) and (13.38), the relations valid for an isentropic process (Chapter II, § 4):

$$\delta(\rho e) - h\delta\rho = \rho(\delta e + p\delta v) = \rho T\delta s = 0 \tag{13.57}$$

$$\delta h = \left(\frac{\partial h}{\partial p}\right)_s \varpi = \frac{1}{\rho}\varpi \tag{13.58}$$

Similar equalities exist between the corresponding gradients. As a result we obtain:

$$\frac{\partial h}{\partial x}\delta\rho\delta T^{-1} = \frac{-\alpha}{\rho^2 c^2\, c_p T}\frac{\partial p}{\partial x}\varpi^2 = \frac{\partial T^{-1}}{\partial x}\delta h\delta\rho \tag{13.59}$$

On the other hand, one has:

$$-\frac{\mathrm{v}}{T}\frac{\partial \delta h}{\partial x}\delta\rho = \frac{-\mathrm{v}}{Tc^2}\varpi\frac{\partial}{\partial x}\left(\frac{\varpi}{\rho}\right) = \frac{\mathrm{v}}{\rho^2 Tc^2}\frac{\partial\rho}{\partial x}\varpi^2 - \frac{\mathrm{v}}{2\rho Tc^2}\frac{\partial\varpi^2}{\partial x} \tag{13.60}$$

Then the last term is differentiated by parts, and the divergence term is neglected as it vanishes by integration. This gives us:

$$-\frac{\mathrm{v}}{T}\frac{\partial\delta h}{\partial x}\delta\rho = \frac{\mathrm{v}}{\rho^2 Tc^4}\frac{\partial p}{\partial x}\varpi^2 + \left[\frac{\partial}{\partial x}\left(\frac{\mathrm{v}}{2\rho Tc^2}\right)\right]\varpi^2 \tag{13.61}$$

Expanding the last term and regrouping the whole expression, it becomes finally:

$$-\frac{\mathrm{v}}{T}\frac{\partial\delta h}{\partial x}\delta\rho = \frac{\mathrm{v}}{2\rho^2 Tc^2}\left[\frac{1}{c^2} - \frac{\alpha}{c_p}\right]\frac{\partial p}{\partial x}\varpi^2 + \frac{1}{2\rho Tc^2}\left[\frac{\partial \mathrm{v}}{\partial x} - \frac{\mathrm{v}}{c^2}\frac{\partial c^2}{\partial x}\right]\varpi^2 \tag{13.62}$$

Now introducing (13.57), (13.59) and (13.62) into the last bracket of (13.44) written as:

$$\mathrm{v}\left\{[\delta(\rho e) - h\delta\rho]\frac{\partial}{\partial x}(\delta T^{-1}) - \frac{\partial h}{\partial x}\delta\rho\delta T^{-1} - \frac{\partial T^{-1}}{\partial x}\delta\rho\delta h - T^{-1}\frac{\partial\delta h}{\partial x}\delta\rho\right\}$$

we get the following expression:

$$\left[\frac{\mathrm{v}}{\rho^2 Tc^2}\left(\frac{3}{2}\frac{\alpha}{c_p} + \frac{1}{2c^2}\right)\frac{\partial p}{\partial x} + \frac{1}{\rho Tc^2}\left[\frac{1}{2}\frac{\partial \mathrm{v}}{\partial x} - \frac{\mathrm{v}}{c}\frac{\partial c}{\partial x}\right)\right]\varpi^2 \tag{13.63}$$

The remaining terms in the r.h.s. of (13.44) may be reduced as follows:

$$-[u\delta(\rho e) + u\varpi + \tfrac{1}{2}\rho \mathrm{v} u^2]\frac{\partial T^{-1}}{\partial x} + u\delta\rho\frac{\partial}{\partial x}(hT^{-1})$$

$$= \frac{1}{T^2}[\rho u\delta h + \tfrac{1}{2}\rho \mathrm{v} u^2]\frac{\partial T}{\partial x} + \frac{u}{\rho T c^2}\frac{\partial p}{\partial x}\varpi$$

$$= \frac{1}{\rho T}\left[\frac{\alpha}{c_p} + \frac{1}{c^2}\right]\frac{\partial p}{\partial x}u\varpi + \frac{\alpha \mathrm{v}}{2c_p T}\frac{\partial p}{\partial x}u^2 \tag{13.64}$$

and

$$-[\varpi\delta T^{-1} - \rho T^{-1}u^2 - \mathrm{v}T^{-1}u\delta\rho]\frac{\partial \mathrm{v}}{\partial x} + \tfrac{1}{2}u^2T^{-1}\frac{\partial}{\partial x}(\rho \mathrm{v})$$

$$= \left[\frac{\alpha}{\rho c_p}\varpi^2 + \frac{3}{2}\rho u^2 + \frac{\mathrm{v}}{c^2}u\varpi\right]\frac{1}{T}\frac{\partial \mathrm{v}}{\partial x} + \tfrac{1}{2}\frac{u^2 \mathrm{v}}{c^2 T}\frac{\partial p}{\partial x} \tag{13.65}$$

Taking now (13.63), (13.64) and (13.65) into (13.44) we recover the form (13.45).

This closes our discussion on the stability of finite amplitude waves. It is very satisfactory to observe that such simple results could be derived in spite of the basic non-linearity

PART III
Chemical Processes

CHAPTER XIV

Time Order in Chemical Reactions

1. INTRODUCTION

In Chapters XI to XIII, we applied the stability theory to some typical problems of Hydrodynamics. In Chapters XIV to XVI, we consider applications to open chemical systems far from equilibrium.

The main purpose is to get the conditions for the emergence of time order or space order in a chemical system under the influence of dissipation. The situation is similar to the onset of free convection (Chapter XI) or the transition from laminar to turbulent flow (Chapter XII). Here also we have to determine the critical value of the constraints (Chapter III, § 4), corresponding to the limit of stability of the thermodynamic branch. Indeed it is only beyond this transition that one may expect situations entirely different from those obtained by simple extrapolation of the behaviour which prevails near equilibrium.

Unfortunately, the variety of non-linear chemical mechanisms is so large, that a systematic unified approach of this subject is not yet possible. The only unifying features at our disposal are the stability condition (7.40) and the condition (9.57) for the occurrence of oscillations.

In the present chapter we investigate the problem of chemical oscillations in a homogeneous medium. This problem has been studied extensively. The conservation and the perturbation equations are both ordinary autonomous differential equations, involving only the time as independent variable.

We first consider the behaviour of oscillations on the *thermodynamic branch* (Chapter XI, § 5). This situation is realized in the Lotka (1920)–Volterra (1931) model.

We then study the oscillations belonging to a new non-thermodynamic branch. We discuss the difference with the behaviour observed in the first case. The main feature is the advent of new types of chemical oscillating processes, the 'limit cycles' first introduced by Poincaré in the study of the three-body problem (1892). It is very likely that these processes are of great importance for the occurrence of 'chemical clocks', which are typical examples of 'time order' generated by dissipation.

2. THERMODYNAMIC THRESHOLD FOR CHEMICAL OSCILLATIONS

We have shown in (9.56) and (9.57) that the behaviour of a dissipative system, in the neighbourhood of a steady state, is determined by the two inequalities:

$$\omega_r^2 \delta_m^2 S = \omega_r \delta_m P \leqslant 0 \tag{14.1}$$

and

$$\omega_i^2 \delta_m^2 S = \omega_i \delta_m \Pi \leqslant 0 \tag{14.2}$$

These inequalities refer respectively to the real and imaginary parts of each normal mode. They determine the essential features of the time evolution of the normal mode: $\delta_m \Pi$, gives through its sign, the direction of rotation around the steady state while $\delta_m P$ decides the stability of the latter. For an aperiodic system ($\omega_i = 0$) we have

$$\delta_m \Pi = 0 \tag{14.3}$$

and *vice versa*.

In Chapter XI, § 7 we used this condition to prove the principle of the exchange of stabilities for the Bénard problem.

The object of this section, and of the following, is to investigate the conditions under which either the condition (14.3) for aperiodic behaviour, or the stability condition $\delta_m P \geqslant 0$ can be transgressed.

We consider the following scheme of reactions:

$$\begin{aligned} A + X &\underset{k_{-1}}{\overset{k_1}{\rightleftharpoons}} 2X \qquad &\text{(a)} \\ X + Y &\underset{k_{-2}}{\overset{k_2}{\rightleftharpoons}} 2Y \qquad &\text{(b)} \\ Y &\underset{k_{-3}}{\overset{k_3}{\rightleftharpoons}} E \qquad &\text{(c)} \end{aligned} \tag{14.4}$$

where the values of initial and final products A and E are maintained constant in time, so that only two independent variables, X and Y are left. As in (3.45), the k denote kinetic constants.

The thermodynamic state of the system is then characterized by the value of the overall affinity†

$$\mathscr{A} = \mathscr{A}_1 + \mathscr{A}_2 + \mathscr{A}_3 = \mathscr{R}T \ln \left(\frac{k_1 k_2 k_3}{k_{-1} k_{-2} k_{-3}} \cdot \frac{A}{E} \right) \tag{14.5}$$

corresponding to the global reaction

$$A \rightleftharpoons E \tag{14.6}$$

We use here the symbol $\mathscr{A}$ defined in (8.11) for the chemical affinity. On the other hand we use the symbols A, X, Y, E for the chemical species and their concentrations.

At equilibrium one has (e.g. (3.45)):

$$\left(\frac{A}{E}\right)_{eq} = \frac{k_{-1} k_{-2} k_{-3}}{k_1 k_2 k_3}; \quad X_{eq} = \frac{k_1}{k_{-1}} A; \quad Y_{eq} = \frac{k_1 k_2}{k_{-1} k_{-2}} \cdot A \tag{14.7}$$

The time behaviour of the system in the neighbourhood of a non-equilibrium steady state ($\mathscr{A} \neq 0$), can be studied easily for the two limiting cases already discussed in (3.47) and (3.51), where $\mathscr{A}$ is either small ($|\mathscr{A}| \ll \mathscr{R}T$), or very large ($|\mathscr{A}| \to \infty$). We shall investigate these situations first.

(a) Steady state close to equilibrium

If the ratio (A/E) is only slightly different from its equilibrium value (14.7), we obtain as in Chapter III, § 5 linear laws between affinities and reaction rates. We find in this way, using the subscript zero for steady-state values:

$$w_{01} = k_1 A X_{eq} \frac{\mathscr{A}_1}{\mathscr{R}T}; \quad w_{02} = k_2 X_{eq} Y_{eq} \frac{\mathscr{A}_2}{\mathscr{R}T}; \quad w_{03} = k_3 Y_{eq} \frac{\mathscr{A}_3}{\mathscr{R}T} \tag{14.8}$$

The stability condition (9.25) takes the form:

$$P[\delta S] = \delta P = k_1 A X_{eq} (\delta \mathscr{A}_1)^2 + k_2 X_{eq} Y_{eq} (\delta \mathscr{A}_2)^2 + k_3 Y_{eq} (\delta \mathscr{A}_3)^2 \geqslant 0 \tag{14.9}$$

In agreement with our general conclusions, the excess entropy production is positive. The thermodynamic branch is then stable. Moreover no oscillations are possible as the condition (14.3) is satisfied for each normal mode. Therefore an arbitrary fluctuation regresses in an aperiodic way to the steady state.

† To avoid confusion with the parameter defined in (14.25), we use the symbol $\mathscr{R}$ for the gas constant in this chapter.

(*b*) *Steady state far from equilibrium*

When the reverse reactions in scheme (14.4) become negligible ($k_{-i} = 0$; $i = 1, 2, 3$), the overall affinity (14.5) tends to infinity as pointed out in (3.51). The kinetic laws are then:

$$\frac{\mathrm{d}X}{\mathrm{d}t} = k_1 AX - k_2 XY \tag{14.10}$$

$$\frac{\mathrm{d}Y}{\mathrm{d}t} = k_2 XY - k_3 Y \tag{14.11}$$

They admit a single non-vanishing steady-state solution:

$$X_0 = \frac{k_3}{k_2}; \quad Y_0 = \frac{k_1}{k_2} A \tag{14.12}$$

This scheme is isomorphic to the model originally introduced by Lotka (1920) and Volterra (1931), to describe the predator-prey behaviour in interaction. Henceforth, systems such as (14.10), (14.11) will be called a Lotka–Volterra model. It has been used recently to discuss quite fundamental biological phenomena such as biological clocks (see Bunning, 1958, 1960), or time-dependent properties of neural networks (Cowan, 1969).

Let us first study the stability of solution (14.12) by normal mode analysis. In its neighbourhood, $X(t)$ and $Y(t)$ may be written as:

$$X(t) = X_0 + x\,\mathrm{e}^{\omega t}; \quad Y(t) = Y_0 + y\,\mathrm{e}^{\omega t} \tag{14.13}$$

with

$$\left|\frac{x}{X_0}\right| \ll 1; \quad \left|\frac{y}{Y_0}\right| \ll 1 \tag{14.14}$$

Replacing (14.13) into the kinetic equations (14.10–14.11), and neglecting higher order terms in the perturbations, we get the set of linearized equations:

$$\omega \delta X + k_3 \delta Y = 0 \tag{14.15}$$

$$-k_1 A \delta X + \omega \delta Y = 0 \tag{14.16}$$

The corresponding dispersion equation

$$\omega^2 + k_1 k_3 A = 0 \tag{14.17}$$

indicates that small fluctuations around the steady state (14.12) are now periodic with the frequency:

$$\omega_i = \pm(k_1 k_3 A)^{1/2}; \quad \omega_r = 0 \tag{14.18}$$

In accordance with the thermodynamic stability criterion (9.25), we observe that $\delta_X P$ vanishes here identically around the steady state (14.12). Indeed one deduces from (14.10), (14.11) that

$$\delta_X P = \left(k_2 \frac{Y_0}{X_0} - k_1 \frac{A}{X_0}\right)(\delta X)^2 + \left(\frac{k_3}{Y_0} - k_2 \frac{X_0}{Y_0}\right)(\delta Y)^2 = 0 \tag{14.19}$$

Then, using (9.25), (9.27) and (7.40) we deduce

$$\mathrm{d}_t\, \delta^2 S = P[\delta S] = 0$$

As a result $\delta^2 S$ appears here as a constant of motion for an arbitrary disturbance. One recognizes the *non-asymptotic* or *weak* stability condition of Liapounoff (Chapter VI), already met in Chapter XI, § 12, when we studied the stability of a vertical column of fluid. Likewise, according to (14.1) and to the second relation (14.18), one has separately for each normal mode:

$$\delta_m P = d_t \delta_m^2 S = 0 \tag{14.20}$$

As a result $\delta_m^2 S$ is also a constant of motion (for a given normal mode). Therefore, the two quadratic forms $\delta^2 S$ and $\delta_m^2 S$ neither decrease, nor increase along the perturbed motion (§ 3). However, since only the real part ω_r of ω vanishes (cf. 14.18), the perturbed state cannot be interpreted as another steady state infinitely close to (14.12). Indeed, according to (14.18), the frequency ω_i can never vanish. This also results directly from $\delta_m \Pi$. One finds for each separate mode:

$$\delta_m \Pi = ik_2(\delta X^* \delta Y - \delta X \delta Y^*) \tag{14.21}$$

whence

$$\omega_i \delta_m \Pi = -2k_2 \left[k_3(\delta Y_r)^2 + \frac{\omega_i^2}{k_3}(\delta X_r)^2\right] \leqslant 0 \tag{14.22}$$

that is a negative quantity as it must be, according to (14.2). Besides, it is obvious that ω_i can never vanish.

Let us now consider the behaviour of the system for intermediate values of the overall affinity ($1 \ll |\mathscr{A}|/\mathscr{R}T \leqslant \infty$). It is indeed clear from our discussion that the transition point where $\delta_m \Pi$ becomes different from zero lies in this domain. Putting for simplicity the direct kinetic constants equal to unity, and the inverse ones equal to k, it is easily found that the time independent solutions now satisfy the equations:

$$X_0 = 1 + kY_0 - \frac{kRA}{Y_0} \qquad \text{(a)} \quad (14.23)$$

$$k^3Y_0^4 + (1 - kA + 2k^2)Y_0^3 + (k - A - kRA - 2k^3RA)Y_0^2 + (kRA^2 - 2k^2RA)Y_0 + k^3R^2A^2 = 0 \qquad (14.24) \quad \text{(b)}$$

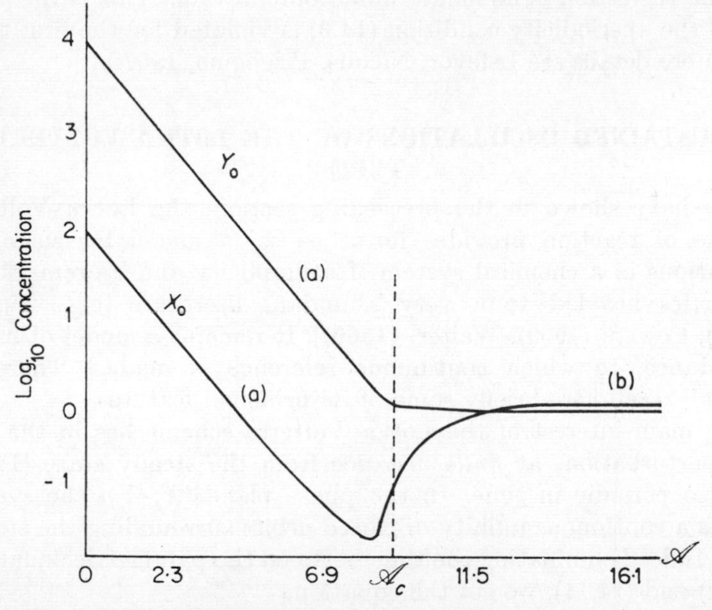

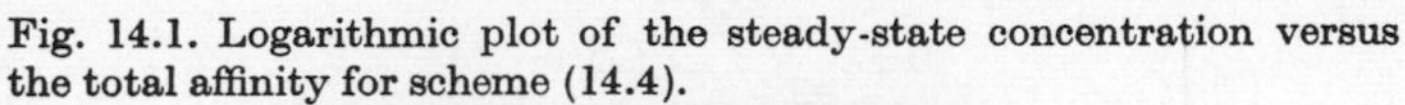

Fig. 14.1. Logarithmic plot of the steady-state concentration versus the total affinity for scheme (14.4).
(a) domain of monotonic behaviour.
(b) domain of oscillatory behaviour.

We have introduced here the parameter:

$$R = \frac{E}{A} \qquad (14.25)$$

which measures the deviation of the system from equilibrium. In Figure 14.1 the steady state solutions for X and Y, corresponding to the thermodynamic branch, have been plotted versus the overall affinity $\mathscr{A}$, for the set of numerical values

$$k = 10^{-2}, \quad A = 1 \qquad (14.26)$$

The dispersion equation for ω is:

$$\omega^2 + (Y_0 - X_0 + 2kX_0 + 2kY_0)\omega + X_0 + 2kX_0 - 1 - 2kX_0^2 - 2kY_0 + Y_0 + 4k^2X_0Y_0 = 0 \quad (14.27)$$

For the whole range of the overall affinity one has the inequality:

$$\omega_r < 0 \quad (14.28)$$

As a result, the thermodynamic branch is stable, and fluctuations will die out. However for values of $\mathscr{A} > 9{\cdot}2\,\mathscr{R}T$, ω becomes complex, and the regression is no longer monotonic in time. This is the point where the aperiodicity condition (14.3) is violated for the first time. (For more details see Lefever, Nicolis, Prigogine, 1967.)

3. SUSTAINED OSCILLATIONS OF THE LOTKA–VOLTERRA TYPE

As we have shown in the preceeding section, the Lotka–Volterra scheme of reaction provides for $\mathscr{A} \to \infty$, a model for sustained oscillations in a chemical system. Its simplicity and its remarkable properties has led to a very abundant literature [e.g. Nicolis, (1969), Cowan, (1969), Walter, (1969)]. It remains a model of major importance, to which continuous reference is made.† Therefore we shall recall here briefly some of its principal features.

The main interest of the Lotka–Volterra scheme lies in the fact that perturbations at *finite* distance from the steady state (14.12) are also periodic in time. In the phase plane (X, Y), the system admits a continuous infinity of closed orbits surrounding the steady state. Indeed eliminating the time between the parametric equations (14.10) and (14.11), we get the equation

$$\frac{\mathrm{d}Y}{\mathrm{d}X} = -\frac{Y(X-1)}{X(Y-1)} \quad (14.29)$$

$$(k_i = 1;\ i = 1, 2, 3;\ A = 1)$$

for the trajectories in the (X, Y) plane. Integrating, we obtain:

$$X + Y - \ln X - \ln Y = K \quad (14.30)$$

or equivalently:

$$X^{-1}\,\mathrm{e}^X = CY\,\mathrm{e}^{-Y};\quad C = \mathrm{e}^K \quad (14.31)$$

where K is an arbitrary constant. This constant is fixed by the initial conditions. Therefore the transcendental equation (14.30) (for more

† For good discussions see the monographs by Davis (1962), Minorsky (1962) or Cesari, (1963).

details see Davis, *loc. cit.*) represents a family of closed curves, corresponding each to a given value of K.

On the other hand, we further deduce from equations (14.10–14.11), that:

$$(X - 1)\frac{dY}{dt} - (Y - 1)\frac{dX}{dt} = (X - 1)^2 Y + (Y - 1)^2 X \quad (14.32)$$

Introducing polar coordinates (ρ, α), by means of the transformation

$$X = 1 + \rho \cos \alpha; \quad Y = 1 + \rho \sin \alpha \quad (14.33)$$

we obtain:

$$\frac{d\alpha}{dt} = (1 + \rho \cos \alpha) \sin^2 \alpha + (1 + \rho \sin \alpha) \cos^2 \alpha \quad (14.34)$$

This equation yields the period of rotation around a particular cycle as:

$$T = \int_0^{2\pi} \frac{d\alpha}{(1 + \rho \cos \alpha) \sin^2 \alpha + (1 + \rho \sin \alpha) \cos^2 \alpha} \quad (14.35)$$

We see that for a Lotka–Volterra type system, there exists a continuous spectrum of frequencies due to the infinite number of possible cycles depending on the initial condition (for further details see §§ 5 and 7). Each cycle appears as a state of marginal stability where even a small perturbation is sufficient to change the motion of the system to a new cycle corresponding to a *different frequence.* In other words, in a Lotka–Volterra type system, there is no mechanism for the decay of fluctuations and therefore there exists no average orbit in the neighbourhood of which the system is maintained. This situation is illustrated in Figure 14.2 in the (X, Y) plane.

One sees that the closer the curves get to the origin, the further they spread away from it in the opposite direction. A small fluctuation, while the system is in the neighbourhood of the origin, produces a very large deviation of the concentrations from their preceeding value, when the system comes back in the opposite portion of the phase plane. Certainly oscillations of this type cannot be expected to lead to reproducible observations as far as the amplitude and frequency of oscillations are simultaneously concerned.†

† In terms of Liapounoff's theory of stability, only the orbits infinitesimally close to the steady state may be considered as stable, since all these trajectories are described with the same universal frequency. On the contrary, at finite distance from the steady state, two neighbouring points belonging to two distinct cycles, tend to go away from each other due to the difference in period. Such motions are therefore unstable in the Liapounoff's sense but are called stable in the extended sense of *orbital stability* (Pars, 1965, section, 23.8).

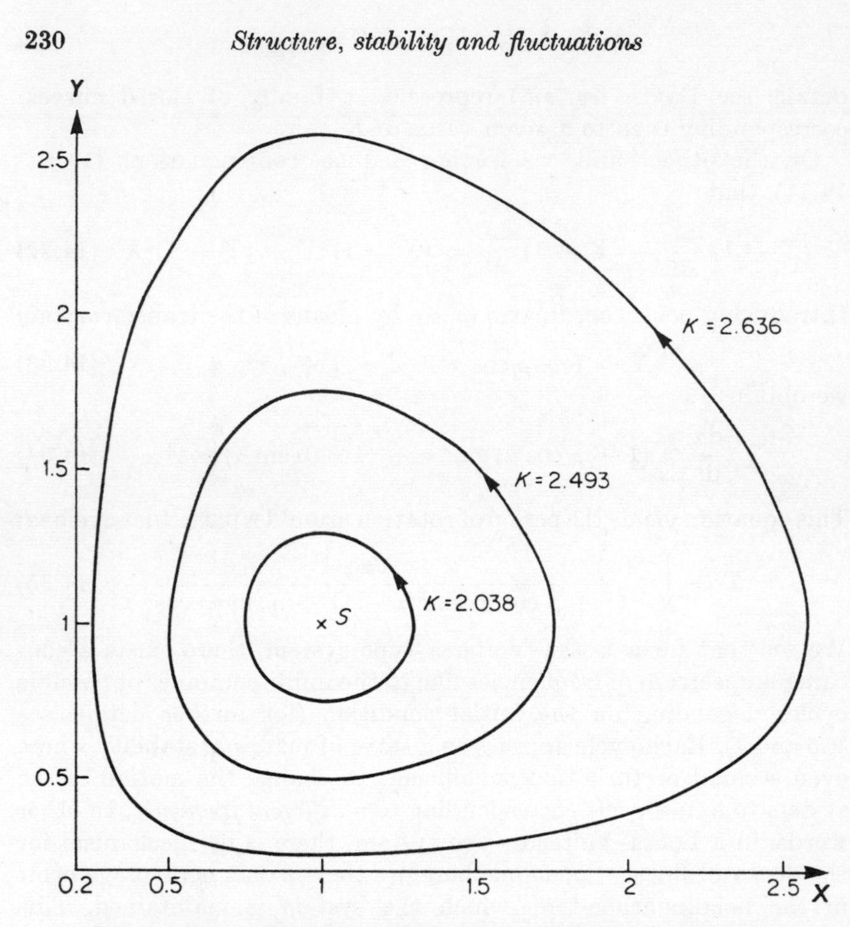

Fig. 14.2. Irreversible orbits in the X, Y plane, for different values of the constant of motion K . S = Steady State.

We shall come back to this problem in § 6. Nevertheless one should notice that the average concentrations have the same value independently of the periodic trajectory considered. Indeed writing for example equation (14.10) in the form ($k = 1$):

$$\frac{\mathrm{d} \ln X}{\mathrm{d}t} = A - Y \tag{14.36}$$

we obtain by averaging over the period T of any arbitrary cycle:

$$\frac{1}{T} \int_0^T \frac{\mathrm{d} \ln X}{\mathrm{d}t} \, \mathrm{d}t = 0 = A - \frac{1}{T} \int_0^T Y \, \mathrm{d}t \tag{14.37}$$

or according to (14.12) and keeping $k_i = 1$:

$$(\overline{Y})_T = \frac{1}{T}\int_0^T Y(t)\,\mathrm{d}t = A = Y_0$$

Likewise (14.38)

$$(\overline{X})_T = 1 = X_0$$

In other words, the average concentrations of X and Y, over an arbitrary cycle remain equal to their steady-state values (14.12). It is interesting to notice that these conditions insure also that the average entropy production over one period remains equal to the steady state entropy production. Starting from:

$$\sigma = AX \ln \frac{A}{X} + XY \ln \frac{X}{Y} + Y \ln \frac{Y}{E} = AX \ln A - Y \ln E$$

$$+ (XY - AX) \ln X + (Y - XY) \ln Y \quad (14.39)$$

Let us calculate the steady state value σ_0 and the average $(\bar{\sigma})_T$ of σ over a period. Because of (14.38) we obtain directly from (14.10), (14.11):†

$$(\overline{\Delta\sigma})_T = (\bar{\sigma})_T - \sigma_0 = -\left\langle \frac{\mathrm{d}X}{\mathrm{d}t} \ln X + \frac{\mathrm{d}Y}{\mathrm{d}t} \ln Y \right\rangle_T \quad (14.40)$$

Integrating (14.40) by parts we indeed verify that

$$(\overline{\Delta\sigma})_T = -\left\langle \frac{\mathrm{d}}{\mathrm{d}t}(X \ln X + Y \ln Y) \right\rangle_T$$

$$+ \left\langle \frac{\mathrm{d}X}{\mathrm{d}t} + \frac{\mathrm{d}Y}{\mathrm{d}t} \right\rangle_T = 0 \quad (14.41)$$

Having discussed the time average properties of the system, let us finally devote some more attention to the constant of motion K defined in (14.30). One has:

$$\frac{\mathrm{d}K}{\mathrm{d}t} = 0 \quad (14.42)$$

† In this section, the bar and the broken brackets denote the same type of average.

Let us introduce the velocities ω_X and ω_Y defined as (see (14.10), (14.11), (14.30)):

$$\omega_X = \frac{\mathrm{d} \ln X}{\mathrm{d}t} = -\frac{\partial K}{\partial \ln Y}$$

$$\omega_Y = \frac{\mathrm{d} \ln Y}{\mathrm{d}t} = \frac{\partial K}{\partial \ln X} \tag{14.43}$$

These equations are reminiscent of the Hamiltonian equations of classical dynamics (e.g. Kerner, 1957, 1959, 1964).

In other words, $K(X, Y)$, appears as an 'energy-like' constant of motion, playing the role of the Hamiltonian. The occurrence of such an invariant which satisfies the existence condition:

$$\frac{\partial^2 K}{\partial X\, \partial Y} = \frac{\partial^2 K}{\partial Y\, \partial X}$$

or

$$\frac{\partial \omega_X}{\partial X} + \frac{\partial \omega_Y}{\partial Y} = 0 \tag{14.44}$$

is quite exceptional and characterizes really the Lotka–Volterra model.

According to (14.19), $\delta^2 S$ is also a constant of motion. Therefore $\delta^2 S$ and K cannot be linearly independent. Otherwise X and Y would be separately determined and all motion would be prohibited. Near the steady state we derive from (14.30), by expansion up to the second order terms, and using (14.12):

$$K = \frac{x^2 + y^2}{2} = -\delta^2 S > 0 \tag{14.45}$$

This provides us with a simple thermodynamic interpretation of K.

4. CHEMICAL INSTABILITIES

We have seen that in the case of the Lotka–Volterra model, the critical point of marginal stability is reached in the limiting situation of an infinite overall affinity (14.18). We want now to study a case of instability for which the excess entropy production first vanishes and then changes its sign for a finite value of the overall affinity. The thermodynamic branch then becomes unstable.

From the general theory discussed in Chapter VII, § 4, it results that we need at least an autocatalytic step to obtain instability. Let us therefore consider the following scheme of reactions:

$$\begin{aligned} A &\rightleftharpoons X && \text{(a)} \\ 2X + Y &\rightleftharpoons 3X && \text{(b)} \\ B + X &\rightleftharpoons Y + D && \text{(c)} \\ X &\rightleftharpoons E && \text{(d)} \end{aligned} \tag{14.46}$$

where the initial and final products are A, B, D, E; only two independent variables X and Y are left. The autocatalytic step (14.46(b)) involves here a trimolecular reaction. This model is certainly not realistic, but still very useful due to its simplicity. In sections 6, 7 and 9 of Chapter XV, more realistic examples, describing some well known biochemical reactions, will be investigated.

The overall reaction corresponding to (14.46) is:

$$A + B \rightleftharpoons E + D \tag{14.47}$$

We have in fact two distinct reactions between the initial and the final products:

$$A \rightarrow E \tag{14.48}$$

and

$$B \rightarrow D \tag{14.49}$$

The equilibrium conditions are easily formulated. We have (e.g. (3.45)):

$$X_{eq} = \frac{k_1 A}{k_{-1}}; \quad Y_{eq} = \frac{k_{-2} k_1}{k_2 k_{-1}} A \tag{14.50}$$

together with:

$$\frac{E}{A} = \frac{k_1 k_4}{k_{-1} k_{-4}}, \quad \frac{D}{B} = \frac{k_2 k_3}{k_{-2} k_{-3}} \tag{14.51}$$

Putting for simplicity all direct kinetic constants k_+ equal to unity, and all inverse ones k_-, equal to k, we obtain the following kinetic equations:

$$\frac{dX}{dt} = A + X^2 Y - BX - X + k(YD + E - X - X^3) \tag{14.52}$$

$$\frac{dY}{dt} = BX - X^2 Y + k(X^3 - YD) \tag{14.53}$$

In the steady state:

$$X_0 = \frac{A + kE}{1 + k}; \quad Y_0 = \frac{kX_0^2 + RD}{X_0^2 + kD} X_0 \tag{14.54}$$

where as in (14.25) we have introduced the parameter

$$R = \frac{B}{D} \tag{14.55}$$

We know from the general theory described in Chapter VII, that steps such as a and d in (14.46) cannot compromise the stability or the asymptotic stability. Therefore we may adopt for A and E, values corresponding to the law of mass action, i.e. $A = k^2E$ (14.51). The corresponding steady-state solution (14.54) is then given by:

$$X_0 = \frac{A}{k}; \quad Y_0 = \frac{A(A^2 + kRD)}{A^2 + k^3D} \tag{14.56}$$

On the other hand, by normal mode analysis, we derive from the linearized perturbation equations around the steady state (14.54) or (14.56), the following dispersion relation:

$$\omega^2 + [X_0^2 + RD + 1 - 2X_0Y_0 + k(3X_0^2 + D + 1)]\omega + X_0^2 + k(X_0^2 + D) = 0 \tag{14.57}$$

Obviously, the values of R for which the coefficient of ω in (14.57) vanishes, correspond to a transition point. Beyond this state, the real part ω_r of the roots ω_1 and ω_2 changes of sign and the system becomes *unstable*. At the marginal state one then has:

$$\omega_1 + \omega_2 = 0 \tag{14.58}$$

Using (14.56), the marginal stability condition can be written in the explicit form:

$$R_c = \frac{k}{(A^2 - k^3D)} \left\{ k^3D + k^2A + (1 + k)\left[2A^2 + k^2 + \frac{A^2(A^2 + k^2)}{k^3D}\right] \right\} \tag{14.59}$$

or equivalently in terms of the corresponding affinity (in units $\mathscr{R}T$)

$$\mathscr{A}_c = \ln \frac{1}{k^2R_c} \tag{14.60}$$

In Figure 14.3, $\mathscr{A}_c$ has been plotted as a function of D for different values of k and for $A = 1$.

As a consequence of (14.59), $R_c > 0$ implies:

$$0 < D < \frac{A^2}{k^3}$$

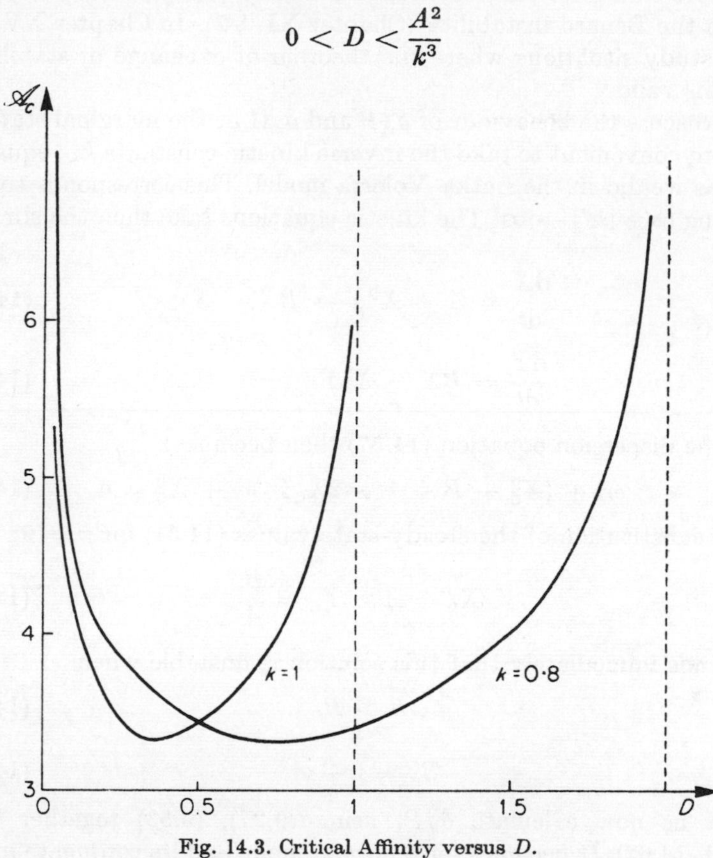

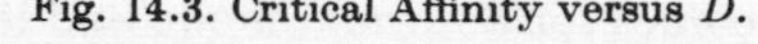

Fig. 14.3. Critical Affinity versus *D*.

For $D = 0$ or ∞, R_c tends towards infinity, while for

$$D = D_M = \frac{1}{k^3(3A^2 + k^2)} \times \{A^2\sqrt{[2(A^2 + k^2)(2A^2 + k^2)]} - A^2(A^2 + k^2)\}$$

R_c reaches its minimum value.

For values of $\mathscr{A}$ smaller than $\mathscr{A}_c$, the steady states lie on the continuation of the law of mass action, along the thermodynamic branch.

It can be verified that in this example the aperiodicity condition (14.3) is violated before the stability condition (cf. 14.1–14.2). The theorem of exchange of stabilities is no longer valid. The situation is therefore similar to the Reynolds instability (Chapter XII, § 2) and not to the Bénard instability (Chapter XI, § 7). In Chapter XV, we shall study situations where the theorem of exchange of stabilities remains valid.

To discuss the behaviour of $\delta_m P$ and $\delta_m \Pi$ at the marginal state, it is again convenient to take the inverse kinetic constants k_-, equal to zero as we did in the Lotka–Volerra model. This corresponds to the limiting case $|\mathscr{A}| \to \infty$. The kinetic equations take then the simpler form

$$\frac{\mathrm{d}X}{\mathrm{d}t} = A + X^2 Y - BX - X \tag{14.61}$$

$$\frac{\mathrm{d}Y}{\mathrm{d}t} = BX - X^2 Y \tag{14.62}$$

and the dispersion equation (14.57) then becomes:

$$\omega^2 + (X_0^2 + B + 1 - 2X_0 Y_0)\omega + X_0^2 = 0 \tag{14.63}$$

After substitution of the steady-state values (14.54) for $k = 0$:

$$X_0 = A; \quad Y_0 = \frac{B}{A} \tag{14.64}$$

one finds immediately that this solution is unstable when

$$B > B_C \tag{14.65}$$

with

$$B_c = 1 + A^2 \tag{14.66}$$

Let us now calculate $\delta_m P$, using (9.27), (9.53) together with (14.61–14.62). It becomes for a normal mode and its conjugate in the neighbourhood of the steady state (14.64):

$$\begin{aligned}
\delta_m P = \tfrac{1}{2} \Big\{ & (2X_0 Y_0 \delta X + X_0^2 \delta Y - B\delta X) \left(\frac{\delta Y^*}{Y_0} - \frac{\delta X^*}{X_0}\right) \\
& + \frac{2}{X_0} \delta X \delta X^* \\
& + (2X_0 Y_0 \delta X^* + X_0^2 \delta Y^* - B\delta X^*) \left(\frac{\delta Y}{Y_0} - \frac{\delta X}{X_0}\right) \Big\} \\
= & \frac{1}{AB} \{B(1 - B)\delta X \delta X^* + A^4 \delta Y \delta Y^*\}
\end{aligned} \tag{14.67}$$

In agreement with our general discussion we find a negative contribution: $-B\delta X\delta X^*$, due to the autocatalytic action of X. For

$$B \leqslant 1 \tag{14.68}$$

$\delta_m P$ is definite positive. This is a sufficient condition for stability. When B increases and reaches finally the critical value (14.66), the excess entropy production vanishes, the negative term compensating then exactly the positive contributions. It is interesting to verify this in detail. Indeed one has:

$$(\delta_m P)_{B=B_c} = \frac{1}{A(1 + A^2)}\{-A^2(1 + A^2)[(\delta X_r)^2 + (\delta X_i)^2] + A^4[(\delta Y_r)^2 + (\delta Y_i)^2]\} \tag{14.69}$$

The perturbations are related through the linearized kinetic equations, written for the marginal state ($\omega_r = 0$) in the form:

$$(1 - B)\delta X_r - \omega_i\delta X_i - A^2\delta Y_r = 0 \tag{14.70}$$

$$(1 - B)\delta X_i - A^2\delta Y_i + \omega_i\delta X_r = 0 \tag{14.71}$$

$$B\delta X_r + A^2\delta Y_r - \omega_i\delta Y_i = 0 \tag{14.72}$$

$$B\delta X_i + \omega_i\delta Y_r + A^2\delta Y_i = 0 \tag{14.73}$$

These equations yield for $B = B_c$:

$$\delta Y_i = \omega_i\delta Y_r; \quad \delta X_i = -\omega_i\delta Y_r; \quad \delta X_r = 0; \tag{14.74}$$

Replacing in (14.69) we get finally:

$$(\delta_m P)_{B=B_c} = \frac{A}{1 + A^2}(A^2 - \omega_i^2)(\delta Y_r)^2 \tag{14.75}$$

yielding back the frequency $\omega_i = \pm A$, as the marginal condition of stability, in agreement with the dispersion equation (14.63). Moreover one sees that ω_i has to be different from zero (overstability).

At the same time, we have for $\delta_m \Pi$:

$$\delta_m \Pi = -4[\delta X_r\delta Y_i - \delta X_i\delta Y_r] \tag{14.76}$$

which, according to (14.74), reduces to the negative definite expression:

$$\omega_i\delta_m\Pi = -4\omega_i^2(\delta Y_r)^2 \leqslant 0 \tag{14.77}$$

We verify here that the sign of the rotation is determined by $\delta_m\Pi$, as prescribed by equation (14.2).

5. TIME BEHAVIOUR BEYOND THE INSTABILITY

It is very fortunate that some insight into the behaviour of the model we studied in Section 4 can be obtained thanks to the remarkable theory of nonlinear oscillations originated by Poincaré, (e.g. Minorsky, *loc. cit.*).

We first convert the system of equations (14.61–14.62) into a single, second order differential equation for X, by differentiating both sides of the equation (14.61) and eliminating Y and $\mathrm{d}Y/\mathrm{d}t$, using the original equations (Lefever and Nicolis, 1970). We also set

$$X(t) = A + x(t) \tag{14.78}$$

In this way, we obtain the following equation for x:

$$\frac{\mathrm{d}^2x}{\mathrm{d}t^2} + \frac{1}{x+A}\left[x^3 + 3Ax^2 + (3A^2 - B - 1)x + A(A^2 - B + 1) - 2\frac{\mathrm{d}x}{\mathrm{d}t}\right]\frac{\mathrm{d}x}{\mathrm{d}t} + x(x+A)^2 = 0 \tag{14.79}$$

This non-linear equation can be further simplified by introducing a new variable ξ through:

$$x = -\frac{A^2\xi}{1 + A\xi}; \quad \xi > -\frac{1}{A} \tag{14.80}$$

We then obtain:

$$\frac{\mathrm{d}^2\xi}{\mathrm{d}t^2} + \left[\frac{A^2}{(A\xi+1)^2} + 2A\xi - B + 1\right]\frac{\mathrm{d}\xi}{\mathrm{d}t} + \frac{A\xi}{1 + A\xi} = 0 \tag{14.81}$$

This equation belongs to a type of nonlinear equation which has been extensively studied by Liénard (1928):

$$\frac{\mathrm{d}^2z}{\mathrm{d}t^2} + f(z)\frac{\mathrm{d}z}{\mathrm{d}t} + g(z) = 0 \tag{14.82}$$

(e.g. Minorsky, 1962, p. 102). For this equation there exists a theorem due to Levinson and Smith (Minorsky, *loc. cit.*) which enumerates the conditions under which the Liénard equation has at least one periodic solution. These conditions imply

$$f(0) < 0$$

In the present case (cf. 14.81):

$$f(0) = A^2 - B + 1 < 0 \tag{14.83}$$

Therefore, beyond the instability corresponding to (14.66) this condition is satisfied.

Further important information about the location of this solution comes from the so-called Bendixon negative criterion (cf. Minorsky, 1962, Chapter III). This theorem states that any periodic solution necessarily must cross the curve

$$\frac{\partial w_X}{\partial X} + \frac{\partial w_Y}{\partial Y} = 0 \tag{14.84}$$

with:

$$w_X = \mathrm{d}X/\mathrm{d}t; \; w_Y = \mathrm{d}Y/\mathrm{d}t$$

For the system (14.61–14.62) this yields the curve:

$$Y = \frac{X}{2} + \frac{B+1}{2X} \tag{14.85}$$

We see that this curve can be in the neighbourhood of the steady state values (14.64), only for B close to the critical value (14.66). Indeed, we verify easily that the sum of the roots of the dispersion equation is:

$$\omega_1 + \omega_2 = \frac{\partial w_{0X}}{\partial X} + \frac{\partial w_{0Y}}{\partial Y} \tag{14.86}$$

where the ω's are the frequencies of the normal modes; the subscript zero in the r.h.s. denotes the steady state. At the marginal state one has

$$\omega_1 + \omega_2 = 0 \tag{14.87}$$

since $\omega_r = 0$. Therefore the curve (14.85) passes through the steady-state point. For $B > B_c$, the curve (14.85) is at finite distance from the steady state. As a result the periodic solution can never be in the immediate neighbourhood of the steady state, except at the point of marginal stability. In this last case there exists an infinite number of periodic trajectories in the immediate neighbourhood of the steady state. This result can be generalized to all models involving two variables (X, Y) and having a point of 'overstability'. According to (14.86) and (14.87) the curve div $\boldsymbol{w} = 0$, then contains the steady-state point. Beyond the marginal state, this curve and therefore also the periodic solution are necessarily at finite distance from the steady state.

6. LIMIT CYCLE

The theory of non-linear oscillations leads to important informations with regard to the periodic solution which appears beyond the instability of the steady state.

Before the instability we have a stable steady-state and the perturbations correspond to complex values of the normal mode frequencies. The situation is described schematically in Figure 9.2. In the usual terminology we have here a stable 'focus'. Beyond the instability the steady state becomes unstable and a stable periodic process called a *limit cycle* occurs. Whatever the initial state, the system approaches in time the same periodic solution whose characteristics, i.e. period and amplitude, are uniquely determined by the non-linear differential equation.

For the chemical examples considered, the characteristics of the periodic processes are therefore uniquely determined by the kinetic constants as well as by the concentrations of the initial and final products.

The marginal stability considered above corresponds to a so-called 'bifurcation point' described by the scheme:

stable limit cycle
↗
stable focus
↘
unstable focus.

We have performed a detailed numerical analysis of our model (Lefever and Nicolis, 1970). It confirms entirely the theoretical predictions.

In Figure 14.4, the trajectories obtained by numerical integration of the kinetic equations have been represented in the phase plane (X, Y), for a large variety of initial conditions corresponding to $A = 1$, $B = 3$.

It is observed that starting from the neighbourhood of the steady state as initial condition, the system attains asymptotically, a closed orbit or limit cycle in (X, Y) space. This implies that for long times, $X(t)$ and $Y(t)$ exhibit periodic undamped oscillations. The characteristics of these oscillations, including their frequency are *independent of the initial conditions*. Indeed starting with initial states as different as

$$(X = Y = 0), (X = Y = 1), (X = 10, Y = 0), (X = 1, Y = 3),$$

we find that for t sufficiently large, the system approaches always the same asymptotic trajectory referred to above. It is also interesting to observe that for the prescribed values of A and B, the approach to the limit cycle is very fast. The general tendency is that, the further a system is in the unstable region, the faster it approaches the limit cycle.

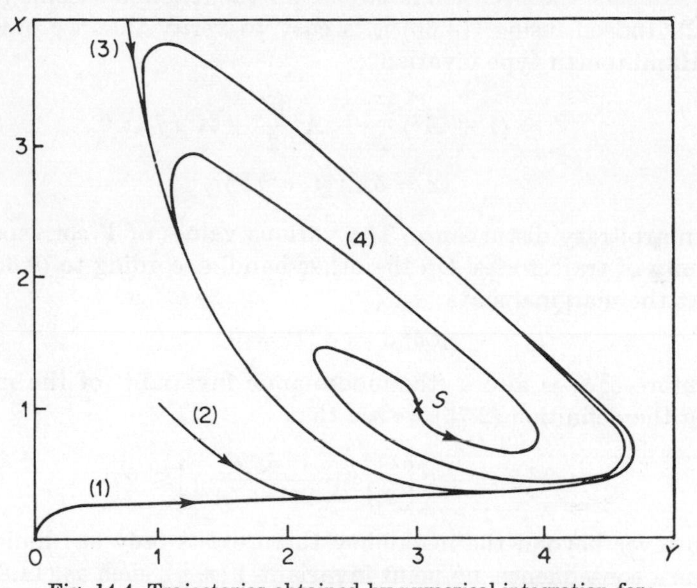

Fig. 14.4. Trajectories obtained by numerical integration for: (1) $X = Y = 0$; (2) $X = Y = 1$; (3) $X = 10$; $Y = 0$; (4) $X = 1$; $Y = 3$.

It is possible to prove that the limit cycle is unique and stable with respect to small fluctuations (Lefever and Nicolis, *loc. cit.*, 1970).

The great conceptual importance of the limit cycle from the thermodynamic point of view comes from its 'ergodicity'. Whatever the initial conditions, the final state is always described by the same periodic trajectory. In this respect, there is a close analogy with the ergodic processes considered in statistical mechanics when the system tends to an equilibrium state, irrespective of the initial condition.

7. COMPARISON BETWEEN THE LOTKA–VOLTERRA MODEL AND LIMIT CYCLE BEHAVIOUR

We have seen that the essential difference between the Lotka–Volterra model considered in §§ 2–3 and the autocatalytic scheme investigated in §§ 4–6, arises from the fact that the Lotka–Volterra mechanism leads to an infinite number of possible periodic motions around the steady state. Such a situation is represented schematically in Figure 9.3. The steady state is then a 'centre'. The trajectories correspond to various values of a Hamiltonian type of invariant (14.30). At the *marginal state* the situation is similar for the reaction scheme (14.61–14.62). Indeed, using (14.66) it is easy to verify that we have also the Hamiltonian type invariant:

$$V = (1 + A^2)\frac{x^2}{2} + A^2\frac{y^2}{2} + A^2xy \geqslant 0 \qquad (14.88)$$

$$(x = \delta X;\ y = \delta Y)$$

for an arbitrary disturbance. The various values of V correspond to a family of trajectories. On the other hand, according to (9.55), one has at the marginal state:

$$d_t\delta_m^2 S = \delta_m P = 0 \qquad (14.89)$$

Therefore $\delta_m^2 S$, is also a 'thermodynamic invariant' of the motion. Using the definition (2.75) we see that

$$\delta_m^2 S = -k\left[\frac{xx^*}{2A} + \frac{Ayy^*}{2(1 + A^2)}\right] \leqslant 0 \qquad (14.90)$$

However beyond the instability there exists only *one* limit cycle and as a consequence, no point invariant $V(x, y)$, such as (14.88) can exist. If such an invariant were to exist, its value would depend on the initial conditions and the evolution could not be ergodic.

The difference between these two types of behaviour corresponding respectively to the Lotka–Volterra model and to the existence of a single limit cycle, is well illustrated by the behaviour of the concentration autocorrelation function as has been shown recently by Lefever and Nicolis (1970).

This function is defined by

$$c(\tau) = \frac{1}{\langle\overline{X^2(t)}\rangle}\langle\overline{X(t)X(t + \tau)}\rangle; \quad c(0) = 1 \qquad (14.91)$$

Here, the brackets $\langle\ \rangle$, correspond to an ensemble average with respect to the initial conditions. On the other hand the bar corresponds to a time average:

$$\overline{f(t)} = \frac{1}{T}\int_0^T f(t)\,dt \qquad (14.92)$$

Lefever and Nicolis (1970) have discussed in detail the corresponding 'spectral function':

$$G(\omega) = \int_0^\infty c(\tau) \cos \omega\tau \, d\tau \tag{14.93}$$

In the case of a limit cycle, one finds, sufficiently far from the instability point, a narrow frequency range around the fundamental frequency. On the contrary, the Lotka–Volterra model provides us with a large dispersion of frequencies. Therefore this model is probably more adequate to represent 'noise' than a well defined time structure.

8. FLUCTUATIONS

In Chapter VIII we have studied the relation between stability theory and fluctuations. It is of special interest to understand how condition (8.1) which expresses that the steady state is 'more probable' than the perturbed state may be reconciled with the growth of fluctuations which characterizes unstable situations (Chapter VIII, § 4).

The study of fluctuations in unstable systems is far from being complete. However interesting results have already been obtained for the Lotka–Volterra model. As we have seen in this chapter, this model has properties very similar to that of unstable systems at the *marginal state*. Therefore, from such a point of view, the study of the fluctuations for the Lotka–Volterra model presents a great interest.

The method used is exactly the same as in Chapter VIII. A partial differential equation can be derived for the reduced moment generating functions (8.19). This leads after a few transformations to (Nicolis, 1970):

$$f(\mathscr{S}) = \exp \frac{(\mathscr{S} - 1)\bar{X}(t)}{1 + (\mathscr{S} - 1)\bar{X}(t)\Delta(t)} \tag{14.94}$$

This formula has to be compared to (8.23); $\bar{X}(t)$ is the time dependent solution for the Lotka–Volterra equations (14.10–14.11). A first difference with (8.23) is that here, the macroscopic chemical equations lead to a time *dependent* solution. But the major difference with (8.23) is in the appearance of the denominator in (14.94). The time dependent quantity $\Delta(t)$ is defined as

$$\Delta(t) = -\int_0^t \frac{A}{\bar{X}(\tau)}\, \mathrm{d}\tau \leqslant 0 \tag{14.95}$$

For small fluctuations ($\mathscr{S} - 1$, small) we recover the usual Poisson distribution around $\bar{X}(t)$. However $\Delta(t)$ increases with time. (In first approximation $\Delta(t)$ is proportional to t.) Therefore, fluctuations *increase*. This is not surprising, as in the Lotka–Volterra model (14.10–14.11), there is no mechanism for the regression of fluctuations. We have here therefore an example where the steady state is stable in respect to small fluctuations, as we obtain (see (8.26)):

$$\rho(X) \sim \mathrm{e}^{-\frac{(\delta X)^2}{2\bar{X}(t)}} \tag{14.96}$$

whereas large fluctuations increase in time. We may expect qualitatively similar results for system presenting a chemical instability.

Because of the existence of the constant of motion (14.30) associated with the Lotka–Volterra model, it has been often suggested that fluctuations may be described in terms of the Gibbs canonical ensemble, valid for equilibrium situations (Kerner, *loc. cit.*). Such an analogy is not confirmed by the Nicolis calculations (*loc. cit.*). Small non equilibrium fluctuations are at least in simple cases correctly represented by the non equilibrium extension (8.6) of Einstein's formula, and are therefore related to thermodynamic quantities. Large amplitude fluctuations require the study of specific models.

9. EXAMPLE OF OSCILLATING SYSTEMS—THE ZHABOTINSKI REACTION

In order to illustrate the conclusions derived in the preceding sections, we shall now consider the behaviour of some well established oscillating systems in organic chemistry and biology. The main purpose is to show that the experimental results are in agreement with the properties of a system functioning *beyond* the marginal stability of the thermodynamic branch.

A particularly interesting class of oscillating systems in organic chemistry, has been described recently by Belusov (1958) and Zhabotinski (1964, 1966). A typical example studied extensively by Zhabotinski, is the oxydation in solution of malonic acid in the presence of cerium sulphate and potassium bromate. The remarkable feature of this process is that it leads to time oscillations in the concentrations of Ce^{3+} and Ce^{4+}, which can easily be followed by spectrometric methods. Such a behaviour is shown in Figure 14.5 (following Degn, 1967).

Although the details of the reaction scheme corresponding to this

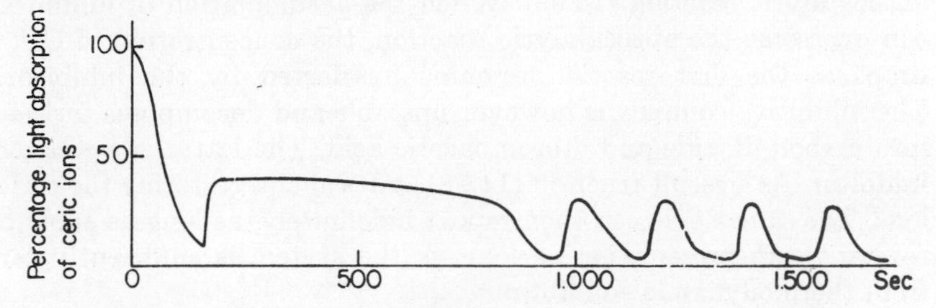

Fig. 14.5. Light absorption of ceric ions (317 mμ) as a function of time. Composition of the mixture: 0·12 m molar ceric sulphate; 0·60 m molar potassium bromate; 48 m molar malonic acid; 3 normal sulphuric acid at 60°C† (m molar = 10^{-3} molar or mole/liter).

behaviour are unknown, it seems that the following three global reactions (Degn, *loc. cit.*, 1967) play an important role:

(*a*) oxydation of the malonic acid

$$CH_2(COOH)_2 + 6Ce^{4+} + 2H_2O \rightarrow 2CO_2 + HCOOH + 6Ce^{3+} + 6H^+ \quad (14.97)$$

(*b*) oxydation of the cerium ions

$$10Ce^{3+} + 2HBrO_3 + 10H^+ \rightarrow 10Ce^{4+} + Br_2 + 6H_2O \quad (14.98)$$

(*c*) transformation of malonic acid into bromomalonic acid

$$CH_2(COOH)_2 + Br_2 \rightarrow CHBr(COOH)_2 \quad (14.99)$$

These reactions permit to understand, at least qualitatively, the basic effects responsible for the oscillations. According to Zhabotinski†, at least one reaction: (14.98), is autocatalytic. During the period of induction (Figure 14.5) this reaction proceeds at the same rate than (14.97). In other words the transformation:

$$Ce^{3+} \rightarrow Ce^{4+}$$

in (14.98) thus compensates the transformation

$$Ce^{4+} \rightarrow Ce^{3+}$$

† See Zhabotinski (1968) for further information about the kinetics and the experimental conditions.

in (14.97), so that the concentration of Ce^{3+} remains constant in the system. Simultaneously however, bromine is formed by reaction (14.98) and combine with malonic acid in (14.99) to give bromomalonic acid and di-bromomalonic acid. Di-bromomalonic acid forms a complex with Ce^{3+}, and therefore acts as an inhibitor of the autocatalytic reaction (14.98). When the accumulation of inhibitor can overcome the autocatalytic reaction, the concentration of Ce^{4+} drops as the first reaction remains unaffected by the inhibitor. The inhibitive complex is however unstable and decomposes further into carbon dioxide and dibromoacetic acid. The latter is a weaker inhibitor. As a result reaction (14.98) starts again, restoring the ceric ions. The same cycle can then repeat indefinitely, as long as enough reactants are present, i.e. as long as the system is sufficiently far from thermodynamic equilibrium.

It has been verified experimentally that the oscillations are completely reproducible under identical experimental conditions in amplitude, form and frequency. Therefore, it seems that they should correspond to a limit cycle around an unstable steady state. In addition we shall see in Chapter XV, § 6, that spatial differentiation can also be observed when this process takes place. This again is an argument in favour of the 'limit cycle' explanation of these oscillations. We shall consider further examples of oscillating chemical reactions in Chapter XV.

In conclusion it may be stated that sequences of chemical reactions operating beyond the instability of the thermodynamic branch may lead to a highly organized and reproducible time structures. In Chapter XV we shall show that a similar conclusion may be drawn for the emergence of space structure under far from equilibrium conditions.

CHAPTER XV

Space Order and Dissipation in Chemical Reactions

1. INTRODUCTION

In the preceding chapter, we have discussed the problem of stability in chemical systems with respect to fluctuations which do not violate spatial uniformity. Both the unperturbed and the perturbed systems we considered were homogeneous. We shall now consider the more general problem of stability with respect to diffusion i.e. to perturbations localized in space.

We shall show that the same effects which were responsible for the occurrence of periodic trajectories such as limit cycles, can also, under almost identical conditions, induce spatial patterns when diffusion is taken into account. Indeed far from thermodynamic equilibrium, the competition between homogeneization of chemical components by free diffusion and spatial localization due to local disturbances of chemical processes involving autocatalytic steps, give rise to instabilities and lead ultimately to the appearance of stable non uniform distributions of matter. We have here an example of *symmetry breaking* transitions, as the state after the transition is less symmetric than the state before.

From our general discussion in Chapter VII, we know that the critical point beyond which spatial differentiation may be possible, corresponds to the situation where the excess entropy production $\delta_m P$ (Chapter IX, § 6) vanishes. However this excess entropy production now contains the contribution of diffusion in addition to that of the chemical reactions.

It is noteworthy to point out that the question of stability with respect to diffusion was first investigated by Turing (1952), in a remarkable paper: 'On the Chemical Basis of Morphogenesis'. This author has in fact proved the existence of symmetry breaking transitions for a number of specific examples. Until recently however, Turing's arguments had only been used in the context of morphogenesis (Maynard-Smith, 1968–69). Our approach shows that the Turing instability is one of the phenomena associated with the breakdown of the thermodynamic branch (Prigogine *et al.*, 1967, 1968, 1969).

2. SYMMETRY BREAKING INSTABILITIES

Let us again consider the scheme (14.46). We neglect in this paragraph all reverse reactions. The kinetic equations become now:

$$\frac{\partial X}{\partial t} = k_1 A + k_2 X^2 Y - k_3 BX - k_4 X + D_X \frac{\partial^2 X}{\partial r^2} \tag{15.1}$$

$$\frac{\partial Y}{\partial t} = k_3 BX - k_2 X^2 Y + D_Y \frac{\partial^2 Y}{\partial r^2} \tag{15.2}$$

The difference with equations (14.61–14.62) is that we now retain diffusion effects. To simplify, we consider a one-dimensional medium with periodic boundary conditions in space.

The time independent homogeneous solution (14.64) corresponding to the thermodynamic branch still exists.

We now consider instead of (14.13), space dependent perturbations. Using r as the geometric coordinate we may write:

$$X - X_0 = x \exp\left(\omega t + \frac{ir}{\lambda}\right) \tag{15.3}$$

$$Y - Y_0 = y \exp\left(\omega t + \frac{ir}{\lambda}\right) \tag{15.4}$$

where λ is the wavelength of the inhomogeneity and:

$$\left|\frac{x}{X_0}\right| \ll 1; \quad \left|\frac{y}{Y_0}\right| \ll 1 \tag{15.5}$$

Taking the kinetic constants equal to unity, we obtain easily as in (14.57) the dispersion equation:

$$\omega^2 + (A^2 + 1 - B + a + b)\omega + A^2(1 + a) + (1 - B)b + ab = 0 \tag{15.6}$$

with

$$a \equiv \frac{D_X}{\lambda^2}; \quad b \equiv \frac{D_Y}{\lambda^2} \tag{15.7}$$

For homogeneous perturbations ($\lambda \to \infty$), we recover the dispersion equation (14.63).

There are now two possibilities for the occurrence of an instability. The first one corresponds to the vanishing of the coefficient of ω, together with a positive value of the independent term. Then the roots are *pure imaginary quantities* and the marginal state corresponds to an *overstability* as in § 4. Condition (14.66) for an instability is replaced by

$$B'_c < B < B''_c; \quad B'_c(\lambda) = 1 + A^2 + a + b \tag{15.8}$$

$$B''_c(\lambda) = (A^2 + b)(1 + a)/b \tag{15.9}$$

The second marginal state corresponds to the vanishing of the independent coefficient in (15.6). Then, one root e.g. ω_1 is necessarily equal to zero, which gives separately

$$\omega_{1r} = 0 \quad \text{and} \quad \omega_{1i} = 0 \tag{15.10}$$

This marginal state corresponds therefore in the present case to an *exchange of stabilities*. The instability condition here becomes

$$B > B''_c \tag{15.11}$$

We then have to look for the critical value of the wavelength λ_c, at which the instability sets in. As in the Bénard problem (see Chapter XI, § 9), let us calculate the wavelength which gives the minimum value of $B'_c(\lambda)$ and $B''_c(\lambda)$, in equations (15.8–15.9). First of all, we observe directly that the minimum value of $B'_c(\lambda)$, is still given by (14.66) ($\lambda \to \infty$). On the other hand, the minimum of $B''_c(\lambda)$, leads immediately to:

$$\lambda_c^2 = \frac{1}{A} (D_X D_Y)^{\frac{1}{2}} \tag{15.12}$$

and upon substitution in (15.9), we get the critical value:

$$B_{cr} = \left[1 + A\left(\frac{D_X}{D_Y}\right)^{\frac{1}{2}}\right]^2 \tag{15.13}$$

The instability which will in fact appear in the system when we increase progressively B, will correspond to the smaller of the two values (14.66) (case of overstability) and (15.13) (case of exchange of stabilities). This depends on the ratio of the diffusion coefficients.

For equal diffusion coefficients, (14.66) will be reached first and we shall obtain a limit cycle. On the contrary for sufficiently small values of D_X/D_Y, we obtain a Turing instability. The importance of unequal diffusion coefficients to obtain symmetry breaking was recently stressed by B. Edelstein (1970). However as pointed out in § 3, there are situations where one finds symmetry breaking instabilities even with equal diffusion coefficient.†

To discuss these expressions it is useful to keep in the equations, the values of kinetic coefficients. Then, (15.12–15.13) have to be replaced by

$$\lambda_c^2 = \left(\frac{k_4}{k_1^2 k_2}\right)^{\frac{1}{2}} \frac{(D_X D_Y)^{\frac{1}{2}}}{A} \tag{15.14}$$

and

$$B_c = \left[\frac{k_1}{k_4}\left(\frac{k_2}{k_3} \cdot \frac{D_X}{D_Y}\right)^{\frac{1}{2}} A + \left(\frac{k_4}{k_3}\right)^{\frac{1}{2}}\right]^2 \tag{15.15}$$

In this way, we see that the instability appears really as a cooperative effect involving both chemical reactions and diffusion. Indeed from equation (15.14), we see that finite inhomogeneities can only arise if the rate of diffusion is comparable with the rate of the chemical reactions. If diffusion as compared to chemical reactions becomes very rapid, the instability occurs for increasing wavelengths, and the system remains practically homogeneous.

Let us now indicate briefly three other typical mechanisms which may also present symmetry breaking transitions. The first is:

$$\begin{aligned} A + X &\to 2X & &\text{(a)} \\ X + Y &\to 2Y & &\text{(b)} \\ Y + V &\to V' & &\text{(c)} \\ V' &\to E + V & &\text{(d)} \end{aligned} \tag{15.16}$$

The overall reaction is

$$A \to E \tag{15.17}$$

Here only mono and bimolecular steps are involved. This scheme is a modification of the Lotka–Volterra model (14.4). We have simply

† Recently Othmer and Scriven, have presented an exhaustive algebraic discussion of all situations which may arise in the case of two interdiffusing and reacting species (1969).

added the 'appendix' (15.16c–d), involving the compound V'. Such an appendix plays also a role in the next scheme:

$$\begin{aligned}
A &\rightarrow X && \text{(a)} \\
X + Y &\rightarrow C && \text{(b)} \\
C &\rightarrow D && \text{(c)} \\
B + C &\rightarrow Y + C && \text{(d)} \\
Y &\rightarrow E && \text{(e)} \\
Y + V &\rightarrow V' && \text{(f)} \\
V' &\rightarrow E + V && \text{(g)}
\end{aligned} \tag{15.18}$$

We have here the two overall reactions:

$$\begin{aligned}
A + B &\rightarrow D \\
B &\rightarrow E
\end{aligned} \tag{15.19}$$

This is a simplified form of Turing's original scheme, which is

$$\begin{aligned}
A &\rightarrow X && \text{(a)} \\
X + Y &\rightleftharpoons C && \text{(b)} \\
C &\rightarrow D && \text{(c)} \\
B + C &\rightarrow W && \text{(d)} \\
W &\rightarrow Y + C && \text{(e)} \\
Y &\rightarrow E && \text{(f)} \\
Y + V &\rightarrow V' && \text{(g)} \\
V' &\rightarrow E + W && \text{(h)}
\end{aligned} \tag{15.20}$$

It differs from (15.18) only by the presence of the intermediate component W. Also the step (b) is supposed to be reversible in (15.20).

One may briefly describe schemes (15.18) and (15.20) as processes which transform the initial products A, B into the final (or 'waste') products D, E, through the intermediate products X, Y, under the catalytic action of C, V, V', W, and according to the scheme:

$$\begin{array}{l}
A \rightarrow X \rightarrow C \rightarrow D \\
\qquad\qquad\quad \nearrow \\
B \rightarrow Y \xrightarrow{\qquad\quad} E
\end{array} \tag{15.21}$$

For more than two variable components X, Y, the problem becomes much more involved.

3. THERMODYNAMIC INTERPRETATION OF SYMMETRY BREAKING INSTABILITIES

As we have shown in (7.62), the explicit condition for stability is:

$$P[\delta S] = \int \sum_{\alpha} \delta J_{\alpha} \, \delta X_{\alpha} \, \mathrm{d}V > 0 \qquad (15.22)$$

This form allows us to analyse the physical mechanism governing the onset of an instability†. We shall discuss two examples.

We first consider the scheme (14.46). As for equations (15.1–15.2), we neglect the reverse reactions but retain diffusion. We then obtain for X and Y, taking into account the steady state value (14.64), the relation (taking $V = 1$):

$$P[\delta S] = \frac{1 - B}{A} x^2 + \frac{A^3}{B} y^2 + \frac{D_X}{\lambda^2 A} x^2 + \frac{D_Y}{\lambda^2 B} y^2 \qquad (15.23)$$

We have taken the kinetic constants as well as $\mathscr{R}T$ equal to unity. Two interesting remarks may be made:

(1) In agreement with our general discussion a negative term $-B/Ax^2$, due to the autocatalytic action of X, appears. This is the *dangerous contribution*, in respect to stability (Chapter VII, § 4).

(2) The explicit contribution of diffusion is positive and proportional to D/λ^2. Therefore, if there is an instability, increasing values of D must give rise to increasing values of the critical wavelength λ_c; if λ_c, were to remain constant, the contribution of diffusion to (15.23) would become dominant and $P[\delta S]$, would be always positive. This is also in agreement with our formula (15.12) for λ_c derived from the dispersion equation. But diffusion has a second role: the manifold of perturbations which we may introduce into (15.23) is now increased by the consideration of inhomogeneous systems.

It is easy to verify that the perturbations x, y, which satisfy the linearized perturbation equations at the marginal state ($\omega_r = \omega_i = 0$), lead to the vanishing of $P[\delta S]$. Indeed in this case

$$y = -\frac{1 + D_X/\lambda^2}{D_Y/\lambda^2} \cdot x$$

and therefore

$$\{P[\delta S]\}_{B=B_{cr}} = 0 \qquad (15.24)$$

The excess entropy production indeed vanishes.

† We ignore here $T_{,j}$ and v_l in (2.21).

Let us now consider briefly as a second example, the scheme (15.18). It is easy to verify that the excess entropy production is now:

$$P[\delta S] = \left(\sqrt{\frac{Y_0}{X_0}}\,x + \sqrt{\frac{X_0}{Y_0}}\,y\right)^2 + \left(\sqrt{\frac{V_0}{Y_0}}\,y + \sqrt{\frac{Y_0}{V_0}}\,v\right)^2$$
$$+ \frac{c^2}{C_0} + \frac{y^2}{Y_0} + \frac{v'^2}{V_0'} - \frac{V_0}{V_0'}\,yv' - \left(\frac{1}{V_0} + \frac{Y_0}{V_0'}\right) vv'$$
$$- \frac{Y_0}{C}\,xc - \left(\frac{B}{Y_0} + \frac{X_0}{C_0}\right) cy$$
$$+ \frac{1}{\lambda^2}\sum_j \frac{D_j}{j_0}(\delta j)^2 \quad (j = X, Y, V, V', C) \tag{15.25}$$

where the last term represents the diffusion contribution. Let us find out in this expression the terms introducing a dangerous contribution to the stability condition. First of all, we may disregard the positive definite quantities. Likewise, it is easy to verify that the group of terms arising from the appendix $Y + V \rightarrow V' \rightarrow E + V$, that is

$$\left(\sqrt{\frac{V_0}{Y_0}}\,y + \sqrt{\frac{Y_0}{V_0}}\,v\right)^2 + \frac{v'^2}{V_0'} - \frac{V_0}{V_0'}\,yv' - \left(\frac{1}{V_0} + \frac{Y_0}{V_0'}\right) vv'$$

is separately a positive definite quadratic form, which may therefore be also disregarded. Finally, at the marginal state ($\omega_r = \omega_i = 0$), the linearized perturbation equations yield the following relations:

$$y = -\left(Y_0 + \frac{D_X}{\lambda^2}\right)\frac{x}{X_0}; \quad c = -\frac{D_X}{\lambda^2}\,x$$

This shows that the remaining terms of (15.25), namely:

$$-\frac{Y_0}{C_0}xc \quad \text{and} \quad -\left(\frac{B}{Y_0} + \frac{X_0}{C_0}\right) cy$$

are respectively positive and negative quantities. Therefore, only this last term is a possible source of instability. In other words, we are here concerned with an instability generated by a cross-catalytic effect between Y and C.

It is clear that there are many reaction schemes which may lead to such negative contributions. We shall consider some other examples of biochemical interest in § 7.

As we have noticed, the introduction of an *appendix* (15.16(c), (d)) may play an important role and may permit the occurrence of instability. However it gives a positive contribution to the excess entropy production. It seems therefore to play a role somewhat similar to diffusion, which also gives a positive contribution to the excess entropy production. It leads to a destabilization of the system by increasing the manifold of possible perturbations.

In biochemical reactions such an appendix is quite common; V would be an enzyme and the appendix then corresponds to the classical Michaelis–Menten mechanism (Mahler and Cordes, 1966).

4. THERMODYNAMIC THRESHOLD FOR SYMMETRY BREAKING INSTABILITIES

Exactly as we did in Chapter XIV, § 4 we may also calculate the critical affinity for the instability with respect to diffusion. Instead of (14.57) we now find:

$$\left(\frac{B}{D}\right)_c = R_c = \frac{1}{k^3\gamma A^2 D}\{2k^2[\gamma(1+k)]^{\frac{1}{2}} \times [2k^9 D^3 + 5k^6 A^2 D^2 + 4k^3 A^4 D + A^6]^{\frac{1}{2}} + k^7 D^2 + k^4[2A^2 + k^2\gamma(1+k) + 3k\gamma A^2]D + kA^2[A^2 + k^2\gamma(1+k) + k\gamma A^2]\} \tag{15.26}$$

with:

$$\gamma = \frac{D_Y}{D_X}$$

The corresponding critical wavelength is:

$$\lambda^4 = \frac{k^3\gamma(2k^3 D + A^2)D_X^2}{(k^7 D^2 + 2k^4 A^2 D + kA^4)(1+k)} \tag{15.27}$$

Contrarily to what we observed in Chapter XIV, § 4, for the critical affinity regarding sustained oscillations, the expression (15.27) is not limited to a given range of values for D. Therefore one may expect that for $k^3 D > A^2$, there can be no interference between the two effects, (see for more details Lefever, 1970). The system can become unstable only for *inhomogeneous perturbations* even if the diffusion coefficients are equal.

The comparison of the results cited here and in Chapter XIV, shows some of the various forms of instability phenomena may take.

One of the two types of instability which may occur in the scheme studied in § 2 corresponds to the bifurcation point as defined in Chapter XIV, § 6. For the other one, we have exchange of stabilities and the imaginary part of the frequency vanishes. The unstable steady state is then a 'node' (Figure 9.1). We have therefore a second type of *bifurcation* point:

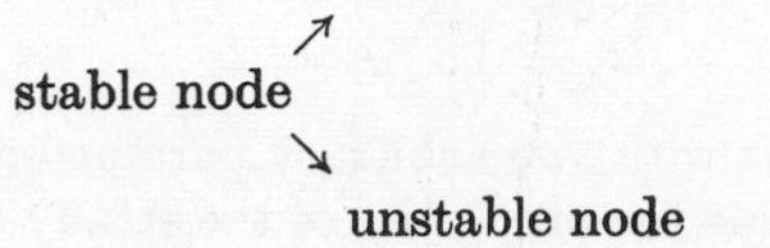

It is the space structure beyond such an instability that we shall study now.

5. DISSIPATIVE SPACE STRUCTURES

The computational apparatus necessary for the study of the time evolution beyond the instability is rather cumbersome. We shall therefore concentrate ourselves again on the scheme (15.1–15.2) which provides us with a simple example. We shall make a further simplification, considering instead of disturbances of arbitrarily wavelength, a model formed by two uniform and identical boxes in the one dimensional representation assumed in § 2. The initial and final products are distributed homogeneously, while X and Y may diffuse freely between the two boxes. Instead of (15.1–15.2) we now have four equations:

$$\frac{dX_1}{dt} = A + X_1^2 Y_1 - BX_1 - X_1 + D_X(X_2 - X_1)$$

$$\frac{dY_1}{dt} = BX_1 - X_1^2 Y_1 + D_Y(Y_2 - Y_1)$$

$$\frac{dX_2}{dt} = A + X_2^2 Y_2 - BX_2 - X_2 + D_X(X_1 - X_2)$$

$$\frac{dY_2}{dt} = BX_2 - X_2^2 Y_2 + D_Y(Y_1 - Y_2) \tag{15.28}$$

The first two refer to box 1; the last two, to box 2. All direct kinetic constants are taken equal to one, while the reverse processes are neglected.

As in (14.64) we then have:

$$X_i = A; \quad Y_i = \frac{B}{A} \quad (i = 1, 2) \tag{15.29}$$

This is the only time independent, homogeneous solution of the system (15.28).

Let us adopt the following numerical values:

$$D_X = 1; \quad A = 2 \tag{15.30}$$

As a result we are left with two arbitrary parameters i.e. D_Y and B, whose values determine the properties of the steady states. Elementary manipulations lead to the following system of equations for the steady state:

$$3X_2^5 - 30X_2^4 + [96 + 2D_Y(B + 3)]X_2^3 - [96 + 12D_Y(B + 3)]X_2^2 + 16D_Y(B + 6)X_2 - 96D_Y = 0 \tag{15.31}$$

$$X_1 = 4 - X_2 \tag{15.32}$$

$$Y_2 = \frac{B(8X_2^2 - 4D_Y - 16X_2 - X_2^3)}{8X_2^3 - 3X_2^2(D_Y + 8) + 8D_Y(X_2 - 2) - X_2^4} \tag{15.33}$$

$$Y_1 = Y_2 + (X_2^2 Y_2 - BX_2)/D_Y \tag{15.34}$$

This system has two types of solutions. On one hand, the homogeneous steady state solution given by (15.29), and on the other hand, an inhomogeneous solution which may be written in two equivalent ways, the model being symmetrical:

$$X_1 > X_2; \quad Y_1 < Y_2 \quad \text{or} \quad X_1 < X_2; \quad Y_1 > Y_2 \tag{15.35}$$

The stability analysis of the steady state solutions of (15.31), (15.34), has been performed both for homogeneous and inhomogeneous fluctuations (Lefever, 1968). One finds that the homogeneous state is unstable in respect to homogeneous perturbations when:

$$B > 5 \tag{15.36}$$

and in respect to inhomogeneous perturbations when

$$B > B_c = 3\,\frac{D_Y + 2}{D_Y} \tag{15.37}$$

The results are represented on Figure 15.1.

Conditions (15.36–15.37) define the domain I where only the stable homogeneous steady state exists. This state becomes unstable with respect to homogeneous perturbations in II and with respect to inhomogeneous perturbations beyond the curve (b). In regions II,

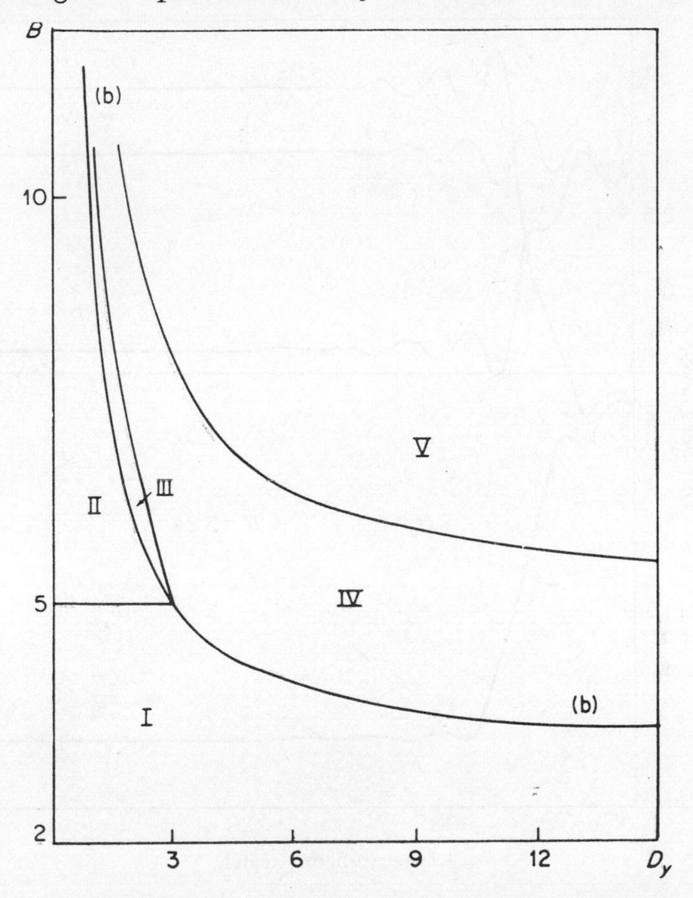

Fig. 15.1. Stability regions and steady states as a function of B and D_Y.

III, V no time independent stable state exists. On the contrary in region IV, the inhomogeneous steady state is stable. This region corresponds therefore to the so-called dissipative space structure.

One may thus see that Figure 15.1 permits us to divide the unstable region into a set of domains where either spatial patterns or a time dependent regime corresponding to a limit cycle may be expected. Observe that the transition from one domain to another may occur by a very slight modification of either the chemical parameter B or the diffusion coefficient D_Y. Figure 15.1 represents a kind of *non equilibrium phase diagram.*

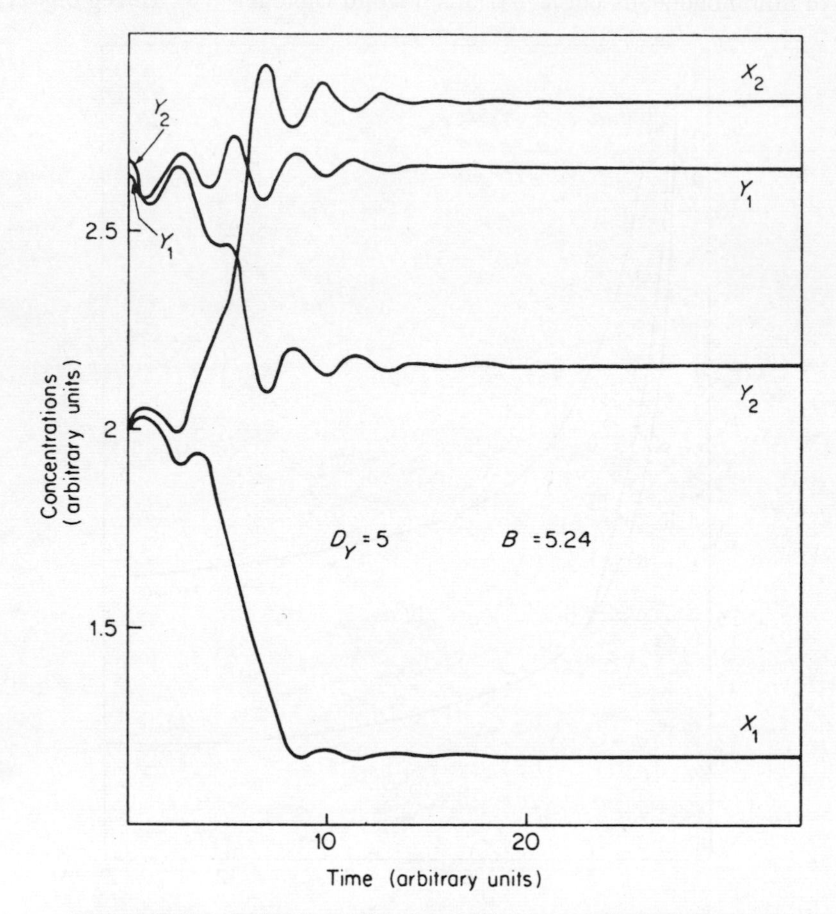

Fig. 15.2. A perturbation of Y in box $2(Y_2)$, around the homogeneous state increases the rate of production of X in that box (X_2), due to the autocatalytic step. This effect grows till the new state on the right.

It is of special interest to investigate how the inhomogeneous state is reached. A typical result is reported in Figure 15.2.

The homogeneous state corresponding to $X_1 = X_2 = 2$, and $Y_1 = Y_2 = 2{\cdot}62$, is destroyed by a small fluctuation $Y_2 - Y_1 = 0{\cdot}04$. It is clearly seen on Figure 15.2 how this initial perturbation is magnified till the inhomogeneous steady state is reached. The configuration (15.35) chosen by the system depends crucially on the nature of the initial perturbation. As in our thermodynamic discussion in Chapter VII, § 4, autocatalysis is here again the determinating factor. Indeed, the concentration of X will be highest in the box where the fluctuation induces an increase of the rate of the autocatalytic step. For example if one admits that the perturbations (x, y) are positive and act in one of the following ways:

$$X_1 + x; \quad Y_1 + y; \quad X_2 - x; \quad Y_2 - y \tag{15.38}$$

it is the configuration $X_1 > X_2$, $Y_1 < Y_2$, which will be realized. In this description it therefore occurs a statistical element. In addition to the 'causal laws' (15.28) we have to know in which box the fluctuation will occur. This in turn determines the choice between the two solutions (15.35) and the subsequent evolution of the system. This choice appears therefore as a primitive type of information which has to be fed into the system and which, in addition to the causal laws, determines the future evolution of the system.

It is also interesting to note that for $D_Y > 3$, the system is unable to find the stable inhomogeneous state. After some time, one observes oscillations of constant amplitude, but the system remains homogeneous.

In Figure 15.3 we see how an initial perturbation around the inhomogeneous state may lead to the exchange of the two inhomogeneous solutions corresponding to the two boxes problem.

An important feature which must be emphasized is that *small fluctuations can no longer reverse the configurations*. Indeed, reversing occurs only if the perturbations of the steady state concentrations are of the same order of magnitude as the difference in concentration between the two boxes. This is far beyond 'spontaneous' thermal fluctuations. As a consequence the possibility of an oscillatory behaviour between the two accessible configurations due to random disturbances, is ruled out.

One can verify that the results obtained with the two boxes model, remain valid when diffusion is represented in a more realistic way. As an improvement, we have studied the steady state distribution for a large number of subsystems. Results are shown in Figure 15.4 and 15.5. The numerical values taken in Figures 15.4–15.5 are:

$$D_X = 0{\cdot}0016; \quad D_Y = 0{\cdot}008; \quad A = 2; \quad B = 5{\cdot}24$$

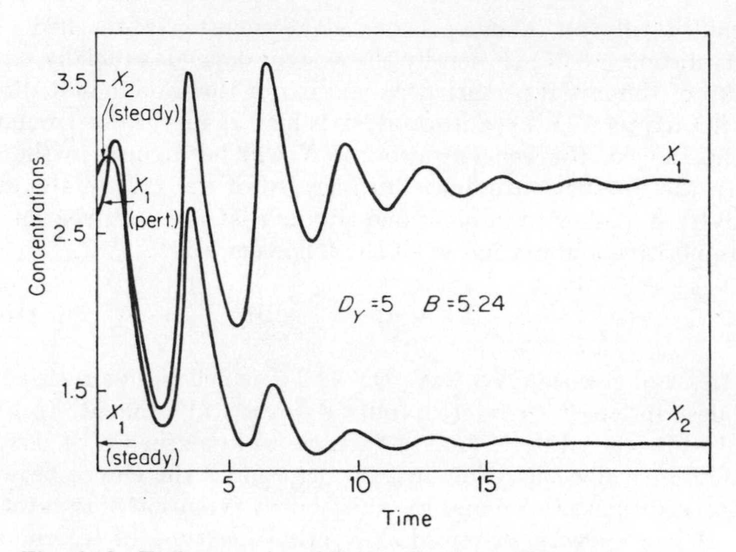

Fig. 15.3. Exchange of stabilities—At the initial time one has $X_1 = 2{\cdot}600$ and $X_2 = 2{\cdot}829$, while the steady state values are given by $X_1 = 1{\cdot}170$ and $X_2 = 2{\cdot}829$.

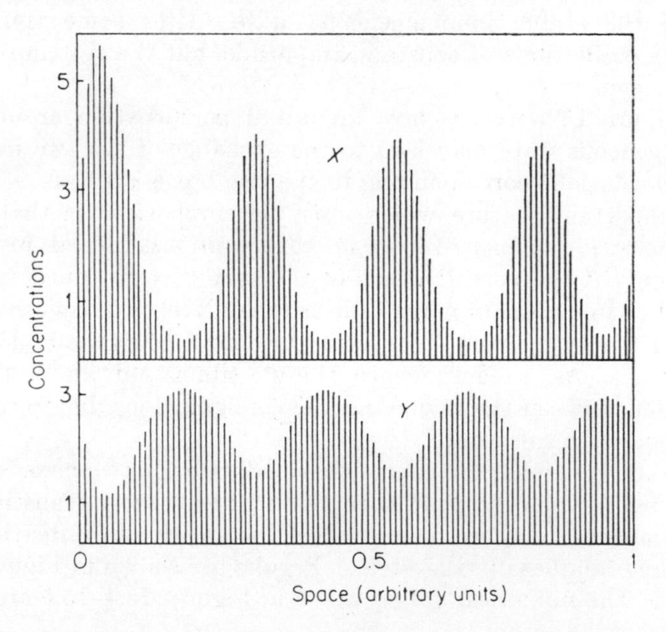

Fig. 15.4. Steady state distributions for a series of 100 boxes with fixed concentrations $X = 2$ and $Y = 2{\cdot}62$ at the boundaries.

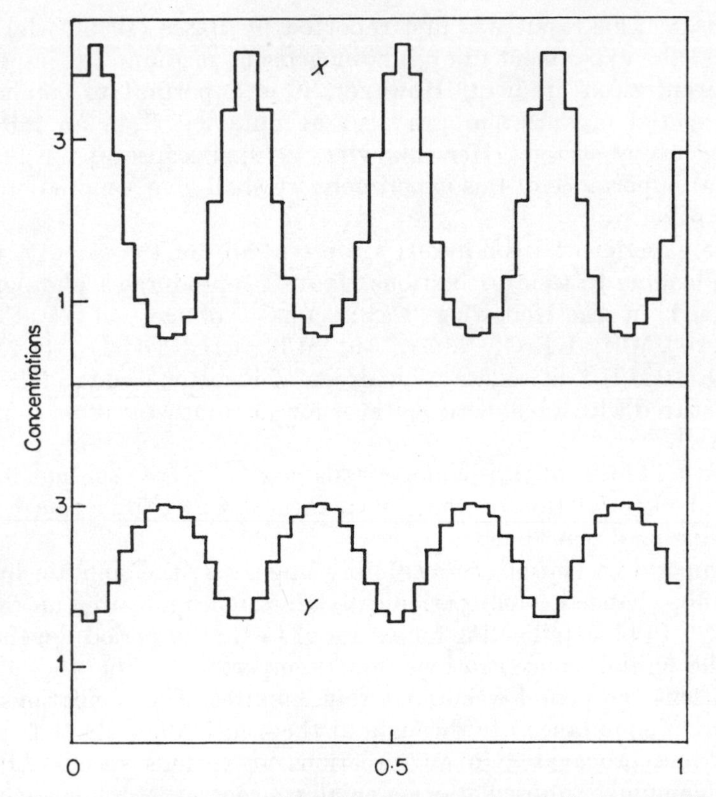

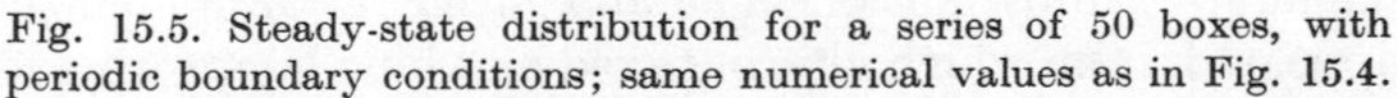

Fig. 15.5. Steady-state distribution for a series of 50 boxes, with periodic boundary conditions; same numerical values as in Fig. 15.4.

In both cases we observe a succession of regions where alternatively X or Y dominate. We really have here a striking example of a dissipative structure created under the action of far from equilibrium conditions. It occurs as a giant non regressing fluctuation, stabilized through the flow of energy and matter exchanged with the outside world.

Let us now consider an example of chemical dissipative structure which has been observed experimentally.

6. EXAMPLES OF A DISSIPATIVE SPACE STRUCTURE. THE ZHABOTINSKI REACTION

Chemical dissipative space structures have been observed experimentally in the case of the Zhabotinski reaction described in Chapter

XIV, § 8. This result was first reported by Busse (1969), who performed the experiment in an inhomogeneous medium, i.e. imposing a concentration gradient. However, it is important to emphasize that spatial organization can also be obtained from an initially homogeneous system (Herschkowitz, 1970). Because of the fundamental importance of this experiment we shall give a more detailed description.

The experimental conditions are essentially of the same type as those leading to time oscillations. Plate I reproduces a photograph obtained in the following way: equal volumes of $Ce_2(SO_4)_3$, $(4 \times 10^{-3}M/l)$; $KBrO_3$, $(3{\cdot}5 \times 10^{-1}M/l)$; $CH_2(COOH)_2$, $(1{\cdot}2M/l)$; $H_2SO_4(1{\cdot}5M/l)$;† as well as a few drops of Ferroïne (redox indicator) were stirred with a magnetic agitator for 30 minutes at room temperature.

Two milliliters of this homogeneous mixture were then put into a test tube kept at the constant temperature of 21°C by a thermostat and stirring discontinued.

Temporal oscillations immediately appeared; the solution in the test tube changed colour periodically from red, indicating an excess of Ce^{3+}, to blue, indicating an excess of Ce^{4+}, the period depending on the initial concentrations and temperature. For the above conditions the period was about four minutes. The oscillations did not occur simultaneously throughout the solution but started at one point and propagated in all directions at various speeds. After a variable number of oscillations a small concentration inhomogeneity then appeared, from which alternate red and blue layers were formed one by one. This evolution is shown in the series of photographs 1 to 6 Plate II. During the formation of these layers, time oscillations continued to be observed in the part of the solution where the structure had not been established.

As the reaction is going on in a closed system the structure is only maintained for a limited time (about 30 minutes) after which the system approaches equilibrium, going back to a homogeneous distribution of matter.

The experiment was repeated at different temperatures. The dissipative structures were always reproduced; only their time of appearance and subsequent life-time varied. It would be of interest to repeat the experiments in steady-state open systems; some experiments of this type have been reported by Zhabotinski, (1968).

† M/l, means: mole per liter.

PLATE I

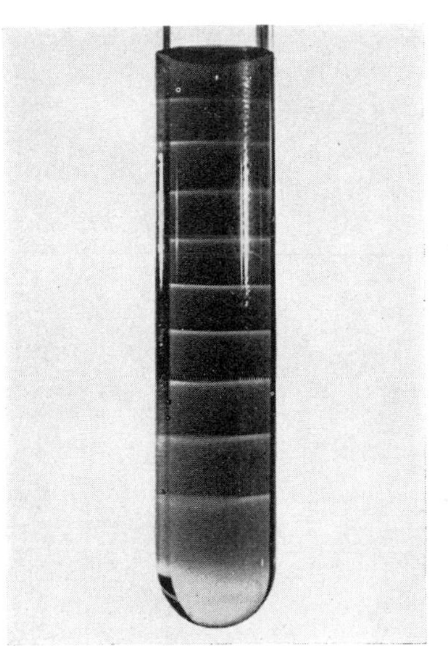

Dissipative structure in the Zhabotinski reaction

(*Facing page 262*)

PLATE II

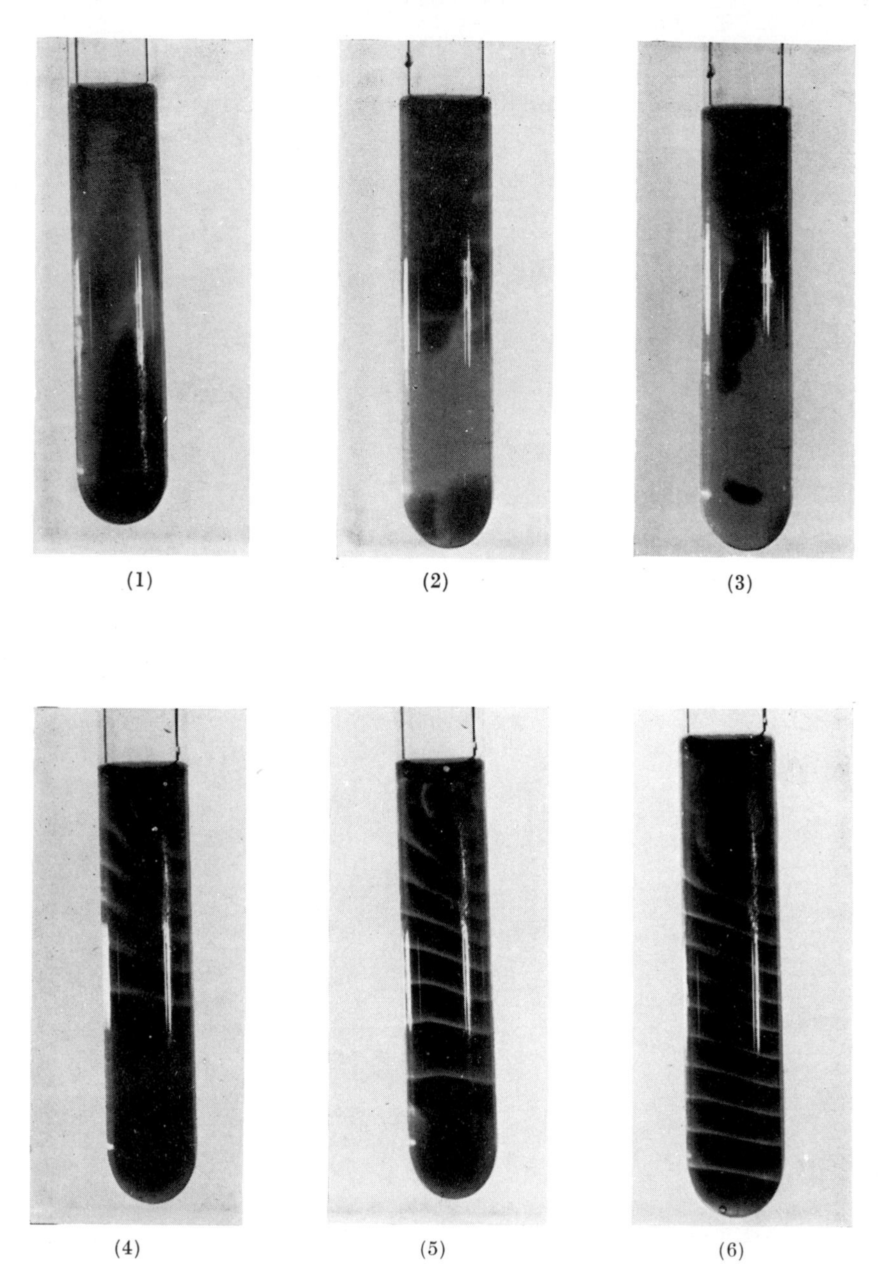

Establishment of the dissipative space structure

The ring pattern observed in the experiment described above was formed starting from a homogeneous medium and always appeared after an oscillatory state. This seems to be supported by the following observation: if one performs the experiment with initial concentrations outside the domain of existence of time oscillations determined by Zhabotinski, the system remains homogeneous and no rings are observed. A range of concentrations where the spatial structure would be established without oscillation, i.e. by exchange of stabilities, as in the case of the Bénard's cells (Chapter XI), is no excluded, but has not yet been observed.

The relation between dissipative space structure and time oscillations is here very intimate. The space structure seems to result from an instability of the limit cycle itself. The following interpretation could be given: far from equilibrium, the homogeneous solution is unstable with respect to small spontaneous fluctuations. These fluctuations do not regress as would be the case near a stable thermodynamic state, but are rather amplified until a new 'quasi-periodic' state is achieved. The latter becomes in turn unstable as soon as diffusion becomes important; then a dissipative space structure appears. Clearly this situation can only be maintained through a subtle balance between reaction rates and diffusion.

7. LIMIT CYCLES AND DISSIPATIVE STRUCTURES IN MULTI-ENZYMATIC REACTIONS

It is of obvious interest to discuss the possiblity of limit cycles and dissipative space structures for biochemical reactions. Oscillations appear in living systems, at widely distinct levels and with widely different properties and frequency values. For example, oscillations have been recorded at the molecular level (oscillations of the concentrations of some metabolites of enzymatic reactions), at the cellular level (they can then be related to the mechanism of genetic induction and repression described by Monod and Jacob; Goodwin, 1963), or at the supracellular level (circadian rythms). These last phenomena have however very long periods and are probably not related exclusively to chemical effects.

Such phenomena are in fact so abundant, that it lies beyond the scope of this book to present a systematic discussion of the properties of biological rythms and their implications as far as space differentiation is concerned. Within the framework of enzymatic reactions however, three characteristic examples presenting widely different catalytic properties have been considered recently from that point of view† (Prigogine, Lefever, Goldbeter, Herschkowitz, 1969):

(*a*) the model of Chernavskaia and Chernavskii (1961) for the *dark part of photosynthesis*;

(*b*) a substrate and product inhibited enzymatic reaction (Sel'kov, 1966);

(*c*) the product activated enzymatic reaction of *phosphofructokinase in the glycolytic cycle* (Higgins, 1964; Sel'kov, 1968).

Each of these processes has been mainly studied, in connection with the problem of *sustained chemical oscillations*. In each case a model yielding satisfactory results has been proposed. We shall consider here these last two examples.

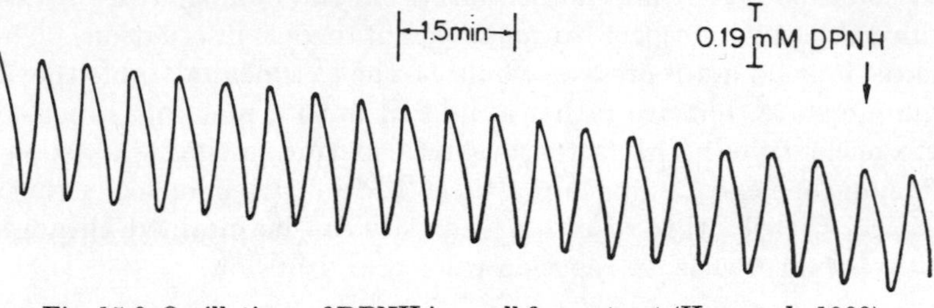

Fig. 15.6. Oscillations of DPNH in a cell free extract (Hess *et al.*, 1966). 1mM = 10^{-3} mole

Damped oscillations in the glycolytic intermediates have been observed by Chance (1964), Duysens (1957), Hommes (1964), Hess (1963) and many others, in yeast cells and muscles. Sustained oscillations have also been described (Chance *et al.*, 1965–1966). A good account of the present situation can be found in the paper of Hess and Boiteux (1968). An example corresponding to the oscillations of DPNH is represented on Figure 15.6 (Hess *et al.*, 1966).

Many enzymes of the cycle could *a priori* be responsible for the observed periodicities. According to most authors, it seems however that the enzyme phosphofructokinase plays the major role. This enzyme is allosteric, i.e., it is made up of equivalent units which possess specific reaction sites for the fixation of substrate and product. Each unit exists in two conformational states: one inactive and one

† The importance of the coupling between diffusion and polymerization processes at the stage of prebiological development has been discussed recently by Katchalsky (1969), who has also stressed the role of non-equilibrium conditions in membrane processes (Katchalsky and Spangler, 1968).

active, having more affinity for the substrate. The conformational equilibrium is concerted, i.e. if one of the units forming a given molecule changes its conformation, then simultaneously the other units do so as well.

Moreover a remarkable property of phosphofructokinase lies in the fact that its products of reaction (FDP or ADP) displace the conformational equilibrium in favour of the active form of the enzyme.† This is clearly a *destabilizing* contribution to the excess entropy production.

The first model for oscillations in the glycolytic cycle, based on the allosteric properties of the enzyme, was proposed by Higgins (1964, 1967). Let us summarize it briefly:

$$\begin{aligned} A &\rightarrow C_1 && \text{(a)} \\ C_1 + D_1 &\rightarrow D_2 && \text{(b)} \\ D_2 &\rightarrow C_2 + D_1 && \text{(c)} \\ C_2 + D_3 &\rightleftharpoons D_1 && \text{(d)} \\ C_2 &\rightarrow F && \text{(e)} \end{aligned} \qquad (15.39)$$

where the symbols denote:

A: initial product: glucose
F: final product: glyceraldehyde $3P$
D_1: active form of the enzyme
D_3: inactive form
D_2: enzymatic complex
C_1: fructose $1P$
C_2: fructose $2P$

The important step is the conformational equilibrium (15.39(d)), where FDP activates the enzyme. With arbitrary values for the kinetic parameters, Higgins obtained for this model, oscillations at the analogic computer.

Further experimental results showed however that under physiological conditions,‡ the enzyme is controlled mainly by the couple (ATP/ADP) and not by ($F1P/FDP$). A revised model taking into

† For a discussion of the allosteric properties of this enzyme see the paper by Monod, Buc and Blangy (1968).

‡ Sel'kov has shown that the Higgins' model can then be reduced to a Lotka–Volterra type scheme.

account most experimental facts has then be presented by Sel'kov (1968).

Let us study the conditions under which this model presents oscillations and thereafter compare with the condition of instability with respect to diffusion:†

$$\xrightarrow{\omega_1} A_3 \qquad \text{(a)}$$

$$A_3 + D_1 \underset{k_{-1}}{\overset{k_{+1}}{\rightleftharpoons}} D_2 \qquad \text{(b)}$$

$$D_2 \xrightarrow{k_{+2}} A_2 + D_1 \qquad \text{(c)} \qquad (15.40)$$

$$\gamma A_2 + D_3 \underset{k_{-3}}{\overset{k_{+3}}{\rightleftharpoons}} D_1 \qquad \text{(d)}$$

$$A_2 \xrightarrow{k_2} \qquad \text{(e)}$$

The reactant A_3 (ATP) enters the system at constant rate ω_1 [step (a)]. D_1, D_2, D_3 have the same meaning as in Higgins' case. A_2 (ADP) is the product of reaction and also the activator (step 15.40(d)). The factor γ is a convenient way to express the fact that the fixation of A_2 on the enzyme activates more than one reacting site because of the concerted conformational equilibrium. The kinetic equations corresponding to the model are:

$$\frac{\partial A_3}{\partial t} = \omega_1 - k_{+1}A_3D_1 + k_{-1}D_2 + D_{A_3}\frac{\partial^2 A_3}{\partial r^2} \qquad (15.41)$$

$$\frac{\partial A_2}{\partial t} = k_{+2}D_2 - k_{+3}A_2^{\gamma}D_3 + k_{-3}D_1 - k_2A_2 + D_{A_2}\frac{\partial^2 A_2}{\partial r^2} \qquad (15.42)$$

$$\frac{\mathrm{d}D_1}{\mathrm{d}t} = -k_{+1}A_3D_1 + (k_{-1} + k_{+2})D_2 + k_{+3}A_2^{\gamma}D_3 - k_{-3}D_1 \qquad (15.43)$$

$$\frac{\mathrm{d}D_3}{\mathrm{d}t} = -\,k_{+3}A_2^{\gamma}D_3 + k_{-3}D_1 \qquad (15.44)$$

$$\frac{\mathrm{d}D_2}{\mathrm{d}t} = k_{+1}A_3D_1 - (k_{-1} + k_{+2})D_2 \qquad (15.45)$$

† (a) means simply: injection of A_3 with the given rate ω_1.

Diffusion of the enzyme is not taken into account since it is much slower than for ATP and ADP. Moreover in agreement with experimental data, the following conditions are assumed:

$$\frac{k_{+1}}{A_3}, \quad k_{-1}, \quad k_{+2}, \quad \frac{k_{+3}}{A_2^{\gamma}}, \quad k_{-3} \gg 1;$$

$$\frac{A_3}{D_0}, \; \frac{A_2}{D_0} \gg 1; \quad D_0 = D_1 + D_2 + D_3 \tag{15.46}$$

where D_0 is a constant. This permits us to transform the three last kinetic equations into algebraic equations, and to eliminate from (15.41–15.42) the enzymatic forms D_1, D_2, D_3.

On the other hand, sustained oscillations are possible in glycolysis at very low rates of entrance of ATP ($\omega_1 \ll 1$). Under these conditions the kinetic equations can be rewritten in the simplified form:

$$\frac{\partial A_3}{\partial t} = \omega_1 - \alpha_1 D_0 \alpha_3^{\gamma} A_3 A_2^{\gamma} + D_{A_3} \frac{\partial^2 A_3}{\partial r^2} \tag{15.47}$$

$$\frac{\partial A_2}{\partial t} = \alpha_1 D_0 \alpha_3^{\gamma} A_3 \, A_2^{\gamma} - k_2 A_2 + D_{A_2} \frac{\partial^2 A_2}{\partial r^2} \tag{15.48}$$

with

$$\alpha_1 = \frac{k_{+1} \,.\, k_{+2}}{k_{-1} + k_{+2}}; \quad \alpha_3^{\gamma} = \frac{k_{+3}}{k_{-3}}; \quad \alpha_5 = \alpha_1 D_0 \alpha_3^{\gamma} \tag{15.49}$$

The steady state is given by:

$$A_2^0 = \frac{\omega_1}{k_2}$$

$$A_3^0 = \frac{\omega_1^{(1-\gamma)}(k_{-1} + k_{+2}) k_2^{\gamma} k_{-3}}{k_{+1} \,.\, k_{+2} \,.\, k_{+3} D_0} \tag{15.50}$$

Linearization of the perturbation equations around this state yields the dispersion relation:

$$\omega^2 + \left[\alpha_5 (A_2^0)^{\gamma} + k_2 - \gamma \alpha_5 A_3^0 (A_2^0)^{\gamma-1} + \frac{D_{A_2} + D_{A_3}}{\lambda^2}\right] \omega$$
$$+ \alpha_5 (A_2^0)^{\gamma} \left(k_2 + \frac{D_{A_2}}{\lambda^2}\right) + \frac{D_{A_3}}{\lambda^2} \left[k_2 + \frac{D_{A_2}}{\lambda^2}\right.$$
$$\left. - \gamma \alpha_5 A_3^0 (A_2^0)^{\gamma-1}\right] = 0 \tag{15.51}$$

The condition for instability with respect to homogeneous and inhomogeneous perturbations are easily found. For homogeneous perturbations it becomes:

$$w_1^\gamma < (w_{1c}')^\gamma = \frac{k_{-1} + k_{+2}}{k_{+1} \cdot k_{+2}} \cdot \frac{k_{-3}}{k_{+3}} \cdot \frac{k_2^{\gamma+1}}{D_0} \cdot (\gamma - 1) \qquad (15.52)$$

which corresponds (Chapter XIV, § 4), to the vanishing of the coefficient of ω in (15.51). The frequency at that point is

$$\omega_i = k_2 \sqrt{(\gamma - 1)} \qquad (15.53)$$

We have a focus followed by a limit cycle. Similarly, the condition for instability in respect to inhomogeneous perturbations is obtained exactly as in § 2 of this chapter. It is

$$w_1^\gamma < w_{1c}^\gamma = \frac{k_{-1} + k_{+2}}{k_{+1} \cdot k_{+2}} \cdot \frac{k_{-3}}{k_{+3}} \cdot \frac{k_2^{\gamma+1}}{D_0} \cdot \frac{D_{A_3}}{D_{A_2}} (\sqrt{\gamma} - 1)^2 \qquad (15.54)$$

together with the critical wavelength (cf. 15.12):

$$\lambda_c^2 = \frac{D_{A_2}}{k_2(\sqrt{\gamma} - 1)} \qquad (15.55)$$

The time behaviour of the system when (15.52) is satisfied, has been followed on an analogic computer (Sel'kov, 1968; Goldbeter, 1969). A typical result is given in Figure 15.7. It indicates clearly that the system tends towards a limit cycle.

A comparison of (15.52) and (15.54) shows that both instabilities will occur under almost identical conditions when:

$$\frac{D_{A_3}}{D_{A_2}} \approx \frac{\sqrt{\gamma} + 1}{\sqrt{\gamma} - 1} \qquad (15.56)$$

This is not excluded since it has been assumed in the model (15.40) that γ takes values larger than 1.

On the other hand, k_2 can be derived from the first relation (15.50). Then using the results of Hess (Sel'kov, 1968, b), which yield for w_1 and A_2^0 the orders of magnitude:

$$w_1 = 6 \times 10^{-6} \text{ mole sec}^{-1}; \quad A_2^0 = 1{\cdot}5 \times 10^{-4} \text{ mole} \qquad (15.57)$$

one obtains for k_2 the approximate value:

$$k_2 = 4 \times 10^{-2} \text{ sec}^{-1} \qquad (15.58)$$

Introducing these values in (15.53) with $\gamma = 2$, one sees that the frequency of oscillations predicted by the model is:

$$\omega = 2{\cdot}4 \text{ minutes}^{-1} \tag{15.59}$$

so that the period is 2·6 minutes, i.e. 2′36″ ($T = 2\pi/\omega$).

This value is in agreement with the gap of 3–5 minutes found

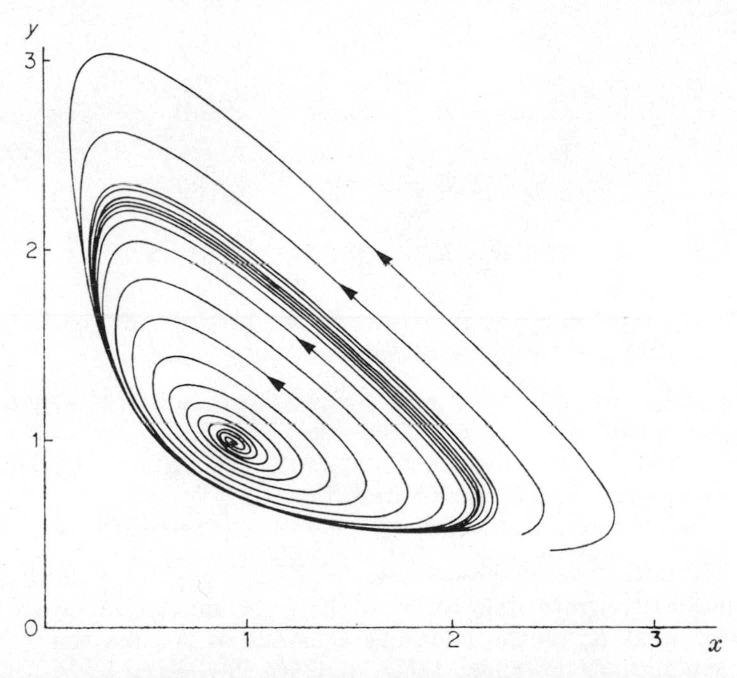

Fig. 15.7. Limit cycle for the phosphofructokinase reaction; x and y are normalized concentrations of ATP and ADP respectively.

experimentally by Hess. Similarly (15.55) shows that spatial differentiation is also closely related to k_2, in this system and in addition to the diffusion constant of ADP. Assuming an order of magnitude of 10^{-6} cm^2 sec^{-1} for D_{A_2}, we see that the model predicts inhomogeneities corresponding to the range:

$$10^{-4} \text{ cm} < \lambda_c < 10^{-2} \text{ cm}$$

i.e. large with respect to molecular dimensions. Our macroscopic treatment is therefore justified *a posteriori*.

As a further illustration concerning the possibility of symmetry breaking instabilities in biological systems, we shall consider Sel'kov's model for an enzyme reaction inhibited by the substrate and product

$$\begin{aligned}
&\xrightarrow{w_i} S_1 + E \underset{k_{-1}}{\overset{k_{+1}}{\rightleftharpoons}} S_1E && \text{(a)}\\
&S_1E \xrightarrow{k_{+2}} E + S_2 \xrightarrow{w_f} && \text{(b)}\\
&S_1 + S_1E \underset{k_{-3}}{\overset{k_3}{\rightleftharpoons}} S_1S_1E && \text{(c)}\\
&S_2 + E \underset{k_{-4}}{\overset{k_4}{\rightleftharpoons}} ES_2 && \text{(d)}\\
&S_2 + S_1E \underset{k_{-5}}{\overset{k_5}{\rightleftharpoons}} S_1ES_2 && \text{(e)}\\
&S_2 + S_1S_1E \underset{k_{-6}}{\overset{k_6}{\rightleftharpoons}} S_1S_1ES_2 && \text{(f)}
\end{aligned} \tag{15.60}$$

where: S_1, represents the substrate, S_2 the product, E, the enzyme and S_1E the *active enzyme-substrate complex*.

ES_2, S_1S_1E, S_1ES_2 and $S_1S_1ES_2$ are inactive enzymatic complexes.

w_i is the rate at which S_1 enters the system and is given by: $w_i = w_0 - k_0S_1$ ($i \equiv$ initial).†

w_f, is the rate at which S_2 disappears ($f \equiv$ final).

It is assumed to be of the form: $w_f = VS_2/(K_m + S_2)$, following an enzymatically irreversible reaction. V is the maximum rate when $S_2 \to \infty$, and K_m is the Michaelis constant of the reaction.‡ The kinetic equations describing the system are then given by:

$$\frac{\partial S_1}{\partial t} = w_i - \frac{k_{+2} e_0 \dfrac{S_1}{K_{S_1}}}{\left(1 + \dfrac{S_2}{K_{S_2}}\right)\left[1 + \dfrac{S_1}{K_{S_1}} + \dfrac{K_{S_1}}{K'_{S_1}}\left(\dfrac{S_1}{K_{S_1}}\right)^2\right]} + D_{S_1}\frac{\partial^2 S_1}{\partial r^2} \tag{15.61}$$

† In the derivation of (15.61–15.62), it is assumed that:

$$S_1 \approx S_2 \approx 1; \quad e_0 \ll 1; \quad k_{+4} = k_{+5} = k_{+6}; \quad k_{-4} = k_{-5} = k_{-6}$$
$$k_{\pm 1}, k_{\pm 3}, k_{\pm 4}, k_{+2} \gg 1.$$

‡ The step $\xrightarrow{w_i} S_1$, corresponding to a given flow w_i, may be interpreted as the usual reversible step $A \underset{k_0}{\overset{k}{\rightleftharpoons}} S_1$, with $w_0 = kA$, related to a given overal affinity.

$$\frac{\partial S_2}{\partial t} = -w_f + \frac{\underset{+2}{k}e_0 \dfrac{S_1}{K_{S_1}}}{\left(1 + \dfrac{S_2}{K_{S_2}}\right)\left[1 + \dfrac{S_1}{K_{S_1}} + \dfrac{K_{S_1}}{K'_{S_1}}\left(\dfrac{S_1}{K_{S_1}}\right)^2\right]} + D_{S_2}\frac{\partial^2 S_2}{\partial r^2} \tag{15.62}$$

where

$$K_{S_1} = \frac{k_{-1} + k_{+2}}{k_{+1}}; \quad K'_{S_1} = \frac{k_{-3}}{k_{+3}}; \quad K_{S_2} = \frac{k_{-4}}{k_{+4}} \tag{15.63}$$

and e_0, is the total quantity of enzyme. Putting

$$\alpha \equiv \frac{K_{S_1}}{K'_{S_1}} \tag{15.64}$$

the instability with respect to diffusion can be expressed in terms of this parameter. The critical value beyond which the system becomes unstable is given by

$$\alpha_c = \frac{D_{S_2}K_{S_2}}{D_{S_1}K_S} \cdot \frac{\xi^2(2\beta - \nu_0)^2}{\beta^2(\nu_0 - \beta)^2} \cdot \frac{1}{\left(1 + \sqrt{\dfrac{(\nu_0 - \beta)}{\beta}}\right)\left(1 + \sqrt{\dfrac{\beta}{(\nu_0 - \beta)}}\right)} \tag{15.65}$$

and corresponds to the critical wavelength:

$$\lambda_c^2 = \frac{\nu_0 - \beta + \sqrt{\{\beta(\nu_0 - \beta)\}}}{\xi(2\beta - \nu_0)} \cdot \frac{D_{S_1}K_{S_1}}{\underset{+2}{k}e_0} \tag{15.66}$$

with:

$$\xi = \frac{k_0 K_{S_1}}{k_{+2}e_0}; \quad \nu_0 = \frac{w_0}{k_{+2}e_0}; \quad \beta = \frac{V}{\underset{+2}{k}e_0} \tag{15.67}$$

Equations (15.65) and (15.66) yield physically acceptable values when

$$\tfrac{1}{2}\nu_0 < \beta < \nu_0$$

In this case, for almost equal diffusion coefficients of S_1 and S_2, the instability is enhanced by product inhibition.

In a very recent paper, Goldbeter and Lefever (1970), have discussed dissipative structures for an allosteric model in connection with the glycolytic oscillations. The main advantage in comparison with the Selkov model (15.40) is that no formal parameter similar to γ, appears in their discussion.

CHAPTER XVI

Multiple Steady States

1. INTRODUCTION

In Chapters XIV and XV we considered situations where the steady-state equations have a single solution corresponding to the thermodynamic branch. For sufficiently large deviations from equilibrium, this branch then becomes unstable. There may, of course, also occur cases where the steady-state equations have more than one stable solution after the thermodynamic branch reaches the instability point.

We discuss first situations characterized by a single independent variable. Then we have always a kinetic potential in the sense defined in Chapter IX. Closely related to this case, is the problem of stability with respect to temperature fluctuations which has been extensively studied by engineers (Aris, 1969, where references to earlier work may be found).

Systems with multiple steady states have attracted much interest in connection with models for biological processes. Bierman (1954) has discussed a model with two stable stationary states. Spangler and Snell (1961, 1967) analysed multienzyme reactions which may present more than one stationary state. More recently, Lavenda and Nicolis (1969) have presented simple reaction schemes and interpreted the transition between steady states as non-equilibrium instabilities arising beyond a critical affinity. An analytically simple scheme has also been presented by Edelstein (1970). We discuss Edelstein's example in § 3.

In § 4 we then consider in more detail a recent model of membrane excitability due to Blumenthal, Changeux and Lefever (1969), which illustrates in a striking way the possible role of multiple steady states and dissipative structures in nerve excitation.

2. SINGLE INDEPENDENT VARIABLE

Let us go back to formula (9.24–9.27). In the case of a single independent variable (cf. 9.29), $d_X P$ can, by definition, be written in terms of a kinetic potential which satisfies inequality (9.44), that is:

$$T\, d_X P = d\Phi \leqslant 0 \tag{16.1}$$

In a steady state Φ will be an extremum in respect to the independent variable, we shall denote by Z (cf. 9.25–9.29). Therefore

$$\frac{\partial \Phi}{\partial Z} = 0 \tag{16.2}$$

and the stability condition for this state implies:

$$\frac{\partial^2 \Phi}{\partial Z^2} > 0 \tag{16.3}$$

Indeed if (16.3) is not satisfied the slightest fluctuations will permit the system to leave this state. Using (9.25) we may also write:

$$d\Phi = \sum_\rho \mathscr{W}_\rho \frac{\partial(\mathscr{A}_\rho T^{-1})}{\partial Z} dZ \tag{16.4}$$

where $\mathscr{W}_\rho$ is the 'relative velocity' which vanishes at the steady state (9.26). The corresponding stability condition is then:

$$\sum_\rho \frac{\partial \mathscr{W}_\rho}{\partial Z} \cdot \frac{\partial(\mathscr{A}_\rho T^{-1})}{\partial Z} > 0 \tag{16.5}$$

We shall see an example in § 3.

The problem of stability in respect to temperature fluctuations is quite similar. The excess entropy production (2.21), (7.61) has now a supplementary contribution due to heat transfer. Let us consider the expression:

$$\sum_\rho \frac{\partial \mathscr{W}_\rho}{\partial T} \cdot \frac{\partial(\mathscr{A}_\rho T^{-1})}{\partial T} \tag{16.6}$$

Because of the Arrhenius factor, the reaction rates increase with temperature. On the contrary for *exothermic* reaction the partial derivative $\partial(\mathscr{A}_\rho T^{-1})/\partial T$, is negative (e.g. Prigogine, Defay, 1954, Chapter IV, § 2). The contribution (16.6) may then become negative and tend to destabilize the system. But we shall not go into a discussion of specific examples here, as they may be found in literature (Aris, *loc. cit.*).

3. MODEL WITH MULTIPLE STEADY STATES

Let us study the following simple scheme (Edelstein, 1970)

$$\begin{aligned} A + X &\rightleftharpoons 2X \\ X + E &\rightleftharpoons C \\ C &\rightleftharpoons E + B \end{aligned} \tag{16.7}$$

The overall reaction is

$$A \rightleftharpoons B \tag{16.8}$$

The mechanism (16.7) consists in the autocatalytic production of X and its 'enzymatic' degradation through a Michaelis–Menten step (see Chapter XV, § 3). We impose the conservation of enzyme condition:

$$E + C = E_T = \text{constant} \tag{16.9}$$

We put again all kinetic constants, corresponding both to the direct and the inverse reactions, equal to one. The kinetic equations are therefore:

$$\begin{aligned} \frac{\mathrm{d}X}{\mathrm{d}t} &= AX - X^2 - XE + C \\ \frac{\mathrm{d}E}{\mathrm{d}t} &= -XE - BE + 2C \end{aligned} \tag{16.10}$$

Using (16.9) the steady-state equations become:

$$\begin{aligned} f(X) = X^3 + (2 + B - A)X^2 \\ + [E_T - A(2 + B)]X - BE_T = 0 \end{aligned} \tag{16.11}$$

$$E = \frac{2E_T}{X + B + 2} \tag{16.12}$$

The steady state values of X for a given A, B and E_T, is easily obtained from (16.11). Figure 16.1 shows a region where multiple steady states occur.

It is easy to verify that (16.4) becomes here:

$$\mathrm{d}\Phi = -\left[\frac{1}{X}(AX - X^2 - EX + C) + \frac{E - C}{CE}(EX - 2C + 2BE)\frac{\mathrm{d}E}{\mathrm{d}X}\right]\mathrm{d}X \tag{16.13}$$

where the derivative $\mathrm{d}E/\mathrm{d}X$, is taken in accordance with (16.12).

The form of $\Phi(X)$, for a given A, corresponding to Figure 16.1, is represented schematically on Figure 16.2.

The upper and lower values of X are stable. The intermediate value is unstable. This can of course be also verified by direct normal mode analysis.

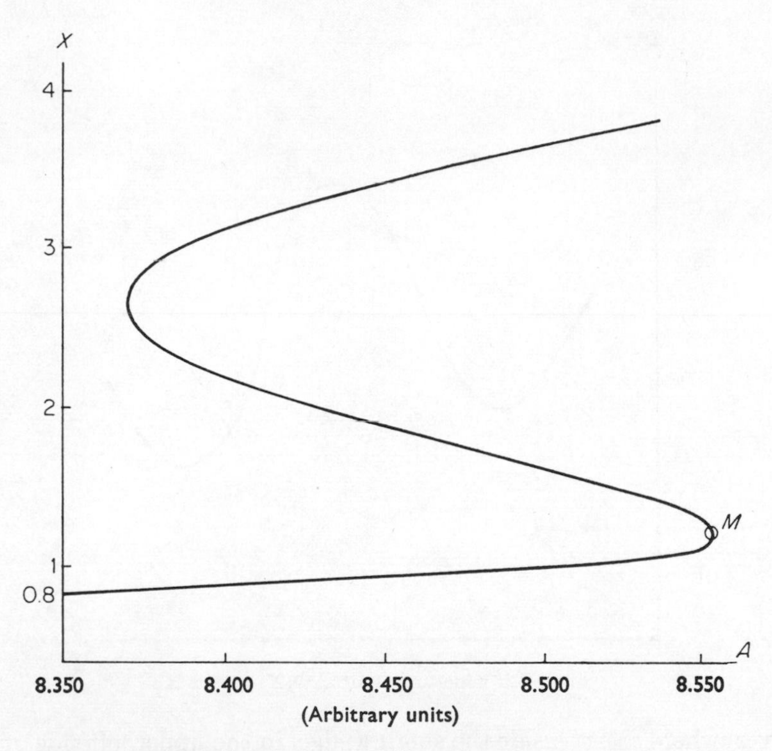

Fig. 16.1. Multiple Steady States of X as a function of A, for $B = 0{\cdot}2$; $E_T = 30$.

The case corresponding to the coalescence of the two extrema in Figure 16.2, was recently investigated by Kobatake (1970). An interesting question would be the meaning of the thermodynamic branch in such an example, where the different steady states are essentially of the same nature. This is in strong contrast with the case of symmetry breaking instabilities studied in Chapter XV. On Figure 16.3 we have represented the projection on the plane AB of the three dimensional diagram ABX (for $E_I = 24{\cdot}8$).

At the thermodynamic equilibrium:

$$A = B = X \tag{16.14}$$

When we go away from this line two situations may occur: either, to given values of A and B corresponds a single steady-state value of X, or, more than one value is possible. The first case is realized

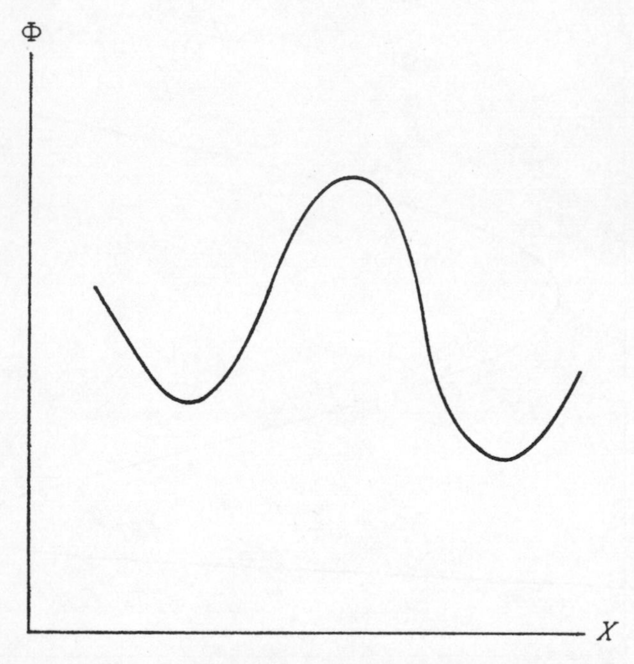

Fig. 16.2. Kinetic Potential $\Phi(X)$ versus X.

everywhere *except* inside the small wedge, in the upper left side of the diagram represented on Figure 16.3. This is the region of multisteady states. Now if we start from state Q and go to state P along the line QRP, we shall have a continuous variation of X on the thermodynamic branch till the final steady state value is achieved. On the contrary if we follow the path QLP, which crosses the wedge, we shall have a discontinuity between the thermodynamic branch and a new non equilibrium branch. We see therefore that C plays in a way, the role of the critical liquid–vapour point in equilibrium thermodynamics. Both stable solutions of Figure 16.1 may be considered as corresponding to the thermodynamic branch, which is here *many valued.*

The situation is completely different from that considered in Chapters XIV and XV. Indeed the onset of a limit cycle or symmetry breaking, always involves an instability and cannot be reached continuously, starting from the thermodynamic branch.

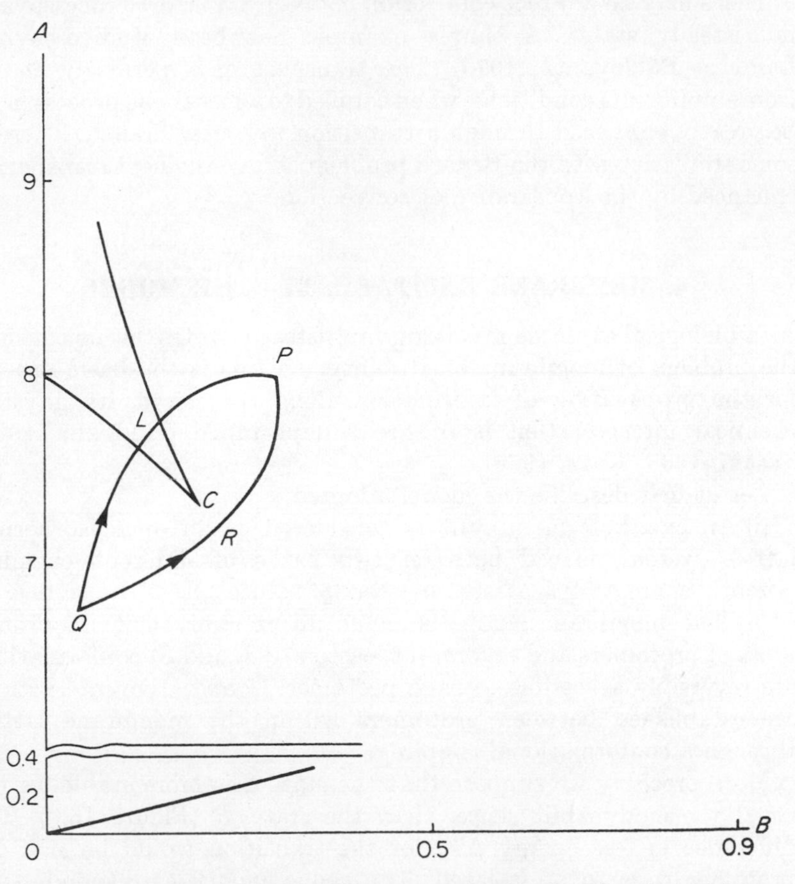

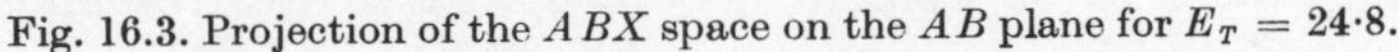
Fig. 16.3. Projection of the ABX space on the AB plane for $E_T = 24{\cdot}8$.

A question of great interest is to know when the system will jump from one branch to another. Do we have in Figure 16.1 to go on the lower branch till the point M, or will the jump occur earlier ? This is a problem of stability in respect to *finite* fluctuations which is not yet fully elucidated.

In a remarkable paper, Kobatake (1970), points out that the transitions between two steady states in a porous charged membrane occurs when the kinetic potential Φ takes the same value in both states. The situation would be similar to that of the Bénard problem (Chapter XI, § 5). This important question would deserve further investigation.

There may be a direct connection between transport processes and multisteady states. A simple example has been studied by A. Lambert–Babloyantz (1970). The transport of a permeant in far from equilibrium conditions, when coupled to a metabolic process, may be greatly enhanced through a transition to a new branch. There is some similarity with the Bénard problem, where the heat transport is enhanced by the appearance of convection.

4. MEMBRANE EXCITABILITY—THE MODEL

As a biological example involving multisteady states, let us consider the problem of membrane excitability. As this is the basic process for the propagation of information along the nerve, its physico-chemical interpretation is of great importance (Hodgkin, 1965; Tasaki, 1968; Katz, 1959).

Let us first describe the model adopted.†

(i) An excitable membrane is considered as an open isothermal lattice system, placed between two baths of different chemical potentials, creating a passive net flux of solute across the lattice.

(ii) The membrane lattice is made up of equivalent lipoproteic units of protomers and several (at least two: R and S) conformations are reversibly accessible to each protomer. Cooperative interactions are established between protomers within the membrane lattice through a conformational coupling.

More precisely we suppose that the state S is more stable, in the usual thermodynamic sense, than the state R (Figure 16.4). The difference in free energy ΔF, for the transition would be ε, if the protomer were to be isolated. The cooperativity corresponds to a decrease of ε with the average number of protomers already in the state R. The simplest quantitative expression of this cooperativity is (Changeux *et al.*, 1967):

$$\Delta F = \varepsilon - \eta\langle r\rangle \tag{16.15}$$

where $\langle r\rangle$ is the average fraction of protomers in the state R and η a positive constant.

† This model is based on a treatment introduced by Hill and Kedem (1966) for transport of non electrolytes through a membrane lattice.

(iii) Each protomer carries at least, two distinct sites for the same ligand, one on the outer face of the membrane, the other on the inner face (Figure 16.5). The ligand, which binds to the receptor sites of the protomer is the permeating species. The ligand therefore *both binds*, (thereby stabilizing a certain conformation) *and permeates* across the membrane. Transport takes place by a 'jump' of the ligand from the outer to the inner site.

(iv) Both the affinity and the permeability of the protomer for the permeating species are altered when a transition occurs from one to another conformation of the protomer.

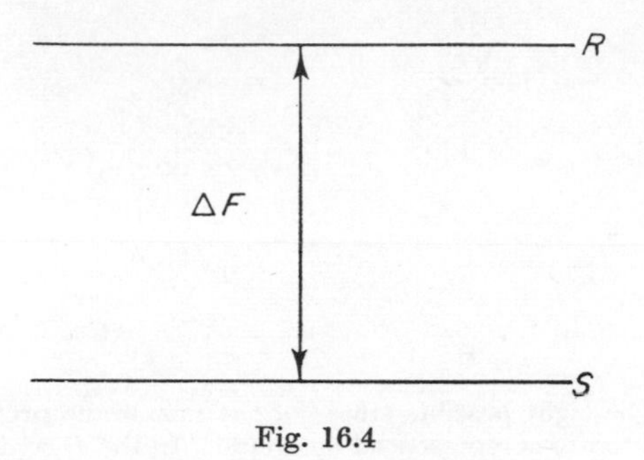

Fig. 16.4

(v) There exists on both sides of the membrane an 'equilibration layer' in which the concentration of the ligand is a function of both diffusion across the membrane, and diffusion between the layer and bulk solution.

Let us now consider the kinetic equations. These equations have been derived for the case of a single permeating species A which, in addition, is a non-electrolyte. Figure 16.5 illustrates the exchanges of matter between the equilibration layers and the membrane.

By adsorption and desorption of the ligand, each protomer conformation (R or S) is assumed to exist under four binding states: binding to the inner side (2 and 6), to the outer side (3 and 7), to both sides (4 and 8), and finally no binding (1 and 5). Translocation through the membrane occurs when a particle jumps from the outer side to the inner side (steps $2 \rightarrow 3$ and $6 \rightarrow 7$). A steady state regime is established through the membrane when the translocation of A is compensated by its diffusion between the equilibration layers on both sides of the membrane and the bulk solution, where the concentration of A is kept constant. For sake of simplicity, the kinetic equations have been written for symmetrical protomers, i.e. the

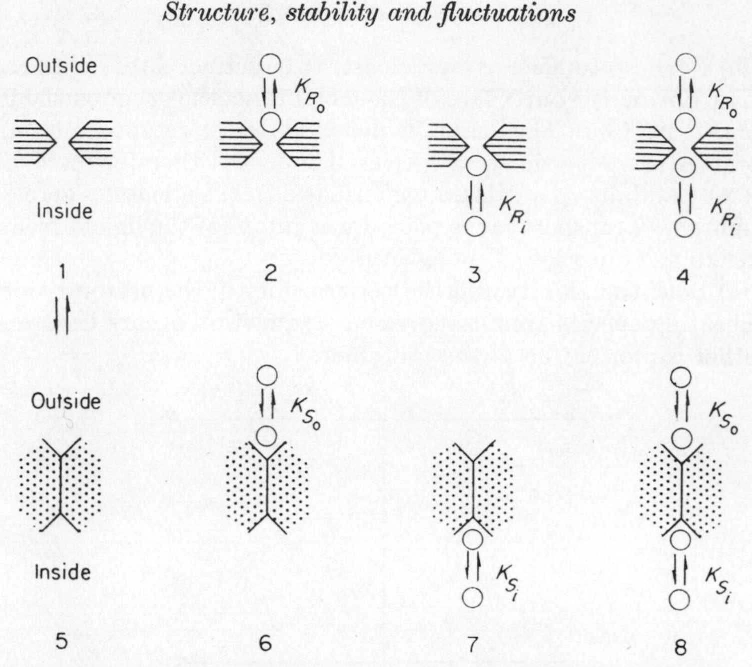

Fig. 16.5. The eight possible states of the membrane protomers. The transported solute is represented by a circle. In the R conformation of the protomer, the solute is both bound and transported. In the S conformation, the solute is bound but not transported. K_{R_0}; K_{R_i}; K_{S_0}; K_{S_i}, are the equilibrium constants for adsorption-desorption of the solute on the outer and inner site of protomers in the R or S state.

dissociation constants of the ligand bound to the R (resp. S) state on its inner (K_{R_i} (resp. K_{S_i})) and on its outer (K_{R_0} (resp. K_{S_0})) sites are equal; one has the relations:

$$K_R = K_{R_0}, \text{ and } K_{S_i} = K_{S_0}. \tag{16.16}$$

We call

$$x_i = \frac{N_i}{N} \tag{16.17}$$

the fraction of protomers in state i. To simplify the equations further, we shall assume that adsorption of the ligand as well as translocation occurs only on the protomers R. As a result:

$$x_6 = x_7 = x_8 = 0$$
$$x_1 + x_2 + x_3 + x_4 + x_5 = 1 \tag{16.18}$$

We shall call B (instead of A) the concentration of the ligand on the *inside* of the membrane.

The kinetic equations now become:

$$\frac{dx_2}{dt} = k_a(Ax_1 - Bx_2) + k_d(x_4 - x_2) + k_m(x_3 - x_2) \tag{16.19}$$

$$\frac{dx_3}{dt} = k_a(Bx_1 - Ax_3) + k_d(x_4 - x_3) + k_m(x_2 - x_3) \tag{16.20}$$

$$\frac{dx_4}{dt} = k_a(Ax_3 + Bx_2) - 2k_dx_4 \tag{16.21}$$

$$\frac{dx_5}{dt} = k_c'x_1 - k_cx_5 \tag{16.22}$$

$$\frac{dA}{dt} = k_p(A_0 - A) + (k_dx_2 + k_dx_4 - k_aAx_1 - k_aAx_3)N \tag{16.23}$$

$$\frac{dB}{dt} = k_p(A_i - B) + (k_dx_3 + k_dx_4 - k_aBx_1 - k_aBx_2)N \tag{16.24}$$

where k_a, k_d, are respectively the constants of adsorption and desorption of the ligand on the protomers ($K_R = K = k_d/k_a$); k_p, the diffusion constant between the bulk solution and equilibration layer, and k_m, the diffusion constant across the membrane. We also introduce the isomerization constant (16.15):

$$\ell' \equiv \frac{k_c}{k_c'} = \exp\left[\beta(\varepsilon - \eta\langle r\rangle)\right] = \ell\Lambda^{\langle r\rangle} \tag{16.25}$$

with

$$\langle r\rangle = \frac{x_1 + x_2 + x_3 + x_4}{x_1 + x_2 + x_3 + x_4 + x_5} \tag{16.26}$$

representing the fraction of protomers in the R conformation. Let us now redefine the parameters and variables in the following way:

$$\alpha_0 = \frac{A_0}{K}; \quad \alpha_i = \frac{A_i}{K}; \quad \alpha = \frac{A}{K}; \quad \beta = \frac{B}{K}; \quad \frac{k_m}{k_d} = \varepsilon;$$

$$k_d' = \frac{k_dN}{K}; \quad \gamma = \frac{k_pK}{k_mN}; \quad x_2 + x_4 = y_0; \quad x_3 + x_4 = y_i;$$

$$x_1 + x_3 = \langle r\rangle - y_0; \quad x_1 + x_2 = \langle r\rangle - y_i \tag{16.27}$$

and rearrange the equations (16.15–16.24) in the following form

$$\frac{dy_0}{dt} = k_d[\alpha\langle r\rangle - (\alpha + 1 + \varepsilon)y_0 + \varepsilon y_i] \tag{16.28}$$

$$\frac{dy_i}{dt} = k_d[\beta\langle r\rangle - (\beta + 1 + \varepsilon)y_i + \varepsilon y_0] \tag{16.29}$$

$$\frac{dx_4}{dt} = k_d[\alpha y_i + \beta y_0 - (\alpha + \beta + 2)x_4] \tag{16.30}$$

$$\frac{d\langle r\rangle}{dt} = k_c[(1 - \langle r\rangle) - l'(\langle r\rangle - y_i - y_0 + x_4)] \tag{16.31}$$

$$\frac{d\alpha}{dt} = k'_d[\gamma\varepsilon(\alpha_0 - \alpha) - \alpha\langle r\rangle + (\alpha + 1)y_0] \tag{16.32}$$

$$\frac{d\beta}{dt} = k'_d[\gamma\varepsilon(\alpha_i - \beta) - \beta\langle r\rangle + (\beta + 1)y_i] \tag{16.33}$$

Assuming fast equilibration of the protomers with respect to the environmental medium, i.e.:

$$k_d,\ k_c > k'_d \tag{16.34}$$

equations (16.28–16.31) may be divided by k_d and k_c, and yield thus a set of algebraic relations. As a result one has:

$$y_0 = \frac{[(1 + \beta)\alpha + \varepsilon(\alpha + \beta)]\langle r\rangle}{[(1 + \alpha)(1 + \beta) + \varepsilon(\alpha + \beta + 2)]} \tag{16.35}$$

$$y_i = \frac{[(1 + \alpha)\beta + \varepsilon(\alpha + \beta)]\langle r\rangle}{[(1 + \alpha)(1 + \beta) + \varepsilon(\alpha + \beta + 2)]} \tag{16.36}$$

Replacing these values into (16.32–16.33), one finds that the behaviour of the membrane in response to alteration of its environment is described by a set of three equations.

Two equations give the time-change of the ligand concentration in the inner and outer equilibration layers:

$$\frac{d\alpha}{dt} = \varepsilon k'_d\left[\gamma(\alpha_i - \alpha) - \frac{(\alpha - \beta)\langle r\rangle}{(1 + \alpha)(1 + \beta) + \varepsilon(\alpha + \beta + 2)}\right] \tag{16.37}$$

$$\frac{d\beta}{dt} = \varepsilon k'_d\left[\gamma(\alpha_0 - \beta) - \frac{(\beta - \alpha)\langle r\rangle}{(1 + \alpha)(1 + \beta) + \varepsilon(\alpha + \beta + 2)}\right] \tag{16.38}$$

A third one yields the fraction of the membrane protomers which are in the R state for given values of these concentrations:

$$\frac{1}{\langle r\rangle} = 1 + \frac{\ell\Lambda^{<r>}\left(1 + \dfrac{4\varepsilon}{\alpha + \beta + 2}\right)}{(1 + \alpha)(1 + \beta) + \varepsilon(\alpha + \beta + 2)} \tag{16.39}$$

Equations (16.37–16.39) are our basic equations for the discussion of the membrane excitability.

5. MEMBRANE EXCITABILITY. STEADY-STATE EQUATIONS

From equations (16.37) and (16.38), one deduces immediately the condition for steady state regime ($d\alpha = d\beta = 0$), between the membrane and its environment:

$$J_d = J_m \quad \text{and} \quad \alpha_0 + \alpha_i = \alpha + \beta \tag{16.40}$$

with

$$J_d = \gamma\varepsilon(\alpha_i - \alpha) = \gamma\varepsilon(\beta - \alpha_0)$$

$$J_m = \frac{(\alpha - \beta)\langle r \rangle}{(1 + \alpha)(1 + \beta) + \varepsilon(\alpha + \beta + 2)} \tag{16.41}$$

A characteristic and important property of the system lies in the possibility of multisteady state regimes for a *given value* of the overall gradient $(\Delta\alpha)_t$ between bulk solutions on both sides of the membrane. Indeed, Figure 16.6 shows that in conditions where the variation of $\langle r \rangle$ with $\Delta\alpha$ is highly cooperative, there might exist three acceptable solutions of equations (16.39) and (16.41). For a given value of the constraint $(\Delta\alpha)_t = \alpha_0 - \alpha_i$, they correspond to distinct values of ligand concentration in the two equilibration layers. Accordingly, three steady states for diffusion through the membrane are *a priori* accessible. However it is easily seen by testing their stability that only the two extreme states (1) and (3) are stable with respect to small disturbances. On the contrary the intermediate one (2) is not and corresponds therefore to some kind of 'threshold state': in this region of maximum cooperativity, the concentration of ligand in the equilibration layers is predominantly controlled by the changes in membrane permeability. Fluctuations around the state (2) increase 'consumption' of ligand by the membrane so abruptly, that its 'production' by diffusion from the bulk solution can no longer compensate this effect. Let us now consider states (1) or (3). Small subthreshold perturbations will let the system relax to the original state, but when the state (2) is reached, the system jumps to the other stable steady state corresponding to a different permeability. An important fact is that, whatever the magnitude of the suprathreshold perturbation may be, the amplitude of the jump is constant since the same steady states are always reached. In addition, once disturbances have been amplified locally by the instability, it is easy to see that this effect might spread over to the neighbouring regions in the membrane, and propagate through the whole nerve. We will not develop here the biological implications of this model, but rather

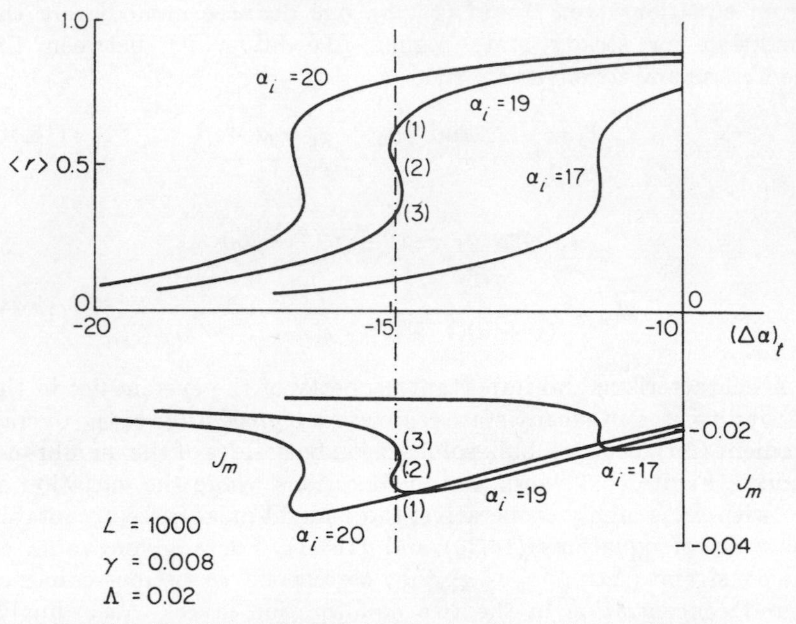

Fig. 16.6. Variation of J_m and $\langle r \rangle$ as a function of the *overall gradient* $(\Delta\alpha)_t$, between the *bulk solutions*. For a given value of $(\Delta\alpha)_t$, chosen in the region of cooperative change, three steady state solutions for J_m and $\langle r \rangle$ are possible.

show how cooperativity is affected by the non equilibrium constraints. Indeed the results reported here indicate how the cooperative *molecular* properties of membrane units can be amplified by energy dissipation and account for the phenomenon of membrane excitability. If we go back to the behaviour of the system under *equilibrium conditions*, equation (16.39) giving the conformational state of the membrane reduces immediately to the one previously derived by Changeux *et al.* (1967):

$$\frac{1}{\langle r \rangle} = 1 + \frac{l\Lambda^{\langle r \rangle}}{(1+\alpha)^2} \tag{16.42}$$

with $\alpha = \beta = \alpha_0 = \alpha_t$.

One then sees immediately that for example, within the framework of numerical values considered here for Λ, no multiple states for $\langle r \rangle$ are possible at equilibrium for a single value of α, and consequently there is no threshold state or instability. As shown on

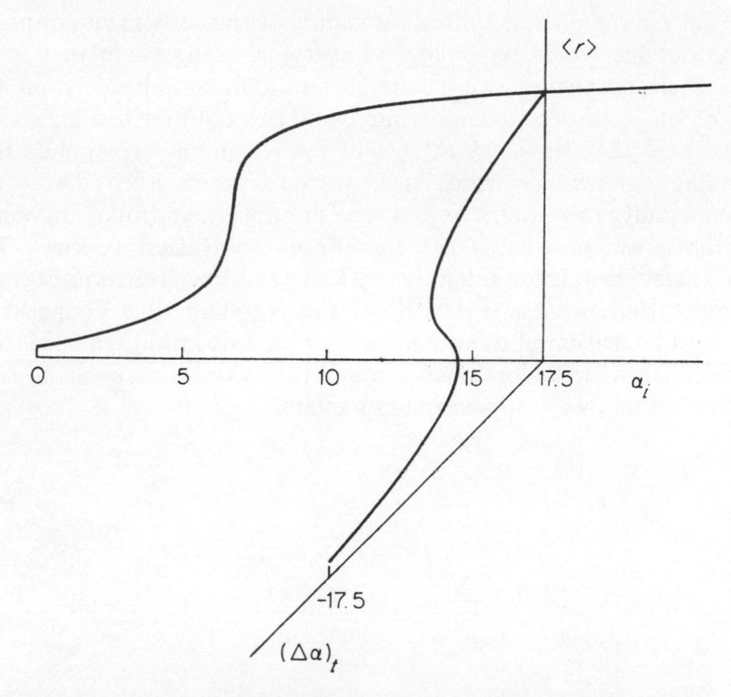

Fig. 16.7. In the plane corresponding to thermodynamic equilibrium (i.e. for $\alpha_0 = \alpha_i$), $\langle r \rangle$ is plotted as a function of α_i, the normalized concentration of A in the inner bulk solution. Only one value of $\langle r \rangle$ is then accessible to the membrane for a single value of α_i. In a plane perpendicular to the plane of the figure $\langle r \rangle$ is plotted as a function of the overall gradient $(\Delta\alpha)_t$ for a given value of α_i. The figure shows how the behaviour of the system is modified when the membrane is brought into environmental asymmetry.

Figure 16.7 a multivalued $\langle r \rangle$ function appears here as a result of the 'functioning' of the system inserted in a concentration gradient.

The system possesses the freedom to leave the equilibrium plane where $\alpha_0 = \alpha_i$, and to find new working conditions more adapted to its environment. As a result we see that when the deviations of the overall gradient from equilibrium are small, the conformational curve $\langle r \rangle$, corresponds to an extrapolation of the equilibrium situation.

One may therefore say that the membrane state lies then on the 'thermodynamic branch'. On the contrary for higher values of the gradient, when the membrane exhibits a low permeability and is in its resting state. The membrane organization of conformation would then lie on the other branch which corresponds to a *dissipative structure*. This again exemplifies the fact that the constraints imposed by the outside world on biological systems, bring them in a state where their properties can no longer be understood solely on the basis of an extrapolation starting from the equilibrium situation. We see here that the state at rest of a membrane corresponds to a molecular organization which is separated from its excited state by a discontinuity; the jump across this discontinuity, following small perturbations, constitutes the membrane excitation process. The excited state itself is obtained by gradual deviation from equilibrium. The excitation process is based on the coupling of a cooperative structural transition of the membrane, with a downhill translocation of ions. We may therefore qualify membrane excitation as an *assisted phase transition*, as it appears in our model.

CHAPTER XVII

Unity of Physical Laws and Levels of Description

1. INTRODUCTION†

It is a rather remarkable coincidence that the idea of evolution emerged in the nineteenth century associated with two conflicting aspects:

> In thermodynamics, the second law is formulated as the Carnot–Clausius principle. It appears essentially as the evolution law of continuous disorganization, i.e. of disappearance of structure, introduced by initial conditions.
>
> In biology or in sociology, the idea of evolution is, on the contrary, closely associated with an increase of organization giving rise to the creation of more and more complex structures.

It is clear that both concepts have deep philosophical implications. The extension of the thermodynamical concept of evolution to the world as a whole, leads to the idea that 'structure' originated in some far distant 'golden age'. Since then, chaos is progressively taking over.

The biological concept of evolution points precisely in the opposite direction. Nobody has expressed this better than the french philosopher Henri Bergson, whose whole metaphysics is essentially a meditation on the biology of his time: 'The deeper we go into the nature of time the more we understand that duration means invention, creation of forms, continuous elaboration of what is absolutely new' (Bergson, 1907, 1963). To some extent the ideas of Spencer are similar. Indeed Spencer believes that the basic principle of evolution in nature is the principle of 'instability of the homogeneous' (for an excellent discussion of Spencer's ideas see Henderson, 1917).

† This chapter follows closely two recent papers (Prigogine, 1970, a, b).

How can one reconcile these two apparently opposite aspects of evolution!

There is no doubt that both correspond to different facets of physical reality. For instance, if we let two liquids mix, diffusion will occur, with a progressive 'forgetting' by the system of its initial conditions. This is a typical example of situations described by an increase of entropy. On the contrary in biological systems heterogeneity is the rule. Inequalities between concentrations are maintained by chemical reactions and active transport. Therefore a 'coherent' behaviour is the characteristic feature of biological systems (Weiss, 1968).

Are there consequently two different irreducible types of physical laws! The difference of behaviour is so drastic that such a question is indeed fully legitimate (e.g. Heitler, 1961).

The point of view considered in this monograph suggests that there is only one type of physical law, but different thermodynamic situations: near and far from equilibrium. Broadly speaking *destruction of structures* is the situation which occurs in the neighbourhood of thermodynamic equilibrium. On the contrary, *creation of structures* may occur, with specific non-linear kinetic laws beyond the stability limit of the thermodynamic branch. This remark justifies Spencer's point of view (1862): '*Evolution is integration of matter and concomitant dissipation of motion*'.

For all these various situations, the second law of thermodynamics still remains valid.

2. BIOLOGICAL STRUCTURES

It is certainly tempting to describe biological structures as open chemical systems operating beyond the stability of the thermodynamic branch. Such models applied to living systems are clearly incomplete, since much more precise information on the type of chemical reactions involved, would be necessary to explain essential features of life such as replication phenomena.

In Chapter XV, § 7, we have seen that important biochemical processes seem indeed to occur beyond the stability of the thermodynamic branch. In more qualitative terms a similar discussion may be presented for biological cycles as a whole such as photosynthesis (Chernavskaya, Chernavskii, 1961; Prigogine and others, 1969). It is even possible to isolate inside such cycles several separate groups of reactions, operating beyond the thermodynamic branch (see Goldbeter, 1969; Rodeyns, 1969). This seems to be one of the ways in which the organism maintains its 'coherence'.

The existence of the biological information would at least be partially governed by the necessity to maintain the state of the biological system beyond the thermodynamic branch.

In spite of the incomplete character of this view, it has a number of attractive features:

(i) It links the occurrence of space-time order to the specific function of the system.

(ii) It combines various ideas and points of view which at first seemed contradictory. Life no longer appears as an island of resistance against the second law of thermodynamics, or as the deed of some Maxwell demons. It would appear now as a consequence of the general laws of physics, appropriate to specific chemical kinetics and to far from equilibrium conditions. These specific kinetic laws are those which permit the flow of energy and matter to build and maintain functional and structural order in open systems.

Some authors conclude that life follows the second law of thermodynamics; others insist that the laws of biology have a somewhat special status. For instance, Elsasser has coined the term biotonic to denote a causal relationships involving an increase of 'information' (Elsasser, 1958).

Both views find a justification in our approach: indeed, the entropy production always remains a positive quantity. However, the type of solution changes radically, before and after the instability.

What would have appeared *a priori* as a tremendously improbable fluctuation, becomes beyond the instability the macroscopic solution of the problem.

In the last chapters we used a physico-chemical language. We spoke about catalytic reactions. Others may prefer to speak about negative feedback, auto-regulation and so on. Of course, this is only a question of terminology.

Again, some authors have insisted on the chemical aspects of life (Calvin, 1969). Clearly, our conclusions are not in contradiction with this point of view, as precise chemical conditions are to be met to produce the chemical instabilities.

Recently, Trintscher (1965) has established that the entropy production per unit mass and unit time (as measured by the metabolism) increases during the first period of ontogenesis, then passes through a maximum and decreases to reach eventually a steady state value. It is at the least very tempting to imagine that the first stage corresponds to the period where the minimum level of dissipation is reached (Prigogine, 1969). Correspondingly, in the prebiological stage, important steps would be associated with the creation of dissipative structures sufficiently far from equilibrium.

The above interpretation is so far sheer speculation. However, it has now been verified that there indeed exists new *dynamic* states of matter induced by a flow of free energy far from equilibrium (Chapter XV, § 1). Such states are governed by a new physical chemistry on a supermolecular level, while all laws referring to the molecular level remain essentially unchanged. In all the cases considered, the coherent behaviour on the supermolecular level corresponds in fact to an amplification of specific *molecular* properties (such as kinetic constants) in far from thermodynamic equilibrium conditions.

3. HIERARCHY OF STRUCTURES

An interesting aspect of chemical instabilities investigated above, is the inter-relation between the system and its boundary conditions. The occurrence of a dissipative structure depends, in general, on the boundary conditions, which in turn may be modified by the former.

We have in fact considered till now only simple cases where the chemical instabilities occur in a homogeneous medium. Let us briefly consider the much more involved situation where the instability appears in a non-homogeneous medium. We again consider the scheme (14.46), but we take into account the diffusion of component A (cf. Herschkowitz and Platten, 1971). In addition to (15.1), (15.2), we have the equation:

$$\frac{\partial A}{\partial t} = -A + D_A \frac{\partial^2 A}{\partial r^2} \tag{17.1}$$

taking for convenience all the kinetic constants k, equal to unity. The solution of (15.1), (15.2) and (17.1), as well as its stability may be obtained e.g. by the local potential method (cf. Chapters X and XII). A typical solution before the instability is represented on Figure 17.1.

In Figure 17.2, we have represented a steady state corresponding to a dissipative structure, arising beyond the instability. For the sake of simplicity we have represented only the unknown function X (r). Note the short wavelength oscillations, characteristic of the dissipative structure over a section of the system.

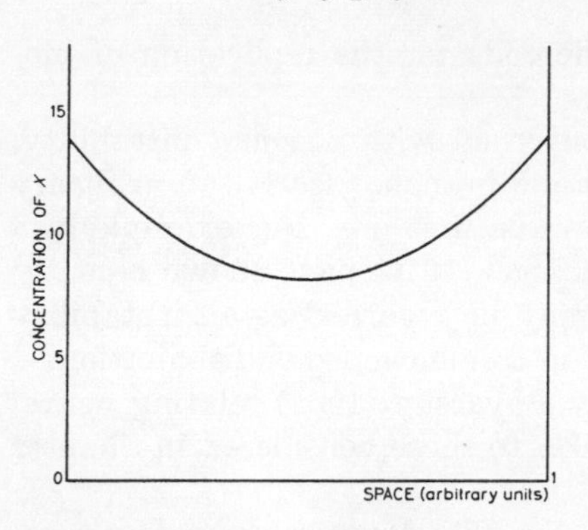

Steady State on the thermodynamic branch corresponding to equations (15.1), (15.2), (17.1), for fixed X, Y and A on the boundaries.
Numerical values used:
$D_A = 1.97 \times 10^{-1}$;
$D_X = 1.05 \times 10^{-3}$;
$D_Y = 5.26 \times 10^{-3}$; $B = 10$

Fig. 17.1

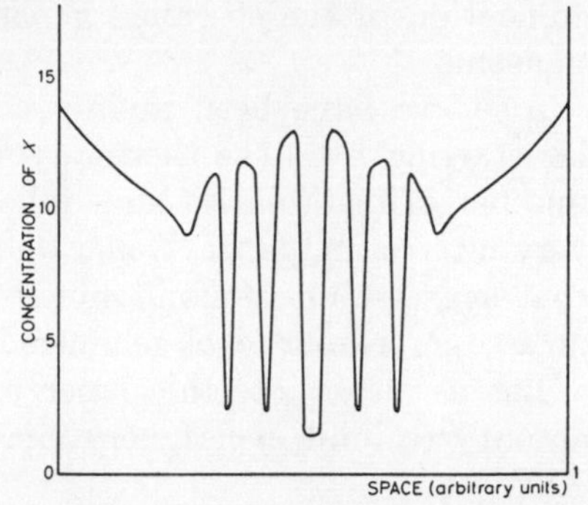

Steady State beyond the critical point corresponding to equations (15.1), (15.2), (17.1), for fixed X, Y and A on the boundaries.
Numerical values used:
$D_A = 1.97 \times 10^{-1}$;
$D_X = 1.05 \times 10^{-3}$;
$D_Y = 5.26 \times 10^{-3}$; $B = 26$

Fig. 17.2

An important feature is that the size of the dissipative structure is now well defined. According to the numerical values used, there is a whole wealth of possibilities; steady dissipative structures, chemical waves corresponding to well defined limit cycles at each point and so on. Again, as in Figure 15.1, we may construct a *phase diagram*, representing all these possibilities.

We find here one more reason to assimilate a biological system to a dissipative structure. Indeed, the boundaries of a living system appear far less arbitrary than for usual systems on the thermodynamic branch. They constitute the specific domain of space-time organization characteristic of the dissipative structure.

The concept of stability applies to a much wider range than could be included in this monograph. On one side, we have the possibility of instabilities on a mere molecular level. Indeed, besides the fluctuations of the macroscopic variables so far considered, we could also be concerned with fluctuations of *internal variables*, i.e. describing the intramolecular state. For example, fluctuations in the kinetics of a polymerization may give rise to fluctuations in the type of polymers produced. This, in turn, may give rise to instabilities in far from equilibrium situations. Possibly such instabilities contribute to the edification of the so called genetic code for the replication of biomolecules.†

Also, we have been mainly concerned with the *first* instability, i.e. starting from the thermodynamic branch. Clearly, an arbitrary number of instabilities may follow the first one. For example in a very interesting paper, Keller and Segel (1970) have shown recently that aggregation of slime molds may be regarded as an instability on a supercellular level generated in a homogeneous distribution.

Let us also quote the paper by Boyarsky (1967) relating neural activity to limit cycles, very similar to those considered in Chapter XIV, § 6.

In more qualitative terms, the paper 'The Measurement of Cultural Development in the Ancient Near East and in Anglo-Saxon England' by Carneiro (1969) uses explicitly the stability concept in a historic context. These are only a few examples chosen between many publications discussing similar problems.

Clearly we perceive a whole hierarchy of structures separated by discontinuities.

Concluding rather optimistically, we feel that the unified description of the macro-world, developed in this monograph, may prove useful for future progress. The concept of stability really reconciles unity of physics, with the existence of well defined levels of description.

† Indeed in a most remarkable communication, Eiger (1970, 1971) has studied a population of 'competing' and autocatalytic species of biological relevance. He has shown that it undergoes an evolution through a series of instabilities leading finally to the appearance of a biological code.

References

R. Aris, *Chem. Eng. Sci.*, **24**, 149 (1969).

A. Babloyantz, *Phys. Fluids*, **12**, 262 (1969).

A. Babloyantz, to be published 1970.

B. P. Belusov, *Sborn referat. radiat. meditsin za* (1958): *(Collection of Abstracts on Radiation Medicine)*, p. 145, Medgiz, Moscow, 1959.

H. Bergson, *Evolution Créatrice*, Alcan, Coll. de Bibl. de Philosophie, 3e Ed., 1907; *Oeuvres*, Presses Universitaires de France, Paris, 1963.

A. Bierman, *Bull. Math. Biophys.*, **16**, 203 (1954).

M. A. Biot, *Variational Principles in Heat Transfer*, Oxford Mathematical Monographs, 1970.

D. Blangy, H. Buc and J. Monod, *J. Mol. Biol.*, **31**, 13 (1968).

R. Blumenthal, J. P. Changeux and R. Lefever, *Compt. Rend.*, **270**, 389 (1970).

L. L. Boyarsky, *Curr. Mod. Biol.*, **1**, 39 (1967).

E. Bun; ing, Cold Spring Harbor Sym. Quant. Biol. (1960); *The Physiological Clock*, Academic Press, Publ. Berlin, Göttingen, Heidelberg, 1964.

H. Büsse, *J. Phys. Chem.*, **73**, 750 (1969).

H. W. Butler and R. L. Rackley, *Intern. J. Heat Mass Transfer*, **10**, 1255–1266 (1967) (Application of a variational formulation to non-equilibrium fluid flow); *Phys. Fluids*, **10**, 2499–2500 (1967) (Variational formulation for non-isothermal viscous liquids).

H. W. Butler and D. E. MacKee, *Intern. J. Heat Mass Transfer*, **13**, 43–54 (1970) (A Variational Solution to the Taylor Stability Problem based upon Non-Equilibrium Thermodynamics).

H. Callen, *Non-Equilibrium Thermodynamics, Variational Techniques and Stability*, University of Chicago Press, Chicago and London, 1965.

M. Calvin, *Chemical Evolution*, Clarendon Press, Oxford, 1969.

L. Carneiro, *Transactions New York Academy of Sciences*, 1013 (1969).

H. S. Carslaw and G. C. Jaeger, *Conduction of Heat in Solids*, 2nd Ed., Clarendon Press, Oxford, 1959.

L. Cesari, *Asymptotic Behaviour and Stability Problems in Ordinary Differential Equations*, Academic Press, 1963.

B. Chance, B. Schoener and S. Alsauer, *J. Biol. Chem.*, **240**, 3170 (1965).

B. Chance, R. W. Estabrook and A. Ghosch, *Proc. Natl. Acad. Sci. U.S.*, **51**, 1244 (1964).

L. C. Chambers, *Quart. J. Mech. Appl. Math.*, **9**, 234 (1956).

S. Chandrasekhar, *Amer. Math. Monthly*, **61**, 32–45 (1945); Max Planck Festschrift 1958, 103–14, Veb Deutscher Verlag der Wissenschaften, Berlin (1958); *Hydrodynamic and Hydromagnetic Stability*, Clarendon Press, Oxford, 1961.

J. P. Changeux, J. Thiery, Y. Tung and C. Kittel, *Proc. Natl. Acad. Sci. U.S.*, **57**, 335 (1967).

S. Chapman and T. G. Cowling, *The Mathematical Theory of Non-Uniform Gases*, Cambridge University Press, Cambridge, 1939.

N. M. Chernavskaia and D. S. Chernavskii, *Sov. Phys. Usp.*, **3**, 850 (1961).

R. Courant and D. Hilbert, *Methods of Mathematical Physics*, Interscience, New York, 1953.

J. D. Cowan, *A Statistictal Mechanics of Nervous Activity*, preprint 1969.

H. T. Davis, *Introduction to Non-Linear Differential and Integral Equations*, Dover, New York, 1962.

Th. De Donder, *L'Affinité*, (Réd. nouvelle par P. Van Rysselberghe), Gauthier-Villars, Paris, 1936.

S. R. De Groot and P. Mazur, *Non-Equilibrium Thermodynamics*, North Holland, Amsterdam, 1962.

H. Degn, *Nature*, **213**, 589 (1967).

Ch. J. de la Vallée Poussin, *Cours d'Analyse Infinitésimale*, Gauthier-Villars, Paris, 1926.

P. Duhem (1893), see P. Duhem (1899).

P. Duhem, *Traité Elémentaire de Mécanique Chimique*, Hermann, Paris, 1899, *Traité d'Energétique*, Gauthier-Villars, Paris, 1911; *Sur la Stabilité de l'Equilibre au sein d'une Enveloppe imperméable à la Chaleur*, Procès-verbaux de la Société des Sciences Physiques et Naturelles de Bordeaux juillet 1904. See also *Sur la Stabilité isentropique d'un Fluide, compt. Rend.*, **132**, 5 février (1901).

L. N. M. Duysens and H. J. Amesz, *Biophys. Biochem. Acta*, **24**, 19 (1957).

B. Edelstein, These de doctorat; Chimie-Physique, Univ. de Bruxelles. (1970).

M. Eigen, Nobel Symposium, Stockholm, Dec. 1970.

M. Eigen, *Naturwissenschaften* (to appear) (1971).

W. N. Elsasser, *The Physical Foundation of Biology*, Pergamon, London, 1958.

D. M. Gage, M. Schiffer, S. J. Kline and W. C. Reynolds, *Non-Equilibrium Thermodynamics Variational Techniques and Stability*, University of Chicago Press, Chicago and London, 1965.

K. S. Gage and W. H. Reid, *J. Fluid. Mech.*, **33**, 21 (1968).

J. W. Gibbs, *Collected Works*, Longman Green, New York, London, Toronto, 1st Ed. 1928, Reprinted 1931. *Equilibrium of Non-Homogeneous Substances*, 1875–1878.

P. Glansdorff and I. Prigogine, *Physica*, **20**, 773 (1954); *Physica*, **30**, 351 (1964).

P. Glansdorff, *Mol. Phys.*, **3**, 277 (1960).

P. Glansdorff and I. Prigogine, *Physica*, **31**, 1242 (1965).

P. Glansdorff, *Non-Equilibrium Thermodynamics, Variational Techniques and Stability*, University of Chicago Press, Chicago and London, 1965.

P. Glansdorff and I. Prigogine, *Physica*, **46**, 344 (1970).

P. Glansdorff, *Physica*, **32**, 1745 (1966).

P. Glansdorff and N. Banai, *Bull. Acad. Roy. Belg.*, **56** (1970).

M. Goche, *Bull. Classe Sci., Acad. Roy. Belg.*, (1971).

A. Goldbeter, Mémoire de Licence Chimie Physique Universite Libré de Bruxelles (1969).

B. C. Goodwin, *Temporal Organization in Cells*, Academic Press, 1963.

R. F. Greene and H. B. Callen, *Phys. Rev.*, **83**, 1231 (1951).

E. A. Guggenheim, *Thermodynamics*, North Holland, Amsterdam, 1949; *Mixtures*, Univ. Press, Oxford, 1952.

D. F. Hays and N. H. Curd, *Bull. Classe Sci., Acad. Roy. Belg.*, **53**, 469 (1967). (A variational formulation for diffusion problems: concentration-dependent diffusivity); *J. Franklin Inst.*, **289**, n° 4, 300 (1967); General Motors Corporation, Research Publication GMR–562 (1966) (Warren, Michigan). (A variational formulation for diffusion problems concentration dependent diffusivity.)

W. Heitler, *Die Wissenschaft Band*, **116**, Der Mensch und die Naturwissenchaftliche Erkenntonis Friedr. Vieweg and Sohn—Braunschweig 1961.

L. J. Henderson, *The Order of Nature*, Harvard Univ. Press, Cambridge, Mass., 1917.

M. Hershkowitz-Kaufman and J. Platten, *Bull. Acad. Roy. Belg.* (to appear) (1971).

M. Herschkowitz, *Compt. Rend.*, **270**, Série C, 1049 (1970).

B. Hess, *Funktionelle und Morphologische Organization der Zelle*, Springer Verlag, Berlin, 1963.

B. Hess, K. Brand and K. Pye, *Biochem. Biophys. Res. Commun.*, **23**, 102 (1966).

B. Hess and A. Boiteux, *Regulatory Functions of Biological Membranes*, Elsevier, 148, 1968.

J. Higgins, *Proc. Nall. Acad. Sci. U.S.*, **51**, 989 (1964); *Ind. Eng. Chem.*, **59**, 19 (1967).

T. L. Hilland and D. Kedem, *J. Theoret. Biol.*, **10**, 339 (1966).

J. O. Hirschfelder, C. F. Curtiss and R. B. Bird, *The Molecular Theory of Gases and Liquids*, J. Wiley, New York, 1954.

A. L. Hodgkin, *The Conduction of the Nervous Impulse*, University Press, Liverpool, 1964.

F. A. Hommes and F. M. Schurmans-Stukhoven, *Biophys. Biochem. Acta*, **86**, 427 (1964).

D. T. J. Hurle and E. Jakeman, *Phys. Fluids*, **12**, 2704 (1969).

A. Hurwitz, see for example L. Cesari, *Asymptotic Behaviour and Stability Problems in Ordinary Differential Equations*, Ed. Springer, 1963.

E. Jouguet, *Etude Thermodynamique des Machines thermiques*, Ed. Doin, Paris, 1909.

E. Jouguet, *Notes de Mécanique Chimique*, Journal Ecole Polytech. (Paris), 2e Série, 21e Cahier (1921).

G. D. Kahl and D. C. Mylin, *Phys. Fluids*, **12**, 11 (1969).

L. V. Kantorovich and V. I. Krylov, *Approximate Methods of Higher Analysis*, translated by C. D. Benster, Groningen, Ed. P. Noordhoff, The Netherlands, 1958.

A. Katchalsky and P. F. Curvan, *Non-Equilibrium Therm. in Biophysics*, Harvard University Press, Cambridge, Mass., 1964.

A. Katchalsky and R. Spangler, *Quart. Rev. Biophys.*, **1**, 1 (1968).

A. Katchalsky, preprint (1969).

B. Katz, *Nerve, Muscle and Synapse*, McGraw-Hill, 1966.

E. F. Keller and L. A. Siegel, *J. Theoret. Biol.*, **26**, 399 (1970).

E. H. Kerner, *Bull. Math. Biophys.*, **19**, 121 (1957); *Bull. Math. Biophys.*, **21**, 257 (1959); *Bull. Math. Biophys.*, **26**, 333 (1964).

Y. Kobatake, *Physica*, to be published (1970).

M. Kruskal, *Non-Equilibrium Thermodynamics Variational Techniques and Stability*, The University of Chicago Press, Chicago, 1965.

H. Lamb, *Hydrodynamics*, 6th Ed., Cambridge University Press, Cambridge, 1932.

L. D. Landau and E. M. Lifshitz, *Statistical Physics*, Addison-Wesley, Reading, Mass., 1958.

L. Landau and E. M. Lifshitz, *Fluid Mechanics*, Pergamon Press, London, 1959.

L. Landau, *Collected Papers*, 'On a Study of the Detonation of Condensed Explosives', Pergamon Press, New York, London, p. 425, 1965.

J. La Salle and S. Lefschetz, *Stability by Liapounov's Direct Method*, Academic Press, 1961.

B. Lavenda, Thése de doctorat, Chimie-Physique. Universite de Bruxelles (1970).

M. Lax, *Rev. Mod. Phys.*, **32**, 25 (1960).

L. H. Lee and W. C. Reynolds, Technical Report F.M.1 for the National Science Foundation, Thermosciences Division, Department of Mechanical Engineering, Stanford University, Stanford, California, U.S.A. (1964).

L. H. Lee and W. C. Reynolds, *Quart. J. Appl. Math.*, **20**, pt. 1, 1 (1967).

R. Lefever, G. Nicolis and I. Prigogine, *J. Chem. Phys.*, **47**, 1045 (1967).

R. Lefever, *Bull. Classe Sci., Acad. Roy. Belg.*, **54**, 712 (1968).

R. Lefever and G. Nicolis, 'Chemical Instabilities and Sustained Oscillations', submitted to *J. Theoret. Biol.*, (1970).

J. C. Legros and J. Platten, submitted to *Phys. Fluids* (1970).

G. N. Lewis, *J. Amer. Chem. Soc.*, **53**, 2578 (1931).

A. Liénard, *Rev. Gen. Elec.*, **23**, 901 (1928).

M. J. Lighthill, "Viscosity Effects in Sound Waves of Finite Amplitude", see *Surveys in Méchanics*, Ed. by G. K. Batchelor and R. M. Davies, Cambridge University Press (1956).

C. C. Lin, *The Theory of Thermodynamic Stability*, Cambridge Univesrity Press, London, 1955.

A. J. Lotka, *J. Amer. Chem. Soc.*, **42**, 1595 (1920).

H. R. Mahler and E. H. Cordes, *Biological Chemistry*, Harper Int., 1966.

W. V. R. Malkus, *J. Fluid Mech.*, **1**, 521–539 (1956); *Nuovo Cimento Suppl.*, **22**, X, 376 (1961).

D. Massignon, *Mécanique Statistique des Fluides*, Dunod, Paris, 1957.

J. Maynard Smith, Second International Conference. "Theoretical Physics and Biology Versailles (1969)". North Holland Publ. Cy, Amsterdam (1970).

D. McQuarrie, *Supplementary Review Series in Applied Probability*, Methuen, London, 1967.

N. Minorsky, *Non-linear Oscillations*, Van Nostrand (1962).

R. Narasimha, *Bull. Classe Sci., Acad. Roy. Belg.*, October, 1970.

G. Nicolis, *Non-Equilibrium Thermodynamics, Variational Techniques and Stability*, University of Chicago Press, Chicago and London, 1965.

G. Nicolis and Ph. Sels, *Phys. Fluids*, **10**, 414 (1967).

G. Nicolis, *Advan. Chem. Phys.*, **13**, 299 (1967).

G. Nicolis and A. Babloyantz, *J. Chem. Phys.*, **51**, *6*, 2632–2637 (1969).

G. Nicolis, 'Stochastic Analysis of the Volterra-Lotka Model', submitted to *J. Math. Phys.* (1970).
G. Nicolis, J. Wallenborn and M. G. Velarde, *Physica*, **43**, 263 (1969).
J. C. Nihoul, *Bull. Soc. Roy. Sci. Liège, 38e année*, **3–4** (1969).
S. Ono, *Advan. Chem. Phys.*, **3**, 267 (1961).
L. Onsager, *Phys. Rev.*, **37**, 405 (1931).
H. G. Othmer and L. E. Scriven, *Ind. Eng. Chem.*, **8**, 302 (1969).
L. A. Pars, *A Treatise on Analytical Dynamics*, Heinemann, London, 1965.
A. Pellew and R. V. Southwell, *Proc. Roy. Soc. A.*, **176**, 312–43 (1940).
J. Platten, Thése de doctrat, Chimie-Physique, Universite de Bruxelles. 1970.
J. Platten, and R. S. Schechter, *Phys. Fluids*, **13**, n° 3, 823–33 (1970).
H. Poincaré, *J. Math.*, (3), **7** (1881) or *Oeuvres T. 1*, Gauthier-Villars, Paris, 1928.
I. Prigogine, *Bull. Classe Sci., Acad. Roy. Belg.*, **31**, 600 (1945); *Etude thermodynamique des Phénomènes Irréversibles*, Desoer, Liège, 1947; *Thermodynamics of Irreversible Processes and Fluctuations*, Third Symposium on Temperature, Washington, 1954.
I. Prigogine and R. Defay, *Chemical Thermodynamics*, English translation by D. H. Everett, Longmans Green, London, 1954.
I. Prigogine and G. Mayer, *Bull. Classe Sci., Acad. Roy. Belg.*, **5**, XLI (1955).
I. Prigogine, *Introduction to Thermodynamics of Irreversible Processes*, New York, Ed. Thomas (1955).
I. Prigogine, (a) "Unity of Physical Laws and Levels of Description", see *Reducibility*, ed. by M. Greene, Study Group on the Unity of Knowledge, University of California (to appear) (1970) (b) "Dissipative Structures in Biological Systems", Communication presented at the Second International Conf, *Theoretical Physics and Biology*, Versailles (1969), North Holland Publ. Co, Amsterdam (to be published) (1970).
I. Prigogine and R. Balescu, *Bull. Classe Sci., Acad. Roy. Belg.*, **41**, 917 (1955) and **42**, 256 (1956).
I. Prigogine, A. Bellemans and V. Mathot, *The Molecular Theory of Solutions*, North-Holland, Amsterdam, 1957.
I. Prigogine, *Introduction to Thermodynamics of Irreversible Processes*, John Wiley, New York, 1967.
I. Prigogine, *Non-Equilibrium Thermodynamics, Variational Techniques and Stability*, University of Chicago Press, Chicago and London, 1965.
I. Prigogine, "Structure, Dissipation and Life". Communication presented at the First International Conference. "Theoretical Physics and Biology". Versailles (1967). North Holland Publ. Cy. Amsterdam, (1969).
I. Prigogine and G. Nicolis, *J. Chem. Phys.*, **46**, 3542 (1967).
I. Prigogine and R. Lefever, *J. Chem. Phys.*, **48**, 1695 (1968).
I. Prigogine, R. Lefever, A. Goldbeter and M. Herschkowitz, *Nature*, **223**, 913 (1969).
K. Pye and B. Chance, *Proc. Natl. Acad. Sci., U.S.*, **55**, 888 (1966).
P. H. Roberts, *Non-Equilibrium Thermodynamics, Variational Techniques and Stability*, The University of Chicago Press, Chicago, 299–302 (1965).
A. M. Rodeyns, *Mémoire de Licence, Chimie Physique II*, Université Libre de Bruxelles (1969).
P. Rosen, *J. Chem. Phys.*, **21**, 1220 (1953); *J. Appl. Phys.*, **25**, 336 (1954).

R. S. Schechter and D. M. Himmelblau, *Phys. of Fluids*, **8,** 1431 (1965).

R. S. Schechter, *Non-Equilibrium Thermodynamics Variational Techniques and Stability*, University of Chicago Press, Chicago and London, 1965.

R. S. Schechter, *The Variational Method in Engineering*, McGraw-Hill, 1967.

R. S. Schechter, J. R. Hamm and I. Prigogine, to be published.

W. Schottky, H. Ulich and C. Wagner, *Thermodynamik*, Springer, Berlin, 1929.

E. E. Sel'kov, 'Oscillations in Biological and Chemical System', *Moscow Akad. Sci.*, **93,** (1967); *Eur. J. Biochem.*, **4,** 79 (1968).

R. A. Spangler and F. M. Snell, *Nature*, **191,** 457 (1961).

R. A. Spangler and F. M. Snell, *J. Theoret. Biol.*, **16,** 381 (1967).

H. B. Squire, *Proc. Roy. Soc. A.*, **142,** 621–8 (1933).

I. Tasaki, *Nerve Excitation*, C. Thomas, Springfield, Illinois, 1968.

G. P. Thomaes, *Non-Equilibrium Thermodynamics, Variational Techniques and Stability*, University of Chicago Press, Chicago and London, 1965.

L. H. Thomas, *Phys. Rev.*, **86,** 812 (1952).

L. Tisza and P. M. Quay, *Ann. Phys.*, **25,** 48 (1963).

E. Tonti, *Bull. Classe Sci.*, *Acad. Roy. Belg.*, **3** and **4,** 55 (1969).

K. S. Trintseher, *Biology and Information*, Consultants Bureau, New York, 1965.

C. Truesdell and W. Noll, *Flügge's Handbuch der Physik*, Vol. III, Part 3, Springer-Verlag, Berlin, 1965.

C. Truesdell, B. D. Coleman and V. J. Mizel, "Existence of Entropy as a consequence of Asymptotic Stability", *Arch. for National Mechanics and Analysis*, **25,** 243 (1967); **29,** 105 (1968); **30,** 173 (1968).

C. Truesdell, B. D. Coleman and J. Greenberg, "Thermodynamics and the Stability of Fluid Motion", *Arch. for National Mechanics and Analysis*, **25,** 321 (1967).

C. Truesdell and F. Hofelich, "On the Definition of Entropy for Non-Equilibrium States", *Zeitschrift für Physik*, **226,** 395 (1969).

A. M. Turing, *Phil. Trans. Roy. Soc.*, *London*, *Ser. B*, **237,** 37 (1952).

J. S. Turner, *Deep Sea Research*, **14,** 599 (1967) and *J. Fluid. Mech.* **33,** 183 (1968).

W. Unno, *Publ. Astron. Soc. Japan*, **20,** n° 4, 356–75 (1968).

P. Van Rysselberghe, *Compt. Rend.*, (1935).

G. Veronis, *Tellus*, **19,** 326 (1967); see also: *J. Fluid. Mech.* **34,** 315 (1968).

V. Volterra, *Théorie mathématique de la Lutte pour la Vie*, Gauthier-Villars, Paris, 1931.

C. Walter, *J. Theoret. Biol.*, **23,** 23 (1969).

N. Wax, *Selected Papers on Noise and Stochastic Processes*, Dover, New York, 1954.

D. Weihs and B. Gal-or, "General Variational Analysis of Hydrodynamic, Thermal and Diffusional Boundary Layers", *Int. J. Eng. Sci.*, **8,** 231 (1970).

P. Weiss, *Dynamics of Development: Experiments and Inferences*, Academic Press, New York and London, 1968.

Ya. B, Zeldovich and Yu. P. Raizer, *Physics of Shock Waves and High-Temperature—Hydrodynamic Phenomena*, Vol. 1, Ed. W. D. Hayes and R. F. Probstein, Academic Press, New York and London, 1966.

A. M. Zhaboutinsky, *Biofizika*, **2,** 306 (1964); 'Oscillations in Biological and Chemical Systems', Acad. Sci. U.S.S.R., Moscow (Nanka) (1967); *Russ. J. Phys. Chem.*, **42,** 1649 (1968).

Glossary of principal symbols

LATIN SYMBOLS

A_ρ, $\mathscr{A}_\rho$	affinity of the chemical reaction ρ
$A, B, \ldots X, Y, \ldots$	species; concentration
c_v, (c_p)	specific heat at constant volume (pressure)
c_γ	molar concentration of γ
c	speed of sound
D_γ	diffusion coefficient of γ
d_t	time derivative
$D = \dfrac{\mathrm{d}}{\mathrm{d}z}$	(z = reduced height)
E	internal energy
e	internal energy per unit mass
$,i$	$= \dfrac{\partial}{\partial x_i}$
$\mathbf{j}$	density of flow
J	generalized flow
K	equilibrium constant of a chemical reaction
k_+, k_-	kinetic constants for a chemical reaction
k	wave number; Boltzmann's constant
$L_{\alpha\beta}$	phenomenological coefficient
$\mathscr{L}$, L	Lagrangian
M_γ	molar mass of γ
N_γ	mass fraction of γ; molar fraction of γ; number of particles γ
$P[I]$	global production of I per unit time
P_{ij}	pressure tensor
p_{ij}	non-equilibrium part of the pressure tensor
$\mathscr{S}$	density of the entropy flow
U	total energy
$u_i = \delta u_i (\delta u_3 = w)$	velocity perturbation
V	total volume
$\mathbf{v}$	barycentric velocity
$v (= \rho^{-1})$	specific volume
v_γ	partial specific volume of γ
W	amplitude factor of the perturbation w
$\mathbf{W}$	heat flow

w_γ	rate of the ρth chemical reaction
*	complex conjugate superscript
$\times$	reference state superscript
$\bar{x}$ or $\langle x \rangle$	average value of x
$\mathbf{X}$	generalized force

GREEK SYMBOLS

α	expansion coefficient; convection coefficient in Newton's law
α_j	direction cosines; parameters in test functions
β	adverse temperature gradient
γ	$= c_p/c_v$; expansion coefficient due to concentration; species symbol
δ, δ'	variational symbol
δ_{ij}	Kronecker symbol
Δ	finite increment
$\mathbf{\Delta}_\gamma$	diffusion velocity of γ
$\mathbf{\nabla}$	gradient operator
∇^2	Laplacian operator
ε	weighting function
$\eta = \nu\rho$	dynamic viscosity
θ	angle; temperature perturbation
Θ	amplitude factor of the perturbation θ
κ	thermal diffusivity
λ	thermal conductivity; wavelength
μ_γ	chemical potential per unit mass of γ; chemical potential per mole of γ
$\nu_{\gamma\rho}$	stoichiometric coefficient of a chemical reaction ρ
ν	kinematic viscosity; frequency
ξ_ρ	degree of advancement or extent of reaction
ξ	fluctuating variable
Π	evolution potential of a perturbation around a steady state; amplitude factor of the perturbation ϖ
$\varpi = \delta_p$	hydrostatic pressure perturbation
ρ	density
σ	non-dimensional eigenvalue
$\sigma[I]$	source (local production) of I
$\sum_\gamma$	summation over γ
τ	weighting function
Φ, Ψ	flow; potential; local potential

X	isothermal compressibility
Ω	surface; potential energy
ω	potential energy per unit mass
$\omega = \omega_r + i\omega_i$	eigenvalue of a complex normal mode

Index